Communications in Computer and Information Science

2824

Series Editors

Gang Li, *School of Information Technology, Deakin University, Burwood, VIC, Australia*
Joaquim Filipe, *Polytechnic Institute of Setúbal, Setúbal, Portugal*
Zhiwei Xu, *Chinese Academy of Sciences, Beijing, China*

Rationale

The CCIS series is devoted to the publication of proceedings of computer science conferences. Its aim is to efficiently disseminate original research results in informatics in printed and electronic form. While the focus is on publication of peer-reviewed full papers presenting mature work, inclusion of reviewed short papers reporting on work in progress is welcome, too. Besides globally relevant meetings with internationally representative program committees guaranteeing a strict peer-reviewing and paper selection process, conferences run by societies or of high regional or national relevance are also considered for publication.

Topics

The topical scope of CCIS spans the entire spectrum of informatics ranging from foundational topics in the theory of computing to information and communications science and technology and a broad variety of interdisciplinary application fields.

Information for Volume Editors and Authors

Publication in CCIS is free of charge. No royalties are paid, however, we offer registered conference participants temporary free access to the online version of the conference proceedings on SpringerLink (http://link.springer.com) by means of an http referrer from the conference website and/or a number of complimentary printed copies, as specified in the official acceptance email of the event.

CCIS proceedings can be published in time for distribution at conferences or as post-proceedings, and delivered in the form of printed books and/or electronically as USBs and/or e-content licenses for accessing proceedings at SpringerLink. Furthermore, CCIS proceedings are included in the CCIS electronic book series hosted in the SpringerLink digital library at http://link.springer.com/bookseries/7899. Conferences publishing in CCIS are allowed to use our online conference service (Meteor) for managing the whole proceedings lifecycle (from submission and reviewing to preparing for publication) free of charge.

Publication process

The language of publication is exclusively English. Authors publishing in CCIS have to sign the Springer CCIS copyright transfer form, however, they are free to use their material published in CCIS for substantially changed, more elaborate subsequent publications elsewhere. For the preparation of the camera-ready papers/files, authors have to strictly adhere to the Springer CCIS Authors' Instructions and are strongly encouraged to use the CCIS LaTeX style files or templates.

Abstracting/Indexing

CCIS is abstracted/indexed in DBLP, Google Scholar, EI-Compendex, Mathematical Reviews, SCImago, Scopus. CCIS volumes are also submitted for the inclusion in ISI Proceedings.

How to start

To start the evaluation of your proposal for inclusion in the CCIS series, please send an e-mail to ccis@springer.com

Biviana M. Suárez · Romario Conto ·
Freddy Hernández · Olga Usuga ·
Carmen E. Patiño · Cristhian Quiroz
Editors

National Meeting of R Users

3rd R Day Colombia
Medellín, Colombia, November 14, 2025
Proceedings

 Springer

Editors
Biviana M. Suárez ⓘ
EAFIT University
Medellín, Colombia

Romario Conto ⓘ
Metropolitan Institute of Technology
Medellín, Colombia

Freddy Hernández
National University of Colombia
Medellín, Colombia

Olga Usuga ⓘ
University of Antioquia
Medellín, Colombia

Carmen E. Patiño ⓘ
University of Antioquia
Medellín, Colombia

Cristhian Quiroz ⓘ
EAFIT University
Medellín, Colombia

ISSN 1865-0929   ISSN 1865-0937 (electronic)
Communications in Computer and Information Science
ISBN 978-3-032-18454-2   ISBN 978-3-032-18455-9 (eBook)
https://doi.org/10.1007/978-3-032-18455-9

This Springer imprint is published by the registered company Springer Nature Switzerland AG
The registered company address is: Gewerbestrasse 11, 6330 Cham, Switzerland

If disposing of this product, please recycle the paper.

Preface

The R Day Colombia initiative has become a central meeting point for researchers, professionals, students, and industry practitioners who work with statistical computing, data science, and open-source technologies. Over the years, the event has contributed to strengthening the national and regional ecosystem of analytics by promoting the adoption of the R programming language, a tool that has proven essential for scientific research, industrial innovation, and pedagogical transformation. Through its emphasis on reproducibility, methodological rigor, and community building, R Day has positioned itself as one of the most relevant academic gatherings for the Colombian R community.

The event was previously held in Medellín, Colombia, in 2019 and 2023, and in 2025 it took place once again in the city on November 14, returning to a fully in-person format. This third edition reaffirmed its mission of connecting the growing network of R users across the country, while simultaneously welcoming newcomers who are beginning to explore its capabilities. The positive reception of R Day Colombia has demonstrated both the demand and the enthusiasm for collaborative spaces where ideas, tools, and applications built in R can be shared, extended, and critically discussed.

For the 2025 edition, the call for contributions received a total of 43 submissions. From these, 33 were accepted as oral presentations and 10 as poster presentations, showcasing a broad spectrum of topics and methodologies. Among the original submissions, 21 manuscripts were selected for full development and inclusion in this post-proceedings volume. All papers underwent a rigorous double-blind peer-review process, with three independent evaluations per submission. The selection reflects the reviewers' assessment of originality, methodological soundness, clarity of exposition, and relevance to the R ecosystem. This process ensured that the published contributions meet the academic standards expected in the CCIS series.

The works included in this volume span a wide range of scientific domains and demonstrate the versatility of R as both a research and professional tool. Several contributions focus on the development of new R packages and computational tools, highlighting innovations that extend the language's functionality and usability. Others explore advanced visualization techniques and time series modeling, two areas in which R continues to be a leading environment for both research and applied analytics. Additional papers address pedagogy and the integration of R into teaching–learning processes, showcasing the language's value in modern statistics and data science education. The volume also includes contributions in biostatistics, environmental studies, agriculture, public health, and applications of R in finance, natural language processing, and industrial processes. Taken together, these works demonstrate the interdisciplinary scope of R Day and its commitment to fostering collaboration across diverse fields of knowledge.

We extend our sincere appreciation to the institutions that organized and supported R Day Colombia 2025: the Institución Universitaria ITM, Universidad EAFIT, Universidad Nacional de Colombia, Universidad de Antioquia, Fundación Universidad de Antioquia, and ConsultING. Their commitment made it possible to host a high-quality academic event and to bring together a vibrant, multidisciplinary community. Their support was instrumental not only for the organization of the conference, but also for the preparation of this post-proceedings volume.

We also express our gratitude to the authors who contributed their work, and to the reviewers and committee members who generously dedicated their time and expertise to ensure the quality of the scientific contributions. Their careful evaluations and constructive feedback were fundamental to the success of this volume and to the integrity of its academic process.

Finally, we acknowledge the broader R community in Colombia and Latin America, whose participation and enthusiasm continue to inspire the growth of R Day. We hope that the contributions gathered in this volume will serve as a reference for future research, teaching, and professional practice, and that they will encourage new collaborations and innovative uses of the R programming language in the years to come.

December 2025

Romario Conto
Freddy Hernández
Carmen E. Patiño
Biviana M. Suárez
Olga Usuga
Cristhian Quiroz

Organization

General Chair

Biviana Suárez | Universidad EAFIT, Colombia

Program Committee Chairs

Romario Conto | Institución Universitaria ITM, Colombia
Freddy Hernández | Universidad Nacional, Colombia
Olga Usuga | Universidad de Antioquia, Colombia
Carmen Patiño | Universidad de Antioquia, Colombia

Production Editor

Cristhian Quiroz | Universidad EAFIT, Colombia

Reviewers

Jairo Ángel | Universidad de Córdoba, Colombia
Dana Arango | Universidad EAFIT, Colombia
Sergio Barajas | CIMAT, Mexico
Benjamín Baumer | Smith College, USA
Fidel Castro | Universidade Federal de Rio Grande do Norte, Brazil
Valentina Clavijo | Politécnico de Milán, Italy
Cristian Correa | Universidad Nacional, Colombia
Yandira Cuvero | Escuela Politécnica Nacional, Ecuador
Sandra Flores | Universidad de Chile, Chile
Diego Fonseca | Universidad EAFIT, Colombia
Isabel García | Pontificia Universidad Javeriana, Colombia
Eniuce Menezes | Universidade Estadual de Maringá, Brazil
Nicolás Moreno | Universidad EAFIT, Colombia
Diana Moreno | Universidad del Rosario, Colombia
Diana Pérez | Universidad Deusto, Spain
Leticia Ramíre | CIMAT, Mexico

Andrés Ríos	Universidad Industrial de Santander, Colombia
Jessica María Rojas	Institución Universitaria ITM, Colombia
Marco Scavino	Universidad de la República, Uruguay
Sergio Sierra	Institución Universitaria ITM, Colombia
Alejandra Tapia	Pontificia Universidad Católica de Chile, Chile
Carlos Taima	Universidad EAFIT, Colombia
Francisco Zuluaga	Universidad EAFIT, Colombia

Contents

Development of Packages and Tools in R

Development of Packages and Tools in R

The New Exponentiated Exponential (NEE) Distribution: Implementation and Simulation Study Within GAMLSS

Juliana García Villada[1(✉)] [iD], Freddy Hernandez-Barajas[2] [iD], and Olga Cecilia Cecilia Usuga-Manco[1] [iD]

[1] Universidad de Antioquia, Medellín, Colombia
`juliana.garciav@udea.edu.co`
[2] Universidad Nacional de Colombia, Medellín, Colombia

Abstract. In this work, we present the implementation of the New Exponentiated Exponential (NEE) distribution within the gamlss package in R, providing users with a flexible and robust tool for statistical modeling. This implementation enables the computation of probabilities, quantiles, and random numbers, as well as parameter estimation and the development of regression models within the GAMLSS framework. To evaluate its performance, we conducted a simulation study to evaluate the estimation process using the mean and mean squared error. The results demonstrate the effectiveness of the NEE distribution in capturing data variability and improving model accuracy. This contribution extends the capabilities of the gamlss package, offering R users an advanced resource for data analysis and regression modeling, especially in contexts where flexibility and precision are crucial.

Keywords: New Exponentiated Exponential distribution · GAMLSS · regression model

1 Introduction

Linear regression, despite being a widely used method for modeling numerical dependent variables, relies on the assumption that model residuals are normally distributed. However, because the distribution of residuals reflects the shape of the numerical dependent variable and they often exhibit non-normal patterns in empirical data Bono et al. (2017), there is a high likelihood that estimates produced by linear models may be biased. For example, in cognitive neuroscience and experimental psychology, the dependent variable known as reaction times typically presents a positive skew (Tejo et al., 2019). Consequently, recent approaches have suggested regression models that can handle positively skewed data by modeling how covariates influence the location parameter of such variables (Schumacher et al., 2021).

In recent years, a wide range of new probability distribution families has been proposed to enhance modeling flexibility, particularly in the analysis of

B. M. Suárez et al. (Eds.): R Day 2025, CCIS 2824, pp. 3–15, 2026.
https://doi.org/10.1007/978-3-032-18455-9_1

real-world data with complex characteristics. Many of these developments are based on the inclusion of one or more additional shape parameters into classical models, thereby extending their capacity to capture various distributional behaviors (Marshall and Olkin, 1997; Mahdavi and Kundu, 2017; Cordeiro et al., 2013). Although these models offer greater adaptability, they often introduce challenges in terms of interpretation, parameter estimation, and model selection due to increased complexity. To address these issues, several researchers have proposed distribution families that prioritize structural simplicity, aiming to preserve or even increase flexibility while minimizing the number of parameters. For example, Mahmood and Chesneau (2019) introduced the sine-G family, and Maurya et al. (2016, 2018) proposed the logarithmic and new logarithmic transformation (LT, NLT) methods, each designed to reduce model complexity without sacrificing performance.

Continuing in this direction, Hassan et al. (2024) introduced a novel class of distributions known as the New Exponentiated Transformation (NET) family. This family was developed to provide flexible yet simple modeling tools that avoid the complexity of multi-parameter models. As part of the NET family, they proposed the New Exponentiated Exponential (NEE) distribution, designed specifically to model positively skewed data with varying hazard rate behaviors. The authors focused on applications in engineering, where data often exhibit asymmetries due to factors such as material properties, structural design, and environmental conditions. The NEE distribution is characterized by two parameters (μ, σ), and its formulation allows for straightforward theoretical derivation and efficient implementation (Hassan et al., 2024). The distribution exhibits notable flexibility in terms of the behavior of the hazard rate, including decreasing forms and bathtub-shaped forms, which are frequently observed in the reliability and failure time data.

According to Hassan et al. (2024), one of the key advantages of the NEE distribution is its ability to combine flexibility with interpretability, making it a suitable alternative to more complex multi-parameter models. This is particularly valuable in fields such as reliability engineering and survival analysis, where understanding the underlying hazard function is critical. In their empirical analysis, the authors applied the NEE distribution to real-world engineering data sets, including variables such as gauge lengths and failure stresses. The results showed that the NEE model not only fits the data well but also outperforms several benchmark models in terms of goodness-of-fit and simplicity, especially when the data exhibit positive skewness or non-monotonic hazard functions.

This study introduces the `RelDists` package for R, which provides functions for computing the density, the distribution function, quantiles, and hazard rate, as well as for generating random samples for the NEE distribution. A notable contribution of this work is that `RelDists` contains a family function for the NEE distribution, which takes advantage of the `gamlss` package to estimate distribution parameters and regression parameters.

This paper is structured as follows: Sect. 2 provides an overview of the NEE distribution and its statistical properties. In Sects. 2.1 and 2.2, we present the parameter estimation and the NEE regression model, respectively. Section 3 describes the `RelDists` package. Section 4 presents some applications of the studied distribution using real datasets. Section 5 shows the simulation study and the results. Finally, Sect. 6 presents the conclusions drawn from the work.

2 New Exponentiated Exponential Distribution

Hassan et al. (2024) proposed the New Exponentiated Exponential (NEE) distribution which offers a practical and flexible tool for modeling continuous, positively skewed data in applied settings. Its probability density function (pdf) and cumulative distribution function (cdf) are defined as follows:

$$f(y; \mu, \sigma) = \log\left(2^{\sigma}\right) \mu e^{-\mu y} (1 - e^{-\mu y})^{(\sigma-1)} 2^{(1-e^{-\mu y})^{\sigma}}, \tag{1}$$

$$F(y; \mu, \sigma) = 2^{(1-e^{-\mu y})^{\sigma}}, \tag{2}$$

where $\mu > 0$, $\sigma > 0$ and $y > 0$.

Figure 1 depicts the shapes of the pdf and cdf for different values of μ and σ. For simplicity we write $f(y)$ instead of $f(y; \mu, \sigma)$ and the same convention is used for the cdf $F(y)$. If a random variable Y follows a NEE distribution, we can write $Y \sim \text{NEE}(\mu, \sigma)$.

The reliability function $R(y)$ (or survival function $S(y)$) of the NEE distribution is defined as $1 - F(y)$ and is given by

$$R(y; \mu, \sigma) = 1 - (2^{(1-e^{-\mu y})^{\sigma}} - 1) = 2 - 2^{(1-e^{-\mu y})^{\sigma}}.$$

The hazard function $h(y)$ or failure rate function of this distribution is given as:

$$h(y; \mu, \sigma) = \log(2^{\sigma})\mu e^{-\mu y} \frac{(1 - e^{-\mu y})^{\sigma-1}}{2^{(1-(1-e^{-\mu y})^{\sigma})}}.$$

The NEE distribution's hazard function has great flexibility allowing different shapes to be modeled (such as a bathtub or a nearly constant curve). When the parameter σ decreases, the function shows a high risk at the beginning, decreases in the middle phase, and shows a slight increase towards the end. In contrast, as the value of μ increases, the bathtub transition smooths out and the risk tends to stabilize. Figure 2 depicts the possible forms for the hazard function. The hazard function measures the likelihood that a unit, which has not failed by time t, will fail in the next instant (Panis et al., 2020).

The quantile function of the NEE distribution has a simple expression and can be used to obtain the p-quantile and to generate random samples.

$$y_p = \frac{-1}{\mu} \log\left[1 - \left(\frac{\log(1+p)}{\log(2)}\right)^{\frac{1}{\sigma}}\right].$$

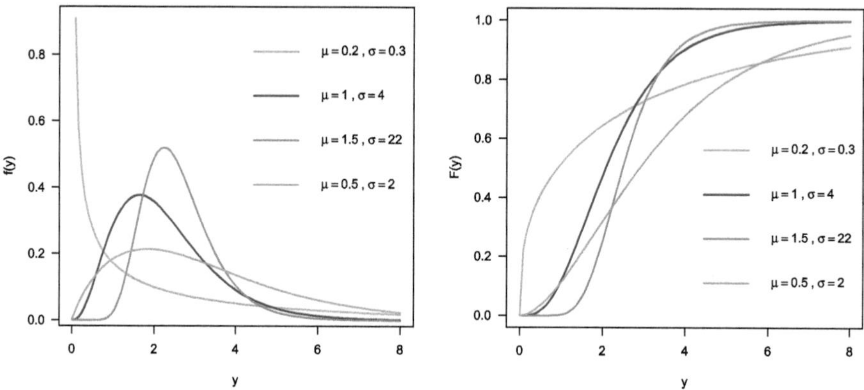

Fig. 1. Probability density function (left) and cumulative density function (right) of the NEE distribution for different values of μ and σ.

The moment generating function (MGF) of Y, defined as $M_Y(t) = E(e^{tY})$, is given by

$$M_Y(t) = \log(2^\sigma) \sum_{r=0}^{\infty} \sum_{j=0}^{\infty} \sum_{k=0}^{\sigma(j+1)-1} \frac{(-1)^k (\log 2)^j}{j!(k+1)^{r+1}} \left(\frac{t}{\mu}\right)^r \binom{\sigma(j+1)-1}{k}, \quad t < \mu.$$

(3)

The MGF uniquely characterizes the distribution and serves as a powerful tool for deriving several distributional properties. In particular, differentiation of $M_Y(t)$ with respect to t allows the direct computation of raw moments of any order. Additionally, the MGF is useful for studying convergence in distribution, obtaining cumulants, and deriving asymptotic properties of estimators.

The r-th raw moment of the NEE distribution is obtained as

$$E(Y^r) = \frac{1}{\mu^r} \log(2^\sigma) \sum_{j=0}^{\infty} \sum_{k=0}^{\sigma(j+1)-1} (-1)^k \frac{(\log 2)^j}{j!} \binom{\sigma(j+1)-1}{k} \frac{\Gamma(r+1)}{(k+1)^{r+1}}. \quad (4)$$

This general expression allows the computation of the mean ($r = 1$), variance ($r = 2$), and higher-order moments such as skewness and kurtosis. These quantities play a fundamental role in describing the location, dispersion, and shape of the distribution.

The moments of the NEE distribution have several important applications. The first moment provides the expected value, which is useful for modeling average behavior in applied problems. The second moment leads to the variance, which quantifies uncertainty and variability. Higher-order moments describe asymmetry and tail behavior, making the distribution suitable for modeling data with skewed or heavy-tailed characteristics.

Moreover, the analytical form of the moments facilitates parameter estimation using the method of moments and supports statistical inference procedures

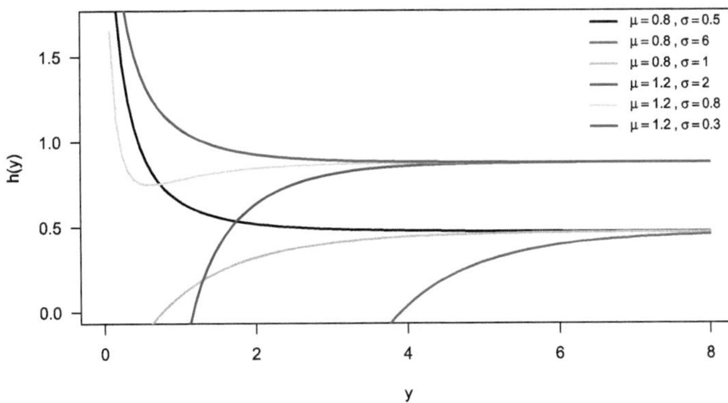

Fig. 2. Hazard function of the NEE distribution for different combinations of the parameter values μ and σ.

such as goodness-of-fit testing and reliability analysis. The availability of the MGF further strengthens the theoretical tractability of the NEE distribution.

2.1 Parameter Estimation

This section presents an estimation method for the NEE distribution based on the maximum likelihood method. Consider a random sample of size n, denoted by $y_1, y_2, \ldots, y_n$, drawn from a NEE population. The log-likelihood function for the parameter vector $\boldsymbol{\theta} = (\mu, \sigma)^\top$ will then be derived in the next paragraph.

$$l(\boldsymbol{\theta}) = \log(L(\boldsymbol{\theta})) = \log\left(\prod_{i=1}^{i=n} f(y_i; \boldsymbol{\theta})\right)$$

$$= \sum_{i=1}^{i=n} \log f(y_i; \boldsymbol{\theta})$$

$$= \sum_{i=1}^{i=n} \log(\log(2^{1-e^{-\mu y}})) + \log(\mu) - \mu y + (\sigma - 1)\log(1 - e^{-\mu y}) + (1 - e^{-\mu y})^\sigma \log(2)$$

By taking the partial derivatives of $l(\boldsymbol{\theta})$ with respect to μ and σ, and letting them be equal to zero, we have the following two equations:

$$\frac{\partial l(\boldsymbol{\theta})}{\partial \mu} = \sum_{i=1}^{i=n}\left[\frac{1}{\mu} - y + \frac{(\sigma - 1)ye^{-\mu y}}{1 - e^{-\mu y}} + \sigma y e^{-\mu y}(1 - e^{-\mu y})^{\sigma-1}\log(2)\right] = 0$$

$$\frac{\partial l(\boldsymbol{\theta})}{\partial \sigma} = \sum_{i=1}^{i=n}\left[\frac{\log(2)}{\log(2^\sigma)} + \log(1 - e^{-\mu y}) + (1 - e^{-\mu y})^\sigma \log(1 - e^{-\mu y})\log(2)\right] = 0$$

The maximum likelihood estimate $\hat{\boldsymbol{\theta}} = (\hat{\mu}, \hat{\sigma})^\top$ for $\boldsymbol{\theta} = (\mu, \sigma)^\top$ is obtained by solving the above nonlinear system of equations.

A contribution of this paper is the development of the RelDists package in the R programming language (R Core Team, 2025), which implements the NEE family and provides a framework for estimating the vector parameter $\boldsymbol{\theta}$ within the GAMLSS framework (Rigby and Stasinopoulos, 2005).

2.2 New Exponentiated Exponential Regression Model

This section introduces the NEE regression model, which employs the method of maximum likelihood (MLE) to estimate its parameters. The aim of the model is to incorporate explanatory variables to describe the behavior of both parameters of the NEE distribution. It is assumed that the observation y_i (for $i = 1, 2, \ldots, n$) are independent and follow the probability density function $f(y; \mu, \sigma)$ as defined in Eq. (1). The formulation of the NEE regression model is given by:

$$g_1(\mu_i) = \eta_1 = \mathbf{X}_{1i}\boldsymbol{\beta}_1,$$
$$g_2(\sigma_i) = \eta_2 = \mathbf{X}_{2i}\boldsymbol{\beta}_2,$$

where each function g_k $(k = 1, 2)$ is a known monotonic transformation that ensures the parameters μ and σ are mapped to their valid domains. Most commonly, the natural logarithm is used as the link function. The design matrices $\mathbf{X}_k$ are specified and include the covariates associated with each parameter, while the vectors $\boldsymbol{\beta}_k$ contain the coefficients representing the fixed effects of these covariates.

3   The RelDists package

The RelDists package implements several distributions, including the NEE distribution. Within a GAMLSS framework, this package allows users to estimate parameters for the NEE distribution or estimate effects for the NEE regression model. The current version of the package is available on CRAN, and users can download it using the following code.

```
install.packages("RelDists")
library(RelDists)
```

Regarding the NEE distribution, the RelDists package provides the functions dNEE() and pNEE() to obtain the probability density function (pdf) and cumulative distribution function (cdf), respectively. They can be used as follows:

```
dNEE(x, mu, sigma, log=FALSE)
pNEE(q, mu, sigma, lower.tail=TRUE, log.p=FALSE)
```

To obtain the reliability function $R(x)$, the user can use the pNEE() function with the argument lower.tail=FALSE. The RelDists package also provides other useful functions. For instance, rNEE() generates random samples, qNEE() obtains quantiles, and hNEE() calculates the hazard function $h(x)$. To use these functions, one can follow the syntax below:

```
rNEE(n, mu, sigma)
qNEE(p, mu, sigma, lower.tail=TRUE, log.p=FALSE)
hNEE(x, mu, sigma, log=FALSE)
```

The function `NEE()` represents the family function that allows the estimation of parameters for the NEE distribution or the estimation of effects for the NEE regression model taking advantange of the `gamlss` package. The help documentation of the `NEE()` function provides five examples that demonstrate how any user can estimate the parameters for the NEE distribution or estimate the effects for the NEE regression model.

As an illustrative example, we present the `R` code used to generate a random sample from the NEE distribution with sample size $n = 500$ and true parameter values $\mu = 2.5$ and $\sigma = 3.5$. The purpose of this example is to demonstrate how the `gamlss` package can be used to estimate the parameters of the NEE distribution based on a random sample.

```
library(RelDists)
library(gamlss)

# Generating random values with known mu and sigma
set.seed(1234)
y <- rNEE(n = 500, mu = 2.5, sigma = 3.5)

# Fitting the model
mod1 <- gamlss(y ~ 1, sigma.fo = ~1, family = NEE)
```

The object `mod1` contains the results of the fitted model. The parameter estimates μ and σ can be extracted from the fitted model and transformed using the inverse link function as follows:

```
exp(coef(mod1, what = "mu"))
exp(coef(mod1, what = "sigma"))
```

From the fitted model, we obtain the estimates $\hat{\mu} = 2.487634$ and $\hat{\sigma} = 3.585429$, which are close to the true parameter values $\mu = 2.5$ and $\sigma = 3.5$. This result illustrates the good performance of the maximum likelihood estimation procedure for the NEE distribution implemented through the `gamlss` framework.

4   Application

In this section, we present two real data sets to illustrate the benefits of using the NEE distribution.

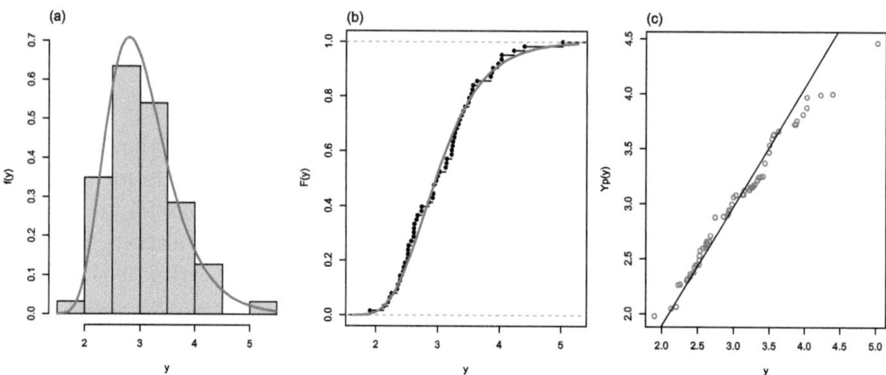

Fig. 3. Histogram (a), empirical cumulative distribution function (b), and empirical quantile-quantile plot (c) for the gauge lengths of 10 mm (y). Red lines correspond to estimated $f(y)$, $F(y)$ and $y_p(y)$ assuming a NEE with $\hat{\mu} = 2.0769$ and $\hat{\sigma} = 255.2289$.

4.1 Application About Gauge Length

Kundu and Raqab (2009) reported a dataset with 63 observations corresponding to gauge lengths of 10 mm. Figure 3 displays, in the left panel, the histogram of the dataset along with a red curve representing the estimated probability density function of the NEE distribution. This curve was obtained using the estimated parameters $\hat{\mu} = 2.0769$ and $\hat{\sigma} = 255.2289$. The estimated density provides a good fit to the histogram, capturing the main features of the observed data. The middle panel of Fig. 3 presents the empirical cumulative distribution function (ECDF) together with the estimated cumulative distribution function (CDF) of the NEE model, computed with the same parameter estimates. It can be seen that the estimated CDF closely follows the empirical pattern. Finally, the right panel of Fig. 3 shows the quantile-quantile (Q-Q) plot for the dataset under the NEE distribution. The points lie approximately along the black reference line, indicating that the NEE model provides an adequate fit to the data.

4.2 Application About Failure Stresses

The data used in this application were originally reported by Bader and Priest (1982), who recorded the failure stresses (in GPa) of 65 individual carbon fibers 50 mm in length. This dataset, referred to as Dataset II in Hassan et al. (2024), has been frequently used to evaluate the performance of various lifetime distributions in reliability analysis. The 65 observations are retained in the analysis presented here. Figure 4 shows the histogram (a), the empirical cumulative distribution function (b), and the quantile-quantile plot (c) comparing the empirical quantiles with those from the fitted NEE distribution.

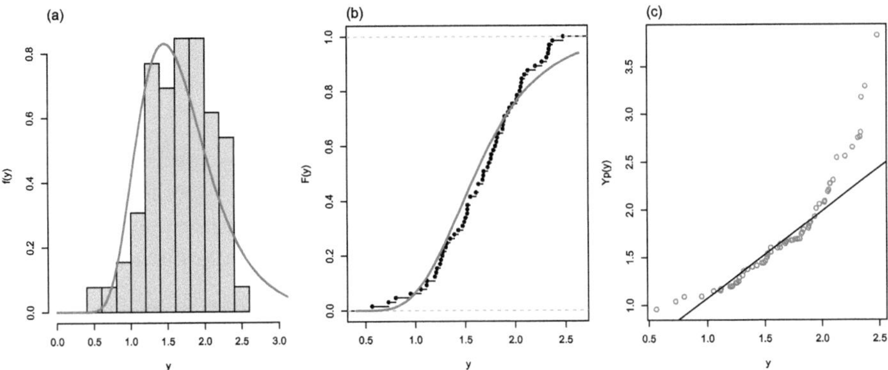

Fig. 4. Histogram (a), empirical cumulative distribution function (b), and empirical quantile-quantile plot (c) for the failure stresses (in GPa) (y) individual carbon fibers 50 mm in length. Red lines correspond to estimated $f(y)$, $F(y)$ and $y_p(y)$ assuming a NEE with $\hat{\mu} = 2.400515$ and $\hat{\sigma} = 25.15236$.

5 Simulation Study

This section shows the findings of a Monte Carlo simulation study based on the guidelines in Friedrich and Friede (2023), with the goal of investigating parameter estimation in models with and without covariates. To evaluate the performance of estimating a generic parameter θ, we used the mean value and the Mean Squared Error (MSE) of the estimator $\hat{\theta}$ defined as,

$$\text{Mean value} = \frac{\sum_{i=1}^{i=k} \hat{\theta}_i}{k}$$

$$\text{Mean Squared Error} = \frac{\sum_{i=1}^{i=k} (\hat{\theta}_i - \theta)^2}{k}$$

5.1 Model Without Covariates

In the first part of the simulation study, we considered data generated from the NEE($\mu = 2.5, \sigma = 25$) distribution to emulate the real data presented in Sect. 4.2. The model used in this part can be summarized as follows:

$$\begin{aligned} y_i &\sim NEE(\mu, \sigma), \\ \mu &= 2.5, \\ \sigma &= 25. \end{aligned} \tag{5}$$

For each sample size $n = 100, 200, 300, \ldots, 800, 900, 1000$, we completed the following steps to simulate the data and estimate the parameters:

1. Generate a sample of size n from a NEE($\mu = 2.5, \sigma = 25$).

12   J. G. Villada et al.

2. Fit a model without covariates using GAMLSS with the NEE family. Estimate $\hat{\mu}$ and $\hat{\sigma}$ by exponentiating the model coefficients.
3. Obtain and store the estimations $\hat{\mu}$ and $\hat{\sigma}$ using the proposed procedure.
4. Repeat steps 1 to 3 for $k = 5000$ times.

Figure 5 illustrates the evolution of the mean estimates for the parameters $\hat{\mu}$ and $\hat{\sigma}$. As the sample size n increases, both estimates progressively converge toward the true parameter values, $\mu = 2.5$ and $\sigma = 25$. Figure 6 shows that, as expected, the mean squared error (MSE) decreases and approaches zero with increasing sample size.

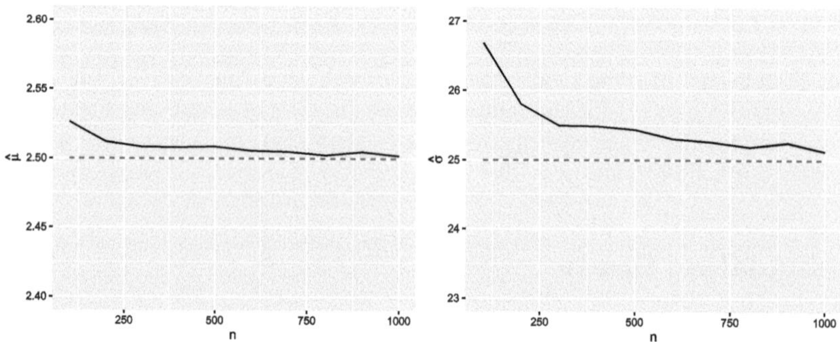

Fig. 5. Mean value for $\hat{\mu}$ and $\hat{\sigma}$ for different sample sizes n. Horizontal red dashed line represents the true parameter values.

5.2   Model with Covariates

In the second part of the simulation study, we considered data generated from the next model:

$$
\begin{aligned}
y_i &\sim NEE(\mu_i, \sigma_i), \\
\log(\mu_i) &= \beta_0 + \beta_1 \times X_{1i}, \\
\log(\sigma_i) &= \gamma_0 + \gamma_1 \times X_{2i}, \\
X_1 &\sim U(0, 1), \\
X_2 &\sim U(0, 1).
\end{aligned}
\tag{6}
$$

The vector parameter was fixed as $\boldsymbol{\theta} = (\beta_0 = -1.4, \beta_1 = 4.6, \gamma_0 = 2.1, \gamma_1 = 2.3)^\top$. These values were selected to ensure a response variable with an approximate $NEE(\mu = 2.5, \sigma = 25)$ to emulate the real data presented in Sect. 4.2. From model (6), we have that $X_1 \sim U(0, 1)$ and it implies that $E(X_1) = 0.5$, so $\mu \approx \exp(-1.4 + 4.6 \times 0.5) \approx 2.5$. The same analysis can be done for the σ parameter to obtain $\sigma \approx 25$.

For each sample size $n = 100, 200, 300, \ldots, 800, 900, 1000$, we followed the next steps to simulate the data and estimate the parameters:

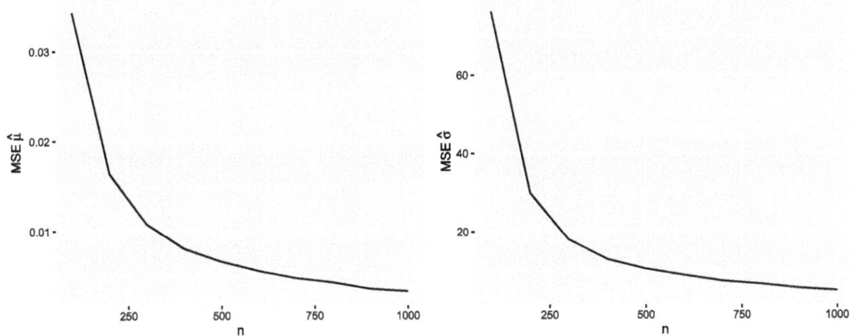

Fig. 6. Mean Squared Error for $\hat{\mu}$ and $\hat{\sigma}$ for different sample sizes n.

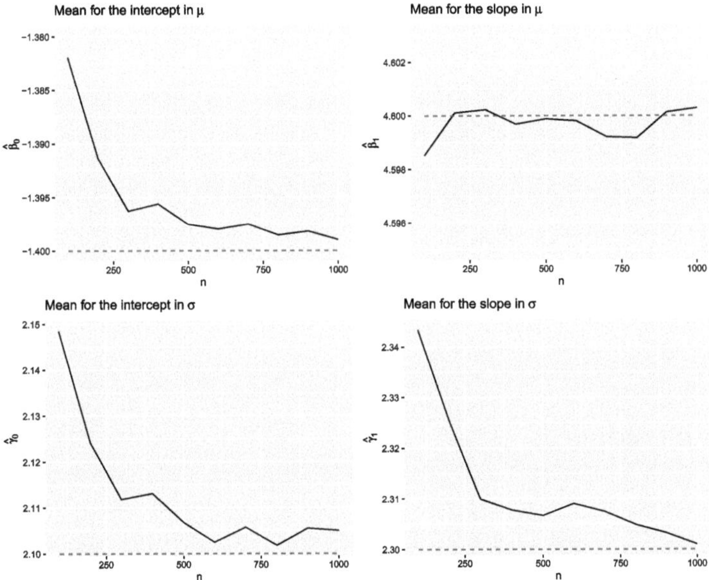

Fig. 7. Mean value for $\hat{\beta}_0$, $\hat{\beta}_1$, $\hat{\gamma}_0$ and $\hat{\gamma}_1$ for different sample sizes n. Horizontal red dashed line represents the true parameter values.

1. Generate a random sample of n values from a NEE distribution given in expression (6).
2. Obtain and store the estimations $\hat{\beta}_0$, $\hat{\beta}_1$, $\hat{\gamma}_0$ and $\hat{\gamma}_1$ using the proposed procedure.
3. Repeat steps 1 and 2 for $k = 5000$ times.

Figure 7 shows the evolution of the mean estimates for the parameters $\hat{\beta}_0$, $\hat{\beta}_1$, $\hat{\gamma}_0$, and $\hat{\gamma}_1$ as a function of the sample size n. As expected, the mean estimates progressively approach the true parameter values (indicated by the red dashed lines) as n increases. Figure 8 presents the corresponding evolution of the mean

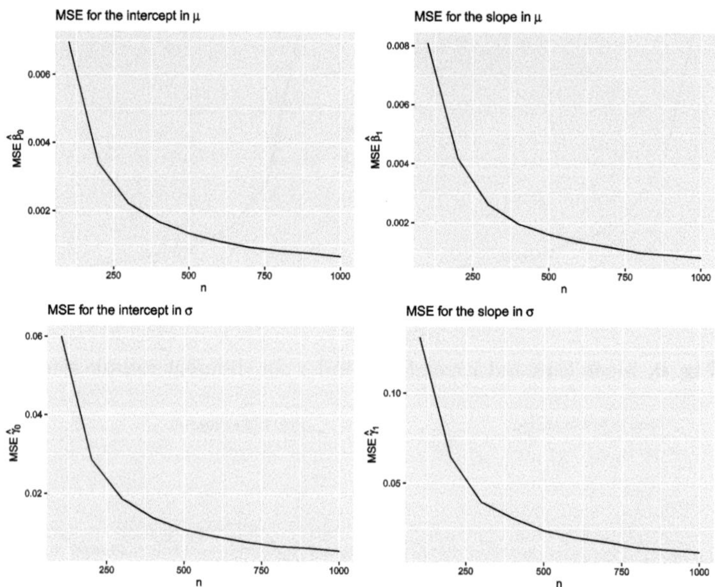

Fig. 8. Mean Squared Error for $\hat{\beta}_0$, $\hat{\beta}_1$, $\hat{\gamma}_0$ and $\hat{\gamma}_1$ for different sample sizes n.

squared error (MSE) for these parameters, which, as anticipated, decreases and tends toward zero with increasing sample size.

6   Conclusions

This study introduced the implementation of the New Exponentiated Exponential (NEE) distribution within the gamlss framework through the RelDists package, expanding the set of flexible tools available for modeling positively skewed data. The NEE distribution combines interpretability with adaptability, allowing users to capture a variety of hazard rate behaviors with only two parameters.

Applications with real datasets demonstrated that the NEE model provides a good fit to empirical data characterized by asymmetry and non-linear patterns, outperforming classical alternatives in both accuracy and parsimony. The Monte Carlo simulation results confirmed the consistency of the maximum likelihood estimators, as the mean and mean squared error of the parameters improved with increasing sample sizes in models with and without covariates.

Overall, this work strengthens the practical use of the NEE distribution in reliability and lifetime analysis, as well as in other applied contexts where flexible modeling is required. Future research may explore extensions of the NEE distribution to multivariate settings, censored data, or Bayesian estimation, and compare its performance with other recently proposed members of the NET family.

References

Bader, M.G., Priest, A.M.: Statistical aspects of fibre and bundle strength in hybrid composites. In: Progress in Science and Engineering of Composites, Proceedings of ICCM-IV, Tokyo, pp. 1129–1136 (1982)

Bono, R., Blanca, M.J., Arnau, J., Gómez-Benito, J.: Non-normal distributions commonly used in health, education, and social sciences: a systematic review. Front. Psychol. **8**, 1602 (2017)

Cordeiro, G.M., Ortega, E.M., da Cunha, D.C.: The exponentiated generalized class of distributions. J. Data Sci. **11**(1), 1–27 (2013)

Friedrich, S., Friede, T.: On the role of benchmarking data sets and simulations in method comparison studies. Biometr. J. (2023)

Hassan, A., Dar, I.H., Lone, M.A.: A new class of probability distributions with an application in engineering science. Pak. J. Stat. Oper. Res. **20**(2), 217–231 (2024)

Kundu, D., Raqab, M.Z.: Estimation of $r = p(y>x)$ for three-parameter weibull distribution. Stat. Probab. Lett. **79**(17), 1839–1846 (2009)

Mahdavi, A., Kundu, D.: A new method for generating distributions with an application to exponential distribution. Commun. Stat. - Theory Methods **46**(13), 6543–6557 (2017)

Mahmood, Z., Chesneau, C.: A new sine-g family of distributions: properties and applications. arXiv preprint arXiv:1903.04338 (2019)

Marshall, A.W., Olkin, I.: A new method for adding a parameter to a family of distributions with application to the exponential and Weibull families. Biometrika **84**(3), 641–652 (1997)

Maurya, S., Kumar, D., Singh, S., Singh, U.: One parameter decreasing failure rate distribution. Int. J. Stat. Econ. **19**(1), 120–138 (2018)

Maurya, S.K., Kaushik, A., Singh, R.K., Singh, S.K., Singh, U.: A new method of proposing distribution and its application to real data. Imp. J. Interdiscip. Res. **2**(6), 1331–1338 (2016)

Panis, S., Schmidt, F., Wolkersdorfer, M.P., Schmidt, T.: Analyzing response times and other types of time-to-event data using event history analysis: a tool for mental chronometry and cognitive psychophysiology. i-Perception **11**(6) (2020)

R Core Team: R: A Language and Environment for Statistical Computing. R Foundation for Statistical Computing, Vienna, Austria (2025)

Rigby, R.A., Stasinopoulos, D.M.: Generalized additive models for location, scale and shape,(with discussion). Appl. Stat. **54**, 507–554 (2005)

Schumacher, F., Lachos, V., Matos, L.: Scale mixture of skew-normal linear mixed models with within-subject serial dependence. Stat. Med. **40**(7), 1790–1810 (2021)

Tejo, M., Araya, H., Niklitschek-Soto, S., Marmolejo-Ramos, F.: Theoretical models of reaction times arising from simple-choice tasks. Cogn. Neurodyn. **13**(4), 409–416 (2019). https://doi.org/10.1007/s11571-019-09532-1

Computational Optimization and Theoretical Foundations of Sample Size in Experimental Designs

Juan Sebastián Ramírez-Ayala, Diana Catalina Hernández-Rojas, Wilmer Pineda-Ríos[(✉)], and Ana María Gomez-Lamus

Escuela Colombiana de Ingeniería Julio Garavito, Bogotá, Colombia
`{ju.ramirez,diana.hernandez-r}@mail.escuelaing.edu.co`,
`{wilmer.pineda,ana.gomez}@escuelaing.edu.co`

Abstract. This work presents a comprehensive framework for determining the optimal number of replications in Completely Randomized Designs (CRD) using both classical and modern computational approaches. Traditional methods based on statistical power, such as those developed by Harris, Hurvitz, and Mood (HHM), and Tukey's method for confidence interval control, are revisited and automated through Monte Carlo simulations in R. The methodology emphasizes the use of the non-central F-distribution and random effects modeling to estimate the empirical power of the test and the required replication size for different design configurations. Additionally, operating characteristic (OC) curves are implemented to visualize the relationship between statistical power and variance ratios under fixed and random effects models. The study also incorporates a budget-constrained optimization approach to allocate replications efficiently across treatments with different costs and variances, using a Lagrangian minimization framework. Overall, this integrated approach enhances experimental design efficiency by combining theoretical rigor with computational simulation, providing practical tools for determining replication numbers under diverse experimental and resource conditions.

Keywords: Experimental design · Completely Randomized Design · statistical power · Monte Carlo simulation · Harris-Hurvitz-Mood method · Tukey method · random effects model · operating characteristic curves · budget constraint · R programming

1 Introduction and Motivation

Determining the optimal sample size, specifically the number of experimental replicates (r), is the cornerstone of any robust experimental design. In the current context, marked by a "reproducibility crisis" in the biological and social sciences, the statistical justification for sample size has shifted from a formal requirement to an imperative for ensuring inferential validity. A design with

© The Author(s), under exclusive license to Springer Nature Switzerland AG 2026
B. M. Suárez et al. (Eds.): R Day 2025, CCIS 2824, pp. 16–42, 2026.
https://doi.org/10.1007/978-3-032-18455-9_2

insufficient statistical power (under-replicated) dramatically increases the false negative rate (Type II error), wasting resources and missed opportunities for discovery. Conversely, an over-replicated design incurs unnecessary economic and ethical costs, particularly in industries where experimental units are expensive or destructible.

Historically, classical approaches to resolving this problem in Completely Randomized Designs (CRDs) have relied on analytical approaches. Seminal methodologies such as those proposed by Harris, Horvitz, and Mood (HHM), or Tukey's confidence interval control, provided the first closed-form solutions to this challenge. However, their implementation typically relied on static tables and operating characteristic curves (OC curves) developed by Pearson and Hartley, which, while theoretically sound, have significant limitations in the modern research environment. These limitations include the difficulty in obtaining precise values, the restriction to fixed significance levels (usually $\alpha = 0.05$), and, crucially, the inability to adapt to complex variance structures or nonlinear budget constraints.

To address these shortcomings, this report presents a theoretical and technical consolidation of the development of the R package dexp and its graphical interface in Shiny. This work responds to the need to integrate classical General Linear Model (GLM) theory with the flexibility of Monte Carlo simulations. Unlike existing tools in the R ecosystem, such as pwr or WebPower, which focus primarily on post-hoc power calculations for standard fixed effects, dexp introduces economic optimization algorithms based on Lagrange multipliers. This allows for calculating the number of replicates under strict budget constraints, considering variable costs per treatment—a critical feature for applications in engineering and agronomy. Throughout this document, the preliminary findings and package documentation are restructured to meet rigorous peer-review standards. The formal notation of GLM for fixed and random effects is explored in greater depth, the computational accuracy of the simulations is validated against analytical solutions, and the software's usability is standardized for an international audience. The ultimate goal is to provide a comprehensive framework that guides researchers from the mathematical foundations to practical decision-making in experimental design.

2   Implemented Methods

2.1   Sample Size Calculation Based on Power

One of the classical methodologies for determining the optimal number of replicates in a Completely Randomized Design (CRD) is based on the calculation of the test's statistical power $(1 - \beta)$, defined as the probability of detecting significant differences among treatments when they truly exist. To apply this method, it is necessary to specify in advance the significance level α, the desired power $(1 - \beta)$, the error variance σ^2, and the minimum detectable difference Δ between treatments [1,3,4,6,7].

The procedure relies on the noncentrality parameter ϕ of the F distribution, which is calculated as:

$$\phi^2 = \frac{r\Delta^2}{2t\sigma^2},\tag{1}$$

where t denotes the number of treatments and r the number of replicates. Because r is initially unknown, an iterative process is required in which successive values of r are proposed, ϕ is computed, and the associated power is evaluated until the desired level is reached.

Traditionally, this calculation was performed using tables or operating characteristic curves derived from the noncentral F-distribution, such as those published by Pearson and Hartley [8] and used in classical experimental design texts [3,6,7]. These tables allow the estimation of the type II error probability (β) and, consequently, the power $1 - \beta$ for different experimental configurations.

In contrast to these tabular methods, the present work automates the process via Monte Carlo simulations, following the theoretical foundations presented by Melo, López, and Melo [4]. Two functions in **R** were implemented that combine the principles of the random-effects model with the empirical estimation of power.

The first function estimates empirical power under the model:

$$Y_{ij} = \tau_i + \varepsilon_{ij}, \qquad \tau_i \sim \mathcal{N}(0,\sigma_\tau^2), \quad \varepsilon_{ij} \sim \mathcal{N}(0,\sigma^2).\tag{2}$$

The second function performs a sequential search to determine the minimum number of replicates r required and also generates a power curve that visualizes the relationship between sample size and achieved power.

2.2 Harris–Hurvitz–Mood (HHM) Method for Determining the Number of Replicates

The Harris–Hurvitz–Mood (HHM) method determines the number of replicates required to obtain significance in a specified proportion of experiments when there are differences greater than a preset value d. It is assumed that the values within each population are normally distributed with a common variance across populations, estimated by S_1^2 with df_1 degrees of freedom.

Given S_1^2, df_1, and d, the required number of replicates at significance level α is obtained from:

$$r = 2 \cdot (df_2 + 1) \cdot \left(\frac{K \cdot S_1}{d}\right)^2,\tag{3}$$

where K is the value from Table A.9 in the appendix of [4], and df_2 is the estimated degrees of freedom of the second estimator of the population variance S_2^2. If r is too small to provide an estimate of df_2, the smallest feasible value of df_2 should be used.

The Constant K as a Standardized Difference. The constant K is defined as the ratio between the specified difference a (between the true mean and the null hypothesis) and the preliminary estimate of the standard deviation s_1:

$$K = \frac{a}{s_1}. \tag{4}$$

In the appendix of Harris et al. (1948) [1], the determination of K is based on solving:

$$P\left[t > t_{2\gamma}(n) - \sqrt{n+1}\,\frac{a}{s_2}\right] = \beta, \tag{5}$$

where t follows a Student's t distribution with n degrees of freedom, s_2 is the standard deviation of the new sample, and $t_{2\gamma}(n)$ is the critical t value for significance level γ.

In their original work, the authors implemented a numerical transformation to evaluate this probability, integrating directly to obtain values for $m = 1, 2, 4, 8, 16, 32$ and $n = 1, 2, 4, 16, 32, 64, 128$, and completing the remaining entries by interpolation.

Computation of k. The function `k_table_mc` uses Monte Carlo simulation. Rather than applying analytical transformations to decouple the dependent distributions of t and $F = s_1^2/s_2^2$, it directly and independently simulates the distributions of s_1 and s_2 from their respective chi-square distributions, assuming unit population variance due to scale invariance.

For each candidate value of K, samples of the sample mean $\bar{x}$ under the alternative hypothesis $\mu = a = K \cdot s_1$ are simulated, and the statistic $t = \sqrt{n+1}\,\bar{x}/s_2$ is computed. The power is empirically estimated as the proportion of replications in which t exceeds the critical value $t_{1-\alpha}(n)$. Finally, a root-finding method (the `uniroot` function in R) is used to find the value of K such that the simulated power equals β (Table 1).

Table 1. Calculated values of the constant k for different degrees of freedom n and m, obtained using a maximum of 1000 iterations.

n	$m=1$	$m=2$	$m=3$	$m=4$	$m=5$	$m=6$	$m=8$	$m=12$	$m=16$	$m=24$	$m=32$
1	57.176	19.442	14.348	12.521	11.610	11.044	10.420	9.814	9.572	9.320	9.192
2	23.975	7.745	5.580	4.771	4.381	4.128	3.859	3.611	3.500	3.389	3.336
3	17.512	5.569	3.975	3.389	3.099	2.924	2.729	2.544	2.456	2.380	2.340
4	14.438	4.587	3.261	2.789	2.543	2.396	2.232	2.081	2.009	1.943	1.910
5	12.638	4.004	2.842	2.433	2.214	2.086	1.942	1.807	1.748	1.690	1.662
6	11.396	3.597	2.561	2.186	1.992	1.875	1.744	1.622	1.569	1.515	1.489
7	10.462	3.297	2.344	2.003	1.826	1.720	1.595	1.487	1.438	1.389	1.366
8	9.720	3.062	2.179	1.861	1.693	1.594	1.484	1.381	1.334	1.292	1.267
9	9.110	2.877	2.043	1.742	1.587	1.495	1.392	1.297	1.252	1.208	1.188
10	8.631	2.715	1.933	1.648	1.502	1.415	1.316	1.222	1.185	1.144	1.125

2.3 Tukey's Method

Tukey's method focuses on controlling the length of the confidence interval for differences between treatment means in CRDs. The goal is to determine the number of replicates r such that a $(1 - \alpha) \times 100\%$ confidence interval has a maximum length of $2d$ for any pair of means.

For t treatments, all pairwise comparisons are considered. The standardized difference between the maximum and minimum mean follows a studentized range distribution:

$$(\bar{Y}_{\max} - \bar{Y}_{\min})/\sqrt{\mathrm{CME}/r} \sim q(t; df_2), \tag{6}$$

where $q(t; df_2)$ denotes the studentized range with t means and df_2 error degrees of freedom.

The probability that the interval length covers all differences less than or equal to $2d$ is given by:

$$P_0 = P\left(\frac{2S\, q_{(1-\alpha)}}{\sqrt{r}} \leq 2d \right), \tag{7}$$

where $S = \sqrt{\mathrm{CME}}$ and $q_{(1-\alpha)}$ is the corresponding percentile of the studentized range distribution. Rearranging:

$$P_0 = P\left(S^2 \leq \frac{d^2 r}{q_{(1-\alpha)}^2} \right). \tag{8}$$

Introducing a prior estimate of the variance S_1^2 with df_1 degrees of freedom yields an F distribution:

$$P_0 = P\left(\frac{S^2}{S_1^2} \leq \frac{d^2 r}{q_{(1-\alpha)}^2 S_1^2} \right), \tag{9}$$

with $S^2/S_1^2 \sim F(df_2; df_1)$. Solving for r gives:

$$r = \frac{F(df_2; df_1; 1 - \alpha)\, S_1^2\, q^2(t; df_2; 1 - \alpha/2)}{d^2}. \tag{10}$$

2.4 Number of Replicates in a Random-Effects Model

Determining the number of replicates in an experimental design with a random-effects model is fundamental for ensuring adequate statistical power in ANOVA tests [2]. The objective is to determine how many observations per treatment are necessary to detect significant differences among levels of a random factor, given the experiment's inherent variability.

Although the test power is based on the F distribution, its determination can be simplified using operating characteristic (OC) curves, which visualize the probability of committing type II errors (β) as a function of different design configurations.

This approach is particularly useful when factors are not directly controlled but randomly sampled from a broader population (e.g., batches, subjects, locations), and one must ensure the design has sufficient sensitivity to detect real effects among treatments or experimental units.

For one-way ANOVA:

$$Y_{ij} = \mu + T_i + \varepsilon_{ij}, \tag{11}$$

where:

- μ: overall mean,
- T_i: treatment effect,
- ε_{ij}: observational error.

Previously, T_i was estimated as if fixed; under random effects:

$$\varepsilon_{ij} \sim \mathcal{N}(0, \sigma_\varepsilon^2), \tag{12}$$

and

$$T_i \sim \mathcal{N}(0, \sigma_T^2). \tag{13}$$

Thus, there are two sources of variability. In a CRD from the random-effects perspective, we no longer test the same hypothesis as in fixed effects.

Hypotheses in Fixed Effects

$$H_0 : T_1 = T_2 = \cdots = T_K = 0. \tag{14}$$

Hypotheses in Random Effects

$$H_0 : \sigma_T^2 = 0 \quad \text{(the variance of the random effects is zero).} \tag{15}$$

The calculation is analogous; only the hypothesis changes. Note that $T_k = 0 \Rightarrow \sigma_T^2 = 0$, but not conversely. If $\sigma_T^2 = 0$, then the T_k are equal, though not necessarily equal to zero.

Decision Between Fixed and Random Effects. In fixed effects, the T_i are treated as parameters directly affecting the response. In random effects, they are viewed as random contributions explaining part of the unexplained variabilit [9].

Example. If a study evaluates crop yield with specific fertilizers, the researcher may select three fixed fertilizer types to compare (fixed effects). If, instead, three fertilizers are randomly chosen from a larger set, the effects are random and the interest lies in the variability component associated with fertilizer type.

From a statistical perspective, hypothesis testing is analogous in both cases. In fixed effects we compare specific levels (e.g., Fertilizer 1 vs. Fertilizer 2), whereas in random effects we assess how much treatment variability contributes to total response variability.

Calculation of the F Statistic. To test the model—both for fixed and random effects—the F statistic is:

$$F = \frac{MS_{\text{Treat}}}{MS_{\text{Res}}}. \tag{16}$$

An important difference is that the numerator involves both σ_T^2 (treatment variability) and σ_ε^2 (experimental error variability), whereas the denominator involves only σ_ε^2.

The treatment mean square in random effects is based on:

$$SS_{\text{Tot}} = SS_{\text{Treat}} + SS_{\text{Res}}. \tag{17}$$

Because sums of squares follow χ^2 distributions:

$$MS_{\text{Treat}} = \frac{SS_{\text{Treat}}}{K-1} \sim \chi^2 \quad \Rightarrow \quad \mathbb{E}(MS_{\text{Treat}}) = \sigma_\varepsilon^2 + r\,\sigma_T^2, \tag{18}$$

where K is the number of treatments, and for the residual:

$$\mathbb{E}(MS_{\text{Res}}) = \sigma_\varepsilon^2. \tag{19}$$

Thus, under H_0 the F statistic follows a central F distribution, while under the alternative ($\sigma_T^2 \neq 0$) it follows a noncentral F distribution.

Power Simulation. For simulation-based calculation:

1. Fix the number of treatments (t).
2. Fix the number of replicates (r).
3. Specify σ_ε^2 and β.
4. Simulate the model:

$$y_{ij} = \mu + T_j + \varepsilon_{ij}, \tag{20}$$

   with

$$T_j \sim \mathcal{N}(0, \sigma_T^2), \qquad \varepsilon_{ij} \sim \mathcal{N}(0, \sigma_\varepsilon^2).$$

5. Fit the ANOVA:

```
anova(y ~ trt)
```

6. Compute the F statistic and determine whether H_0 is rejected.
7. Repeat to estimate empirical power.

Power is estimated as the proportion of simulations in which H_0 is rejected under the alternative hypothesis, across different r values. The relationship between the number of replicates and achieved power can be visualized using OC curves or Monte Carlo simulations.

2.5   Replications Under Budget

It is a method that efficiently allocates experimental resources when treatments differ in variability and cost per observation. It takes into account a budget constraint, aiming not to exceed the available limit, and optimizes the experiment's precision by assigning more replications to treatments that are more variable and less expensive, and fewer to those that are costlier or less dispersed.

Assuming that treatment costs $(c_i > 0)$ are variable, under the constraint:

$$\sum_{i=1}^{t} c_i r_i = C \quad \text{(Budget constraint)} \tag{21}$$

and considering that:

$$Var(\text{MELI}(L)) = \sum_{i=1}^{t} \frac{\lambda_i^2 \sigma_i^2}{r_i} \quad \text{(Equation 5.25 in Melo.M et al., 2007)} \tag{22}$$

where $L = \sum_{i=1}^{t} \lambda_i \mu_i$ and

$$\text{MELI}(L) = \sum_{i=1}^{t} \lambda_i \bar{y}_i \quad \text{(Equation 5.26 in Melo.M et al., 2007)} \tag{23}$$

Lagrangian Minimization. To minimize $Var(\text{MELI}(L))$ subject to the budget constraint, we construct the Lagrangian:

$$Q = \sum_{i=1}^{t} \frac{\lambda_i^2 \sigma_i^2}{r_i} + \phi \left(\sum_{i=1}^{t} c_i r_i - C \right) \quad \text{(Equation 5.27 in Melo.M et al., 2007)} \tag{24}$$

By solving the above, we obtain:

$$r_i = \frac{|\lambda_i| \sigma_i}{\sqrt{\phi c_i}} \quad \text{(Equation 5.28 in Melo.M et al., 2007)} \tag{25}$$

where the Lagrange multiplier is:

$$\phi = \frac{\left(\sum_{i=1}^{t} |\lambda_i| \sigma_i \sqrt{c_i} \right)^2}{C^2} \tag{26}$$

Recommendation: The coefficients λ_i should be expressed as fractions to simplify the calculation of r_i.

2.6 Replications by Variability

It is a statistical procedure that allocates a fixed total number of observations across different experimental treatments based solely on their variability. Replications are assigned proportionally to the estimated standard deviations: treatments with higher variability receive more replications, while more homogeneous ones receive fewer. Its main purpose is to optimize experimental precision while keeping the total sample size constant, without considering factors such as costs or resource availability.

Building on the procedure from the "Replications under budget" section, the following can be obtained:

2.7 Proportional Allocation (Equation 5.29 in Melo.M et al., 2007)

If, instead of optimizing based on costs, we allocate the observations proportionally to the standard deviations (with a fixed n):

$$\tilde{r}_i = \frac{n\sigma_i}{\sum_{s=1}^{t}\sigma_s}; \quad i = 1,\ldots,t \quad \text{(Equation 5.29 in Melo.M et al., 2007)} \quad (27)$$

Interpretation: Treatments with higher variability (σ_i) receive more replications.

2.8 Assignment of Treatments and Sample Sizes with Cost Function and Maximum Variance

This model aims to determine the optimal combination of treatments and replications, balancing statistical precision and economic efficiency. Its goal is to minimize total costs while maintaining a fixed variance in the mean estimate, taking into account variability both between and within treatments. This approach allows resources to be allocated optimally based on associated costs, providing a quantitative tool for efficiently planning experiments.

In the variance components model, both the number of treatments t and the number of replications r are variables, and their estimates are linked to the control of these variances. A common criterion for choosing r and t is to minimize the costs in estimating the mean μ. A measure of the amount of information available to estimate μ is the variance of the sample mean, given by:

$$v(\bar{y}..) = \frac{\sigma^2}{rt} + \frac{\sigma_A^2}{t} \qquad (28)$$

In the case of a completely randomized (C.R.) design, the problem reduces to finding the values of r and t that minimize the cost function given by:

$$C = C_1 t + C_2 tr \qquad (29)$$

for a fixed variance $v(\bar{y}..)$, where C_1 is the cost per treatment and C_2 is the cost per experimental unit. The mathematical solution, according to [5], is:

$$t = \frac{1}{v(\bar{y}..)} \left(\hat{\sigma}_A^2 + \sqrt{\frac{\hat{\sigma}_A^2 \hat{\sigma}^2 C^2}{C^1}} \right) \tag{30}$$

and

$$r = \sqrt{\frac{\hat{\sigma}^2 C_1}{\hat{\sigma}_A^2 C_2}} \tag{31}$$

3 Examples and Visualizations

3.1 Example of Sample Size Calculation Based on Power

The data used in this function are based on **Example 5.1** presented in [4], where the effectiveness of four diets (D1, D2, D3, D4) on weight loss in people attending a gym is evaluated. In that study, 20 people were randomly assigned (5 per diet) and weight difference was measured after 15 days. From the obtained results (Table 5.2 of the book), a combined variance $\hat{\sigma}^2 = 10.35$ was calculated, derived from the individual variances of each diet: 24.5 (D1), 3.87 (D2), 6.08 (D3) and 6.99 (D4).

Subsequently, in **Example 5.9** of the same book, this variance estimate is used to determine the number of replicates required in a future experiment under similar conditions. Specifying a minimum detectable difference $\Delta = 3$ kg, a desired power of 0.80, and a significance level $\alpha = 0.05$, an iterative calculation was performed that led to $r = 27$ replicates per treatment.

The `find_minimum_r_power()` function implemented in this work generalizes this approach using Monte Carlo simulation, allowing to vary parameters such as the number of treatments (t) and the coefficient ρ (ratio between treatment variance and error variance in random effects models). For $t = 5$ and $\rho = 0.4$, the function estimates that 14 replicates are required to achieve a power of 82.8%, demonstrating the flexibility of the method to adapt to different experimental scenarios.

R code for `find_minimum_r_power()`:

```
library(dexp)
# Find minimum r for desired power of 0.8
results <-
  find_minimum_r_power(t = 4,
                       rho = 0.5,
                       diff = 3,
                       target_power = 0.8,
                       sigma2 = 10.35)
results$optimal_r
[1] 21
results$plot
```

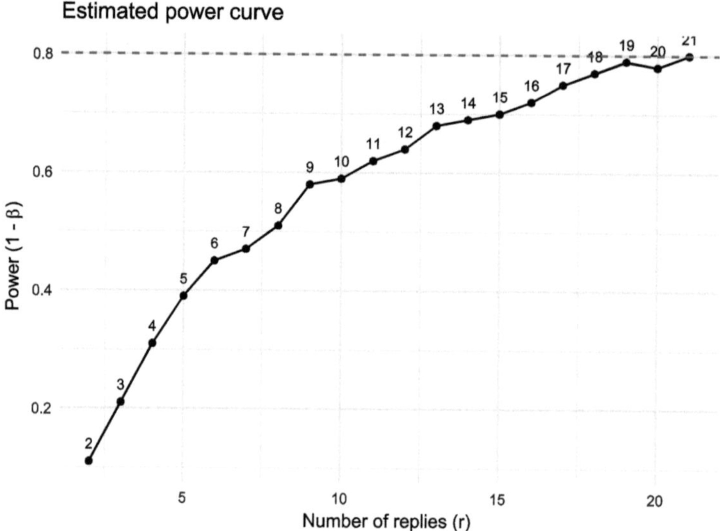

Fig. 1. Power curve.

The output from the `find_minimum_r_power()` function indicates that, under conditions similar to Example 5.1 (four diets with a pooled variance $\hat{\sigma}^2 = 10.35$ and a minimum detectable difference of $3\,\mathrm{kg}$), **21 replicates per treatment** are needed to achieve the target power of 80%. This result aligns with the iterative calculation in Example 5.9 of the textbook, which also yielded a large replicate number ($r = 27$) for the same variance and power. The Monte Carlo simulation provides a more precise and adaptable estimate, confirming that a substantial sample size is required to reliably detect the specified difference given the observed variability. The accompanying power curve (Fig. 1) illustrates how statistical power increases with the number of replicates, aiding in the practical trade-off between experimental effort and sensitivity.

3.2 Example Harris–Hurvitz–Mood (HHM) Method for Determining the Number of Replicates

Example 5.11 from [4] presents an experiment on the amount of fat absorbed by 64 donuts with different types of fat. In this study, an estimate of the experimental variance $S_1^2 = 141.6$ with $df_1 = 40$ degrees of freedom was obtained. Assuming a minimum detectable difference $d = 20$ and a desired power of 80% ($1 - \beta = 0.8$) for $t = 6$ treatments, the manual calculation using the HHM method yields With $df_2 = 60$ we obtain $r = 4.48$, while with $df_2 = 25$ we obtain $r = 4.64$. In both cases, the result is rounded to 5 replicates per treatment.

The implemented `calculate_minimum_r_hhm()` function reproduces this calculation in an automated and iterative manner, obtaining consistent results:

R code for `calculate_minimum_r_hhm()`:

```r
# Find minimum r for desired power 0.8
results <- calculate_minimum_r_hhm(
  t = 6,
  d = 20,
  initial_r = 3,
  s1_sq = 141.6,
  df1 = 40,
  alpha = 0.05,
  beta = 0.8
)


results$r        # estimated number of replicates
[1] 5
results$k        # value of k = a / s1 obtained by simulation
[1] 0.5348283
results$df2      # degrees of freedom of the final error
[1] 22
```

The function confirms that 5 replicates are required to achieve the desired power, matching the conclusion from the book example. The value $K = 0.5348$ obtained through Monte Carlo simulation replaces the tabled values (0.322 and 0.502) used in the manual calculation, providing a more precise estimate tailored to the specific error degrees of freedom ($df_2 = 22$ instead of 60 or 25).

3.3 Example Tukey's Method

Example 5.12 from [4] revisits the parameters from Example 5.11 ($S_1^2 = 141.6$, $df_1 = 40$, $d = 20$) but applies Tukey's method to determine the number of replicates considering confidence intervals for multiple comparisons. With $t = 6$ treatments, $1 - \beta = 0.9$, and initially assuming $df_2 = 30$, the critical values $F(30; 40; 0.10) = 1.54$ and $q(5; 30; 0.10) = 4.30$ are obtained. Using equation (5.24) from the book:

$$r = \frac{S_1^2 \cdot q^2 \cdot F}{d^2} = \frac{141.6 \cdot (4.30)^2 \cdot 1.54}{400} = 10.08$$

Since the df_2 value was underestimated, an iterative adjustment is made taking $r = 9$, which leads to $df_2 = 48$, $q(5; 48; 0.90) = 4.2$, $F(48; 40; 0.10) = 1.48$, and a new calculation of $r = 9.2$. Finally, it is concluded that $r = 10$ replicates are sufficient to obtain a 95% confidence interval that exceeds $2d$ in 90% of the experiments.

The `calculate_r_tukey()` function implements this procedure automatically, performing iterations until converging to the optimal number of replicates:

R code for `calculate_r_tukey()`:

```
# Calculation of the number of replicates using Tukey's method
results <- calculate_r_tukey(
  t = 6,
  d = 20,
  initial_r = 6,
  s1 = sqrt(141.6),
  df1 = 40,
  alpha = 0.05,
  beta = 0.1
)

results$r
[1] 10

results$a
[1] 9.488479

results$r_i
[1] 9

results$a_i
[1] 10.11104

results$r_list
 [1]  5  6  7  8  9 10 11 12 13 14 15

results$a_values
 [1] 15.081301 13.172866 11.857586 10.878156 10.111039
 [6]  9.488479  8.969733  8.528597  8.147330  7.813420
[11]  7.517754

results$chosen_position
[1] 5
```

The value `r` is the number of replicates computed using **Tukey's method**, where `A` is the minimum difference one aims to detect with that sample size. In contrast, `r_i` is the number of replicates whose value most closely achieves the required minimum detectable difference `A_i`. The lists `list_r` and `values_A` contain candidate replicate counts and their corresponding detectable minimum differences, computed with `r` as reference.

According to Tukey's method, to detect a minimum difference of 20 units, a replicate size of 10 is needed. However, this implies that the real detectable difference could be slightly smaller (e.g., 9.49). This can be interpreted as the test being stricter for rejecting H_0 when using $r = 10$. In contrast, using `r_i` provides a slightly more flexible approach that may facilitate detecting the desired difference while balancing type I and type II errors.

3.4   Example Number of Replicates in a Random-Effects Model

Example 5.14 from [4] revisits the data from Example 5.5 (genetic study with bulls) to calculate the power of the F test in a random effects model. With parameters $\alpha = 0.05$, $\nu_1 = 4$, $\nu_2 = 30$, $\hat{\sigma}_A^2 = 158.015$, $\hat{\sigma}^2 = 416.21409$, and $r_0 = 6.97$, a parameter $\lambda = 1.91$ is obtained, corresponding to a power of 0.50. Since this value is considered low, the example proceeds to iteratively determine the number of replicates required to achieve a power of 0.80. Through a trial-and-error process using operating characteristic curves, it is concluded that 12 replicates per treatment are needed.

The `find_minimum_r()` function implements an analogous procedure but using Monte Carlo simulation, generalizing the method for different values of t (number of treatments) and ρ (ratio between treatment variance and error variance). For $t = 5$ and $\rho = 0.4$, the function estimates that 14 replicates are needed to achieve a power of 82.8%, as shown below (Fig. 2):

R code for `find_minimum_r()`:

```
results <-
  find_minimum_r(t = 5, rho = 0.4, target_power = 0.8)
results$optimal_r
[1] 14
results$power
[1] 0.828
results$plot
```

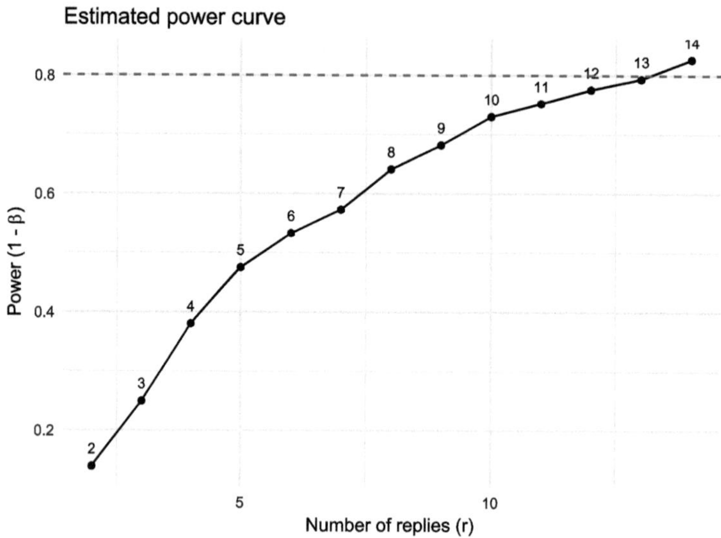

Fig. 2. Power curve.

In an experiment similar to the genetic study with bulls from Example 5.5, but assuming that genetic variability explains 40% of the total variability ($\rho = 0.4$), **14 calves per bull** would be required (70 calves total for 5 bulls) to achieve a statistical power of **82.8%** in detecting significant differences between sires.

3.5   Example Replications Under Budget

Suppose a farmer has the following data on the yield (in tons per hectare) of four sugarcane varieties and their respective costs (taken from [4]) (Tables 2 and 3):

Table 2. Experiment data: observed values, means, standard deviations, and costs per treatment

	V1	V2	V3	V4
1	78.82	56.60	105.126	96.89
2	86.80	63.82	112.940	90.91
3	68.65	58.71	108.118	92.97
4	77.76	70.59	121.105	97.98
5	75.80	81.74	115.870	95.93
Mean	**76.70**	**67.40**	**109.10**	**92.80**
Standard deviation	**6.27**	**9.57**	**12.00**	**3.32**
Costs (COP)	**1000**	**200**	**700**	**1100**

Table 3. Total budget allocated to the experiment

Concept	Value (COP)
Total experiment budget	50.000

We observe that there is a cost for each treatment and a total budget, so according to (5.28 in Melo.M et al., 2007), the appropriate sample sizes for each variety would be:

Starting with the calculation of Φ

$$\Phi = \sqrt{\frac{\left(\frac{1}{4} \cdot 6.27 \cdot \sqrt{1000} + \frac{1}{4} \cdot 9.57 \cdot \sqrt{200} + \frac{1}{4} \cdot 12 \cdot \sqrt{700} + \frac{1}{4} \cdot 3.32 \cdot \sqrt{1100}\right)^2}{50000^2}}$$

$$= 1.448629 \times 10^{-5} \tag{32}$$

$$r_1 = \frac{\frac{1}{4} * 6.27}{\sqrt{1.448629 \times 10^{-05} * 1000}} = 13.0235 \approx 13 \tag{33}$$

$$r_2 = \frac{\frac{1}{4} * 9.57}{\sqrt{1.448629 \times 10^{-05} * 200}} = 44.4486 \approx 44 \tag{34}$$

$$r_3 = \frac{\frac{1}{4} * 12}{\sqrt{1.448629 \times 10^{-05} * 700}} = 29.7915 \approx 30 \tag{35}$$

$$r_4 = \frac{\frac{1}{4} * 3.32}{\sqrt{1.448629 \times 10^{-05} * 1100}} = 6.5751 \approx 7 \tag{36}$$

The replications assigned to each of the treatments are $13, 44, 30, 7$ respectively.

R Code for Allocating Sample Sizes per Treatment Under a Budget Constraint

```
library(dexp)
# Determine the number of replications for four treatments,
given the standard deviations and costs per treatment,
along with the total budget.

sigmas <- c(6.27, 9.57, 12, 3.32)
costs <- c(1000, 200, 700, 1100)
total_cost <- 50000
results<-
```

```
proportionality_with_cost_nor_sample_size(a = 4,
sigmas=sigmas,
costs=costs,
total_cost=total_cost)
[1] 13 44 30 7
```

For a farmer with a fixed budget of 50,000 COP to compare four sugar-cane varieties, the economically optimal experiment requires planting 44 plots of Variety 2, 30 plots of Variety 3, 13 plots of Variety 1, and 7 plots of Variety 4.

3.6 Example Replications by Variability

Suppose a farmer has the following data on the yield (in tons per hectare) of four sugarcane varieties (taken from [4]) (Table 4):

Table 4. Summary of means and standard deviations by variable

	V1	V2	V3	V4
1	78.82	56.60	105.126	96.89
2	86.80	63.82	112.940	90.91
3	68.65	58.71	108.118	92.97
4	77.76	70.59	121.105	97.98
5	75.80	81.74	115.870	95.93
Mean	**76.70**	**67.40**	**109.10**	**92.80**
Standard deviation	**6.27**	**9.57**	**12.00**	**3.32**

It is observed that there is a proportionality in the standard deviations, so according to (5.29), the appropriate sample sizes for each variety would be:

$$\tilde{r}_1 = \frac{(5*4)*(6.27)}{6.27 + 9.57 + 12 + 3.32} = 4.0243 \approx 4 \tag{37}$$

$$\tilde{r}_2 = \frac{(5*4)*(9.57)}{6.27 + 9.57 + 12 + 3.32} = 6.1424 \approx 6 \tag{38}$$

$$\tilde{r}_3 = \frac{(5*4)*(12)}{6.27 + 9.57 + 12 + 3.32} = 7.7021 \approx 8 \tag{39}$$

$$\tilde{r}_4 = \frac{(5*4)*(3.32)}{6.27 + 9.57 + 12 + 3.32} = 2.1309 \approx 2 \tag{40}$$

The replications assigned to each of the treatments are $4, 6, 8, 2$, respectively.

R Code for Sample Size Allocation per Treatment by Variability

```
library(dexp)
# Determine the number of replications for four treatments,
given an initial five replications and the standard deviations
for each treatment.

sigmas <- c(6.27, 9.57, 12, 3.32) #Dev. est. by treatment
results<-
proportionality_without_cost_nor_sample_size(a = 4,
r0 = 5,
sigmas=sigmas)
[1] 4 6 8 2
```

For a farmer comparing four sugarcane varieties without a specific budget constraint but aiming to account for their differing yield variability, the optimal experiment requires planting 8 plots of Variety 3, 6 plots of Variety 2, 4 plots of Variety 1, and 2 plots of Variety 4.

3.7 Example Assignment of Treatments and Sample Sizes with Cost Function and Maximum Variance

A genetic study in cattle involved randomly selecting several males (bulls) and mating them with separate groups of females. At birth, the calves' weights were measured as part of a study on heritable weights (progeny studies). The table below shows the birth weights of the calves from each of the five mating groups (taken from [4]) (Table 5).

Table 5. Weight data (in kg) recorded for different bulls across eight observations

	Bull_85	Bull_113	Bull_134	Bull_158	Bull_165
1	61	75	58	57	59
2	71	102	60	121	46
3	56	95	59	56	120
4	75	103	65	58	115
5	99	98	54	101	93
6	80	115	57	110	105
7	75	–	–	67	75
8	62	–	–	–	115

Generating the ANOVA table (Table 6)

Table 6. ANOVA table for the effect of the bull on the response variable. *Significance codes: 0 '***' 0.001 '**' 0.01 '*' 0.05 '.' 0.1 ' ' 1*

Source of variation	Df	Sum of Squares	Mean Square	F	Pr(>F)
Bull	4	6070	1517.6	3.646	0.0155*
Residuals	30	12486	416.2		

From the previous table, we can obtain the values of $\hat{\sigma}_e^2$ and $\hat{\sigma}_A^2$.

$$\hat{\sigma}_e^2 = \text{CME} = 416.2 \tag{41}$$

$$\hat{\sigma}_A^2 = \frac{\text{CMA} - \text{CME}}{r_0} = \frac{(1517.6 - 416.2)}{6.97} = 158.02 \tag{42}$$

where,

$$r_0 = \frac{\left(35 - \frac{(8^2 + 6^2 + 6^2 + 7^2 + 8^2)}{35}\right)}{4} = 6.97 \tag{43}$$

Assuming a maximum variance $V(\bar{y}_{..}) = 43.49$, with $C_1 = 150,000$ and $C_2 = 50,000$, we find that:

$$t = \frac{1}{43.49}\left(158.015 + \sqrt{\frac{(158.015)(416.21409)(50000)}{150000}}\right) = 7.04 \tag{44}$$

and

$$r = \sqrt{\frac{(416.21409)(150000)}{(158.015)(50000)}} = 3.35 \tag{45}$$

For a sample mean variance not exceeding 43.49, 7 bulls and 3 calves per bull should be selected, assuming an experimental cost of 150,000 per bull and 50,000 per calf.

R Code for Assigning Treatments and Sample Sizes Using a Cost Function and Maximum Variance

```
library(dexp)
# Determine the number of treatments and the sample size per treatment,
considering a cost of 150,000 per treatment, 50,000 per experimental
unit, a within-treatment variance of 416.21, a total variance
proportion of 0.3796, and a maximum tolerable variance for
the sample mean of 43.49.
```

```
results <-
number_of_treatments_and_replicates_with_random_effects (
treatment_cost = 150000,
eu_cost = 50000,
sigma_square = 416.21,
rho = 0.3796,
v_max = 43.49 )
results$num_of_treatments
[1] 7

results$num_of_replicas
[1] 3
```

For a cattle genetics study aiming to estimate heritable birth weight with a maximum tolerable variance for the sample mean of $43.49\,\mathrm{kg}^2$, and given the costs of 150,000 per bull and 50,000 per calf, the optimal experimental design requires sampling 7 bulls and measuring 3 calves per bull.

4   Comparison with Existing Tools

The 'dexp' R package and its accompanying Shiny application represent a specialized advancement in the R ecosystem for experimental design, distinct from general-purpose tools like 'pwr' or 'WebPower'. While these existing packages excel at providing analytical power calculations for standard statistical tests, 'dexp' addresses a crucial gap by modernizing and automating classical, design-specific methodologies—such as the Harris-Horvitz-Mood (HHM) method and Tukey's confidence interval approach—through flexible Monte Carlo simulations. This allows for precise sample size determination beyond the constraints of static tables, adapting to arbitrary significance levels and complex variance structures inherent in Completely Randomized Designs (CRDs) for both fixed and random effects models.

The most significant innovation of 'dexp' is its integration of economic and resource optimization directly into the design process—a feature largely absent from other R packages. Beyond asking "how many replicates for a target power?", 'dexp' can solve "how should I allocate a fixed budget across treatments with differing costs and variabilities to maximize precision?". By implementing Lagrangian optimization.

5   Appendix

Theoretical Foundations: The General Linear Model (GLM)

The foundation of any discussion on sample size must be the formal notation of the General Linear Model. This provides the necessary framework to distinguish between fixed and random effects approaches and to understand how statistical power is derived from variance components.

Matrix Notation for the Completely Randomized Design. The Completely Randomized Design (CRD) is the simplest experimental structure, where treatments are assigned to experimental units in a purely random manner, without blocking constraints. Let $\mathbf{y}$ be an $N \times 1$ observation vector, where $N = \sum_{i=1}^{t} r_i$ is the total number of experimental units, t is the number of treatments, and r_i is the number of replicates for treatment i.

The linear model is defined as:

$$\mathbf{y} = \mathbf{X}\boldsymbol{\beta} + \boldsymbol{\epsilon}$$

where:

- $\mathbf{y}$ is the vector of observed responses.
- $\mathbf{X}$ is the design (or incidence) matrix of dimension $N \times p$, composed of zeros and ones, indicating which treatment was applied to each unit.
- $\boldsymbol{\beta}$ is the vector of unknown parameters of dimension $p \times 1$ (treatment means or effects).
- $\boldsymbol{\epsilon}$ is the vector of random errors of dimension $N \times 1$.

This matrix formulation is crucial because it generalizes power calculations beyond simple balanced cases. Parameter estimation and the sum of squares, which feed the F statistic, depend on the orthogonal projection of $\mathbf{y}$ onto the column space of $\mathbf{X}$.

The Fixed Effects Model vs. Random Effects Model. A critical distinction for sample size calculation, often underestimated in simplified software calculators, is the nature of treatment effects.

Fixed Effects Model. In a fixed effects model, the levels of the treatment are deliberately selected by the experimenter (e.g., three specific fertilizer doses: 0, 50, 100 kg/ha). The scalar model is written as:

$$y_{ij} = \mu + \tau_i + \epsilon_{ij}$$

Under this model, we assume $\sum \tau_i = 0$ (under the zero-sum restriction) and that errors are independent and identically distributed (i.i.d.) with a normal distribution: $\epsilon_{ij} \sim \mathcal{N}(0, \sigma^2)$.

The null hypothesis of interest is:

$$H_0 : \tau_1 = \tau_2 = \cdots = \tau_t = 0$$

Under H_0, the F statistic follows a central F distribution. However, if H_0 is false, the statistic follows a non-central F distribution, whose shape depends on the non-centrality parameter λ. It is this dependency that allows for the calculation of power and, inversely, the required sample size.

Random Effects Model. In a random effects model, the treatment levels constitute a random sample from a larger population of possible levels (e.g., 5 bulls randomly selected from a breed to test weight inheritance, or 3 production batches selected from a year of manufacturing). The model changes conceptually:

$$y_{ij} = \mu + \tau_i + \epsilon_{ij}$$

Here, τ_i is a random variable, not a fixed parameter. We assume:

$$\tau_i \sim \mathcal{N}(0, \sigma_\tau^2), \quad \epsilon_{ij} \sim \mathcal{N}(0, \sigma^2), \quad \text{and } \tau_i \text{ and } \epsilon_{ij} \text{ are independent.}$$

The null hypothesis changes fundamentally. We no longer test whether the means are equal, but whether the treatment variance is null:

$$H_0 : \sigma_\tau^2 = 0$$

Determining the number of replicates in this scenario seeks to ensure sufficient degrees of freedom to estimate the variance component σ_τ^2 with adequate precision. As detailed in the dexp package, this requires different simulation approaches, since the distribution of the test statistic under the alternative hypothesis depends on the variance ratio $\rho = \sigma_\tau^2/\sigma_\epsilon^2$.

The Non-central F Distribution and Statistical Power

To address the methodological request to delve deeper into power calculation mechanisms, this section is dedicated to the non-central F distribution (λ), the mathematical engine behind the `find r minimum Power` function in the dexp package.

The Non-Centrality Parameter (λ). Statistical power $(1 - \beta)$ is defined as the probability of rejecting H_0 given that a specific alternative hypothesis is true. In the context of a one-factor ANOVA (CRD), this probability is the integral of the right tail of the non-central F distribution.

The non-centrality parameter λ encapsulates the effect size and the design size. It is formally defined as:

$$\lambda = \sum_{i=1}^{t} r_i \frac{\tau_i^2}{\sigma^2}$$

This equation reveals three fundamental "levers" that the researcher can manipulate:

1. **Effect magnitude (τ_i):** The greater the discrepancy between treatment means and the grand mean, the larger λ and the higher the power.
2. **Error variance (σ^2):** λ is inversely proportional to experimental variance. Reducing experimental noise (through local control or better instruments) increases power.

3. **Number of replicates** (r_i): Crucially, λ scales linearly with r. This monotonic relationship is what allows the dexp package algorithms to perform an efficient iterative search.

Alternatively, expressed in terms of the minimum detectable difference Δ (the difference between the two most extreme means), and assuming a balanced design $(r_i = r)$, the parameter is often denoted as ϕ in classical texts like Montgomery or Kempthorne:

$$\phi = \frac{1}{2}\frac{\Delta}{\sigma}$$

It is vital to note that the relationship between λ and ϕ varies by author (generally $\lambda = t\phi^2$ or similar). The dexp package internalizes these conversions to avoid the common confusion arising from using old tables.

Sequential Search Algorithm. Since r is discrete and unknown at the start, it cannot be algebraically solved from the cumulative distribution function (CDF) of the non-central F. The package implements a numerical approach:

1. **Initialization:** An initial $r_{initial}$ is proposed (e.g., $r = 2$).
2. **Calculation of λ:** Using the input parameters $(\Delta, \sigma^2, t, r_{actual})$, λ is calculated.
3. **Power Estimation:** $P(F > F_{crit}|\lambda)$ is computed.
4. **Stopping Criterion:** If Power $\geq (1-\beta)_{target}$, it stops. Otherwise, $r \leftarrow r+1$ and it repeats.

This automated cycle eliminates the human error associated with visual interpolation in printed OC curves.

Economic Optimization: The Lagrangian Approach

One of the most novel and relevant contributions of the dexp package, is optimization under budget constraints.

The Resource Allocation Problem. In industrial and agricultural experimentation, the assumption of homogeneous costs is rarely valid. Consider the case study presented in the supplementary data on sugarcane varieties:

- **Variety 2:** High standard deviation ($\sigma = 9.57$), Low cost (200 COP).
- **Variety 4:** Low standard deviation ($\sigma = 3.32$), High cost (1100 COP).

A standard design would assign the same number of replicates (r) to both treatments. This is inefficient: excessive resources are spent measuring a stable and expensive variety (V4), while a cheap and noisy variety (V2) is under-sampled, compromising the overall precision of the comparison.

Mathematical Formulation of the Optimization. The objective is to minimize the variance of the mean estimator (or of a linear contrast), subject to a total budget constraint C.

Objective Function (Minimize Variance):

$$\min_{r_i} \sum_{i=1}^{t} \frac{\sigma_i^2}{r_i}$$

Constraint (Budget):

$$\sum_{i=1}^{t} c_i r_i = C$$

To solve this, we use the method of Lagrange multipliers. We construct the Lagrangian function $\mathcal{L}$:

$$\mathcal{L}(r_1, \ldots, r_t, \phi) = \sum_{i=1}^{t} \frac{\sigma_i^2}{r_i} + \phi \left(\sum_{i=1}^{t} c_i r_i - C \right)$$

Taking partial derivatives with respect to each r_i and setting them to zero:

$$\frac{\partial \mathcal{L}}{\partial r_i} = -\frac{\sigma_i^2}{r_i^2} + \phi c_i = 0$$

Solving for r_i, we obtain the optimal proportionality relationship:

$$r_i \propto \frac{\sigma_i}{\sqrt{c_i}}$$

Interpretation and Application in Dexp. The analytical solution indicates that the optimal number of replicates for treatment i is directly proportional to its variability (σ_i) and inversely proportional to the square root of its cost ($\sqrt{c_i}$).

The `proportionality with cost or sample size` function in the package automates the calculation of the Lagrange multiplier ϕ to satisfy the exact equality budget constraint:

$$r_i = \phi \frac{\sigma_i}{\sqrt{c_i}}$$

Case Study Result: Applying this algorithm to the sugarcane data, the package recommends:

- **Variety 2 (High variance, low cost):** 44 replicates.
- **Variety 4 (Low variance, high cost):** 7 replicates.

This unequal allocation maximizes statistical information per monetary unit invested, a massive strategic advantage over traditional balanced designs.

Appendix B: Notation and Symbols

This appendix provides a comprehensive list of the mathematical symbols and statistical parameters used throughout the document.

Statistical Parameters and Variables

(See Table 7).

Table 7. Mathematical symbols and their definitions.

Symbol	Meaning	Description
α	Significance level	Probability of Type I error (false positive), typically set at 0.05 or 0.10
β	Type II error rate	Probability of failing to reject a false null hypothesis; $1 - \beta$ is the statistical power
γ	Significance level for t tests	Used in the HHM method to denote the significance level for Student's t distribution
Δ	Minimum detectable difference	The smallest difference between treatment means considered scientifically important
d	Half-length of confidence interval	In Tukey's method, the maximum allowable half-length of a confidence interval for mean differences
ε_{ij}	Experimental error	Random variation associated with the j-th observation in the i-th treatment, assumed $\varepsilon_{ij} \sim \mathcal{N}(0, \sigma_\varepsilon^2)$
μ	Overall mean	The grand mean of all observations in the experiment
ρ	Ratio of variances	$\rho = \sigma_T^2/\sigma_\varepsilon^2$; measures the proportion of total variability attributable to treatments in random-effects models
σ^2	Error variance	Variance of the experimental error; denoted σ_ε^2 in random-effects models
σ_A^2, σ_T^2	Treatment variance	Variance component due to random treatment effects in a random-effects model
σ_ε^2	Error variance	Variance component due to experimental error in a random-effects model
σ_y^2	Total variance	Sum of treatment and error variances: $\sigma_y^2 = \sigma_T^2 + \sigma_\varepsilon^2$
τ_i	Treatment effect (fixed)	The effect of the i-th treatment in a fixed-effects model
ϕ	Noncentrality parameter	Parameter of the noncentral F distribution; $\phi^2 = r\Delta^2/(2t\sigma^2)$ in power calculations
a	Specified mean difference	In the HHM method, the difference between the true mean and the null hypothesis value
C	Total budget	Total available resources for the experiment in cost-constrained designs
c_i	Cost per observation	Cost associated with one experimental unit in treatment i
df_1	Degrees of freedom (prior)	Degrees of freedom associated with a prior variance estimate S_1^2
df_2	Degrees of freedom (error)	Error degrees of freedom in the planned experiment; $df_2 = t(r - 1)$ in a balanced CRD
F	F statistic	Ratio of treatment mean square to error mean square in ANOVA
K, k	Standardized difference	$K = a/s_1$; a constant tabulated in the HHM method that depends on α, β, n, and m
n	Sample size per treatment	Number of observations (replicates) per treatment; often denoted r in experimental design

(continued)

Table 7. (*continued*)

Symbol	Meaning	Description
q	Studentized range	Critical value from the studentized range distribution used in Tukey's method
r	Number of replicates	Number of experimental units (observations) per treatment in a balanced design
r_i	Replicates for treatment i	Number of replicates assigned to the i-th treatment in unbalanced designs
r_0	Adjusted replicate number	$r_0 = [n - \sum r_i^2/n]/(t-1)$; a weighting factor for unbalanced data
S_1^2	Prior variance estimate	An estimate of the error variance obtained from previous or pilot experiments
S_2^2	Future variance estimate	The variance estimate that will be obtained from the planned experiment
t	Number of treatments	The number of treatments (or levels of the factor) in the experiment
$t_{1-\alpha}(n)$	Critical t value	The $(1-\alpha)$ quantile of Student's t distribution with n degrees of freedom
T_i	Treatment effect (random)	Random effect of the i-th treatment in a random-effects model; $T_i \sim \mathcal{N}(0, \sigma_T^2)$
Y_{ij}	Response variable	The observed value for the j-th replicate of the i-th treatment
$\bar{Y}_{\max}, \bar{Y}_{\min}$	Maximum and minimum means	The largest and smallest treatment sample means

6 Conclusions and Future Work

Implications for Industry and Research

The transition from sample size calculations based on "rules of thumb" (heuristics) to data-driven optimizations has profound implications:

- **Cost Efficiency (Stakeholder: R&D Managers):** The ability of `dexp` to optimize under budget constraints transforms experimental design from a fixed cost center to a resource optimization problem. In the sugarcane example, reallocating replicates allows maintaining the same statistical precision while reducing total cost, or increasing precision for the same price.
- **Reproducibility:** By facilitating access to rigorous methods such as HHM and power simulations, the tool lowers the barrier for researchers to justify their appropriate N. This directly combats p-hacking and underpowered studies.

Limitations of the Current Study

It is crucial to recognize, as good scientific practice, the boundaries of the current software:

1. **Error Independence:** The current framework assumes errors are independent ($\mathrm{Cov}(\epsilon_i, \epsilon_j) = 0$). In experiments with repeated measures or spatial structures (common in agronomy), this assumption is violated.
2. **Normality:** Although Monte Carlo simulation can handle non-normality, the current default functions generate Gaussian errors.

Future Work and Research Directions

- **Integration with Generalized Linear Models (GLM):** Extend the simulation to handle count (Poisson) or binary (Logistic) responses, vital for clinical and ecological trials.
- **Correlation Structures:** Incorporate covariance matrices for repeated measures designs, using packages such as `nlme` as a simulation backend.
- **Validation with Real Industrial Datasets:** Conduct a meta-analysis comparing the sample sizes suggested by `dexp` with those historically used in high-impact publications to quantify the current "waste" or "risk" in the literature.

References

1. Harris, M., Horvitz, D.G., Mood, A.M.: On the determination of sample sizes in designing experiments. J. Am. Stat. Assoc. **43**, 391–402 (1948)
2. Hinkelmann, K., Kempthorne, O.: Design and Analysis of Experiments, Volume 1: Introduction to Experimental Design, vol. 1. Wiley, New York (2007)
3. Kempthorne, O.: The Design and Analysis of Experiments. Wiley, New York (1952)
4. Melo M., O.O., López P., L.A., Melo M., S.E.: Diseño de Experimentos: Métodos y Aplicaciones, 1 edn. Universidad Nacional de Colombia, Facultad de Ciencias, Bogotá, D.C., Colombia (2007). Primera edición. Diagramación y diseño interior en LATEX. Impreso por Pro-Offset Editorial S.A
5. Mendenhall, W.: Introduction to Linear Models and the Design and Analysis of Experiments. Duxbury Press, Belmont (1968)
6. Montgomery, D.C.: Diseño y Análisis de Experimentos, segunda edn. Grupo Editorial Limusa, S.A., México (2003)
7. Neter, J., Wasserman, W., Kutner, M.H.: Applied Linear Statistical Models, 3rd edn. Richard D. Irwin, Inc., Homewood, IL; Boston, MA (1990)
8. Pearson, E.S., Hartley, H.O.: Biometrika Tables for Statisticians, vol. 1. Cambridge University Press, Cambridge (1996)
9. Raudenbush, S.W.: Random effects models. Handb. Res. Synthesis **421**(3.6) (1994)

Shiny-Based R Framework for Comparative Repeatability and Reproducibility Analysis in Metrology: Integrating DOE and ANOVA Approaches

Juliana Garcia-Villada[ID] and Carmen E. Patiño-Rodríguez[✉][ID]

Universidad de Antioquia, Medellin, Colombia
elena.patino@udea.edu.co

Abstract. The evaluation of measurement systems has evolved toward structured analytical frameworks collectively known as Measurement System Analysis (MSA). Gauge Repeatability and Reproducibility studies (GR&R) provide a framework to quantify and control variability in measurement processes, but their implementation has often depended on closed source or tools with limited adaptability. This study presents an interactive Shiny application developed in R to support systematic comparison, visualization, and interpretation of GR&R, offering an open and adaptable solution consistent with Industry 4.0 principles. The tool includes classical GR&R and statistical approaches based on design of experiments and analysis of variance (ANOVA). Through interactive data selection and visualization, users can identify variability components, analyze error sources, and ensure methodological transparency. The comparison of approaches showed that ANOVA-based methods improve precision in identifying variability sources, providing more robust insights into measurement errors than classical GR&R. The application was tested in industrial and educational settings, demonstrating its effectiveness in improving measurement accuracy, decision-making reliability, and professional skills development.

Keywords: Metrology · GR&R studies · Shiny app

1 Introduction/Background

Measurement System Analysis (MSA) is essential for guaranteeing the reliability of industrial data used in conformity assessment, process capability, and decision-making [1]. When measurement systems introduce uncontrolled variation, both systematic and random errors propagate across analytical stages, compromising quality control and compliance with technical specifications [2, 3].

In this context, Gauge Repeatability and Reproducibility studies (GR&R) help quantify how variability decomposes into instrument, operator, and interaction effects, enabling targeted corrective actions such as calibration, training, or environmental adjustments [4]. GR&R has also gained importance in engineering education, strengthening students' understanding of uncertainty-based decision-making and data integrity [5].

© The Author(s), under exclusive license to Springer Nature Switzerland AG 2026
B. M. Suárez et al. (Eds.): R Day 2025, CCIS 2824, pp. 43–60, 2026.
https://doi.org/10.1007/978-3-032-18455-9_3

Historically, Classical GR&R methods simple, fast, and operationally lightweight were widely adopted for industrial diagnostics and educational contexts [6, 7]. However, they lack statistical robustness and may underestimate complex sources of variability [4]. ANOVA-based approaches extend this framework by explicitly partitioning variance into operator, part, and interaction effects, increasing precision in identifying error sources [8, 9].

Despite their advantages, access to transparent, flexible, and open-source platforms implementing both methods remains limited. To address these gaps, we developed MetR&R, an interactive R Shiny application that integrates descriptive visualization, Classical GR&R, and ANOVA-based GR&R. The tool enhances methodological transparency, supports education, and aligns with the principles of Industry 4.0, where decision-making increasingly relies on real-time data analytics, digital metrology, and traceability [10–12].

The paper is structured as follows: Sect. 1 introduces the conceptual and measurement system foundations and the educational relevance of metrology, highlighting the role of interactive tools. Section 2 presents the `MetR&R` application, including its purpose, R/Shiny platform, interface modules, and the demo data feature. Section 3 describes testing and validation through numerical cross-checks and usability assessment. Section 4 shows results from the app, comparing them with Python and Excel benchmarks. Section 5 covers deployment via Shinyapps.io and GitHub, including accessibility and educational use, and Sect. 6 concludes with key contributions and future research directions.

1.1 Conceptual and Measurement System Foundations

The methodological foundations of MSA are grounded in the statistical principles that determine the reliability, precision, and consistency of data collected in industrial environments. Classical MSA frameworks establish that any decision based on measured data must first validate the system generating those measurements. This evaluation decomposes total variation into three main sources: part-to-part variability, repeatability (instrument-related error), and reproducibility (operator-related error) [1, 13]. The objective is to minimize the latter two so that measured variability accurately reflects intrinsic part differences [4, 13–15]. The statistical basis of GR&R is summarized in Table 1.

Table 1. Statistical framework for variance decomposition GR&R [13, 14, 16]

Equation	Description
$\sigma_{Total}^2 = \sigma_{part}^2 + \sigma_{oper}^2 + \sigma_{oper \times part}^2 + \sigma_{instr}^2$ (1)	Total variance decomposition
$\sigma_{repeat}^2 = \sigma_{instr}^2$ (2)	Repeatability variance
$\sigma_{reprod}^2 = \sigma_{oper}^2 + \sigma_{oper \times part}^2$ (3)	Reproducibility variance
$\sigma_{R\&R}^2 = \sigma_{repeat}^2 + \sigma_{reprod}^2$ (4)	Total measurement system variance

Classical GR&R methods offer intuitive diagnostics using simplified estimators (Table 2), making them ideal for rapid assessments and instructional settings [16].

Table 2. Simplified estimators used in short GR&R [14].

Equation	Description
$ME = k^2 \times \bar{R}$ (5)	Expanded measurement error
$\hat{\sigma}R\&R = \frac{k^2}{5.15} \times \bar{R} = \frac{ME}{5.15}$ (6)	Estimated standard deviation of measurement error
$\frac{P}{T} = \frac{ME}{USL-LSL} \times 100\%$ (7)	Precision to Tolerance Ratio

While Classical GR&R, sacrifice statistical robustness, their reduced cost and simplicity make them attractive tools in contexts where resources are limited, or quick decisions are required. Through this approach, the system's precision is evaluated by comparing measurement variation against process variation, with thresholds typically established at 10% (acceptable), 10–30% (marginal), and above 30% (unacceptable) [3, 13, 14, 18]. These methods remain popular due to their interpretability, especially in training contexts, although they may underestimate complex sources of variability [4].

To address these gaps, statistical approaches ANOVA-based was developed [8, 14, 15]. These methods enable researchers and practitioners to explicitly partition variance components associated with parts, operators, and their interactions (Table 3).

Table 3. GR&R ANOVA-based [14, 15]

Source of Variability	Degrees of Freedom (df)	Sum of Squares (SS)	Mean Square (MS)	F
Operator	$k-1$	SS_O	$MS_O = \frac{SS_O}{(k-1)}$	$\frac{MS_O}{MS_{O\times P}}$
Part	$n-1$	SS_P	$MS_P = \frac{SS_P}{(n-1)}$	$\frac{MS_P}{MS_{O\times P}}$
Operator by Part	$(k-1)(n-1)$	$SS_{O\times P}$	$MS_{O\times P} = \frac{SS_{O\times P}}{(k-1)(n-1)}$	$\frac{MS_{O\times P}}{MS_E}$
Equipment	$nk(r-1)$	SS_E	$MS_E = \frac{SS_E}{nk(r-1)}$	
Total	$nkr-1$	SS_T		

1.2 Normative and Practical Considerations

MSA practices are framed by standards such as AIAG (2010) [19], ISO 22514-7:2022 [20], ISO/IEC 17025:2017 [11], and ISO 9001:2015, which define sampling structures, acceptance thresholds, and validation procedures [14, 16, 17]. Practical implementation varies by sector, with automotive (IATF 16949) [21] and aerospace (AS9100) [22] integrating GR&R into quality assurance routines.

When GR&R results exceed acceptable limits, standards recommend corrective actions including redesign of fixtures, additional operator training, or selecting higher-precision instruments [23]. Recent studies highlight the need for frameworks capable of managing multivariate, nonlinear, or non-normal measurement scenarios [3, 23–25].

1.3 Educational Relevance of Metrology

Metrology education strengthens competencies in uncertainty evaluation, data integrity, and decision-making in industrial environments [10, 26, 27]. Lack of structured training can lead to insufficient understanding of measurement reliability [28, 29]. Integrating MSA into engineering curricula enhances student ability to identify systematic and random errors, interpret variability, and ensure conformity [7, 26, 29, 30].

Interactive digital platforms, particularly open-source tools such as R + Shiny enhance experiential learning by allowing students to manipulate parameters and visualize the impact on variability [16, 31, 32]. These environments also support decision-making in industrial settings through transparent, replicable workflows aligned with Industry 4.0 [5, 11].

1.4 Interactive Data Visualization 38

Interactive visualization is a key component of modern data analysis, allowing users to explore complex information dynamically rather than through static displays. Shiny, an R-based framework, provides a flexible environment for developing such visualizations by integrating statistical computation with real-time interactivity [33].

Shiny is an open-source R package that enables the creation of interactive web applications directly from R without requiring prior knowledge of HTML, CSS, or JavaScript [32, 34]. Its reactive programming architecture automatically updates outputs in response to user inputs, supporting real-time interaction with statistical and computational models [35]. This design extends R beyond statistical analysis, enabling the development of accessible and practical tools for data exploration and visualization. By combining computation and web technologies, Shiny bridges advanced analytical methods with practical usability, expanding access to data analysis while maintaining methodological rigor [36].

In education and research, Shiny applications transform abstract analytical concepts into interactive learning and experimentation tools. In academic contexts, they allow students to adjust parameters, visualize outcomes, and better understand statistical processes and quality control methods [37, 38]. For researchers, Shiny supports transparency and reproducibility by providing platforms where datasets, models, and simulations can be shared interactively, promoting collaborative validation of results [33]. Its alignment with open science and Industry 4.0 principles underscores its value for enhancing accessibility, interactivity, and the integration of analytics into decision-making processes [11, 24, 34, 35].

The integration of Shiny with R builds on a mature computational ecosystem for statistical modeling and quality analysis. Core R functions such as `aov()` and `lm()` support classical ANOVA models, while specialized packages like `lme4` extend these capabilities to mixed-effects modeling, enabling flexible variance decomposition. For quality control, the `qcc` package offers GR&R analysis, capability indices, and control charting [39], while the `metRology` package facilitates the evaluation of measurement uncertainty in compliance with ISO/IEC 17025 standards [40] These computational foundations provide the basis for interactive visualizations that enhance interpretation of measurement system performance [41].

To explore variability across operators and parts, visualization tools were combined with Shiny's interactive framework. Density plots generated with `ggplot2` enabled the representation of continuous distributions as smooth curves, offering clearer insights than traditional histograms into the concentration of measurements and possible overlaps. Comparative boxplots further highlighted interquartile ranges, medians, and outliers, allowing systematic biases between operators to be detected. Scatter plots, enriched with jittering and interactive tooltips through `plotly`, provided a dynamic means to identify patterns and outliers at the operator–part level. Finally, an interactive statistical table built with the `DT` package summarized key metrics such as means, standard deviations, and ranges, complementing graphical exploration with quantitative evidence. Together, these visual tools, when embedded in a Shiny application, transform static statistical outputs into an interactive environment that strengthens both exploratory analysis and decision-making in measurement system evaluation.

2 MetR&R Application Overview

2.1 Motivation and Purpose of the App

`MetR&R` was developed to address the lack of accessible, interactive, and open-source tools for Measurement System Analysis (MSA). The application integrates descriptive visualization, Classical GR&R, and ANOVA-based GR&R analyses, enabling both industrial decision-making and engineering education. It allows users to explore measurement variability, evaluate instruments and operators, and understand uncertainty-based quality control [5, 10–12, 14–16].

The development process emphasized pedagogical value: metrology teaching modules, including measurement uncertainty, repeatability and reproducibility, and instrument calibration, were incorporated to ensure that users not only apply quality control tools but also justify and validate the measurement foundations underlying these analyses. This approach bridges theoretical knowledge and practical application, enhancing comprehension and supporting methodological rigor.

2.2 R/Shiny as a Platform

`MetR&R` was developed using R `shiny` (version 1.11.1 [33, 36]), an open-source framework that enables the creation of interactive web applications directly from R (version 4.5.1 [42]). Shiny's reactive architecture allows outputs to update automatically in response to user inputs, providing real-time interaction with statistical models and ensuring a responsive, dynamic experience. To support the analytical, visualization, and interface requirements of MetR&R, several R packages were integrated into the platform: ggplot2 (v. 3.5.2 [43, 44]) and plotly (v. 4.11.0 [34]) facilitate static and interactive graphics [34, 43, 44]; dplyr (v. 1.1.4 [45]) enables efficient data manipulation [45]; SixSigma (v. 0.11.1 [46]) performs Classical and ANOVA-based GR&R calculations [46]; DT [47] allows interactive table presentation; bslib [48] provides tools for layout management and aesthetic customization; and readxl [49] supports the import of spreadsheet data. This combination of packages allows MetR&R to integrate statistical computations, interactive visualization, and user-friendly interface design in a modular, reproducible, and scalable environment, ensuring that both educational and industrial applications can be effectively addressed.

2.3 Interface and Modules

The interface was designed for clarity, usability, and educational effectiveness. The development followed a structured workflow:

Prototyping. Low-fidelity sketches defined the panel layout, input/output placement, and navigation flow (Fig. 1). Two main panels were initially created to organize key elements, including graphics, textual outputs, and input fields. Although interactivity was not implemented at this stage, the simulated dimensions mirrored the final platform, enabling early detection of potential bottlenecks and redundancies and establishing a clear conceptual foundation for the application.

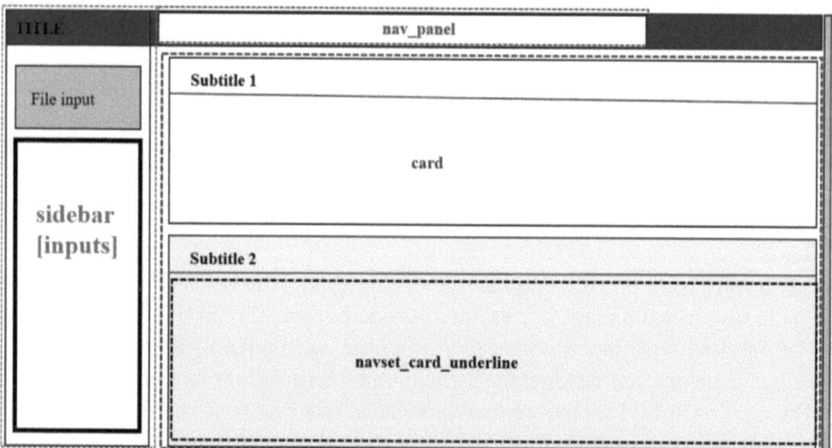

Fig. 1. Initial sketch of the interface with the two main panels

Interaction Logic. User inputs such as operator, part, and measurement repetitions trigger dynamic outputs including variance decomposition, interactive plots, and GR&R

metrics (Fig. 2). The logical framework ensures that dependent and independent variables are correctly mapped to statistical transformations, aligning computational backend operations with research objectives.

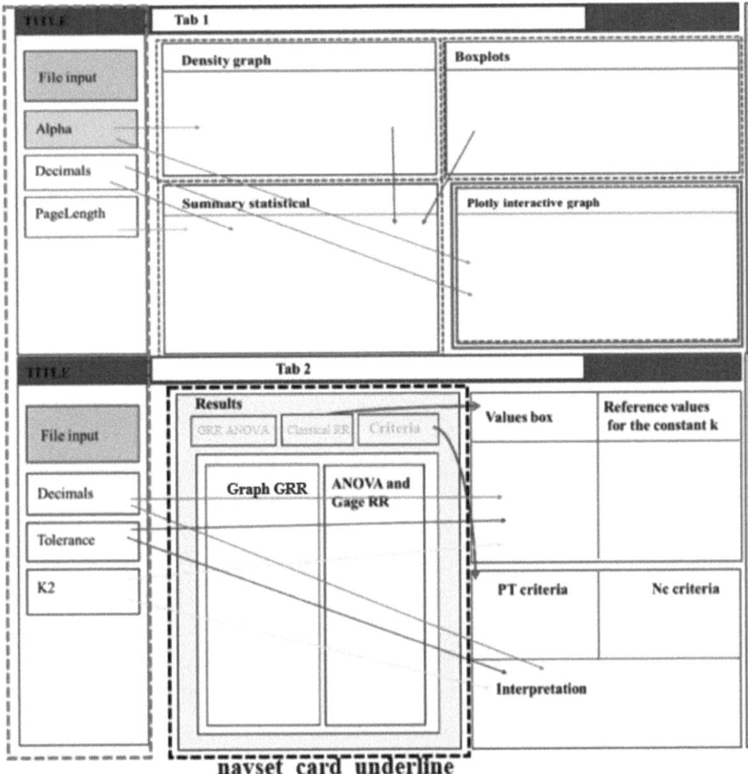

Fig. 2. Interaction diagram between input buttons and visual outputs. The figure shows the correspondence between the input buttons and their associated outputs using directional arrows. The related elements are highlighted in the content, and an interactive graphic is highlighted as part of the visual result.

Reactive Architecture. The UI-server relationship ensures modularity and reproducibility (Fig. 3). The application structure separates preprocessing, analytical routines, and visualization modules, enabling standardized calculations of repeatability, reproducibility, and variance partitioning. This design supports both Classical and ANOVA-based GR&R analyses while maintaining interactivity, performance, and interpretability.

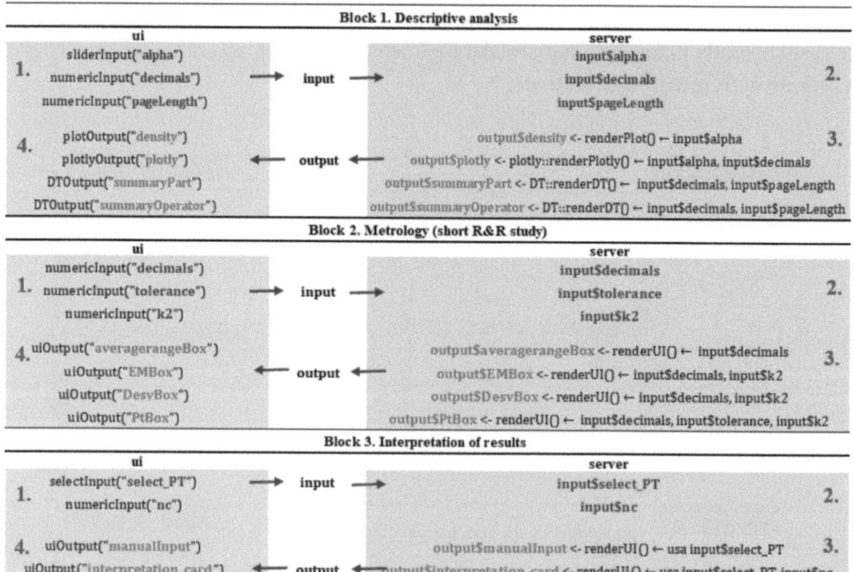

Fig. 3. Reactive interaction architecture between UI and server components in the Shiny application.

Library Integration. Figure 4 illustrates how the R packages referenced in Sect. 2.2 are integrated into the application. `bslib` structures the layout with a fixed sidebar for navigation, while `Shiny` allows buttons to link directly to panels. Descriptive analysis leverages `dplyr` and `DT` for data processing and interactive tables, with `ggplot2` and `plotly` providing dynamic visualization. Metrological analysis uses `card()` and `value_box()` to combine numerical data with visual cues, and GR&R calculations are performed via `ss.rr()` from `SixSigma`, generating both numerical and graphical outputs. Shiny's `HTML()` function integrates explanatory text and acceptance criteria, ensuring a clear, consistent, and pedagogically effective presentation.

2.4 Load Demo Data

`MetR&R` includes a *Load Demo Data* button that enables users to populate the interface with pre-formatted datasets. This functionality allows immediate interaction with visualizations, GR&R calculations, and variance decomposition analyses, facilitating exploration of operator-part interactions, error propagation, and system evaluation in a controlled environment. This feature supports both educational objectives and industrial experimentation by reducing entry barriers for new users. For further details on deployment and dataset access, see Sect. 5.2.

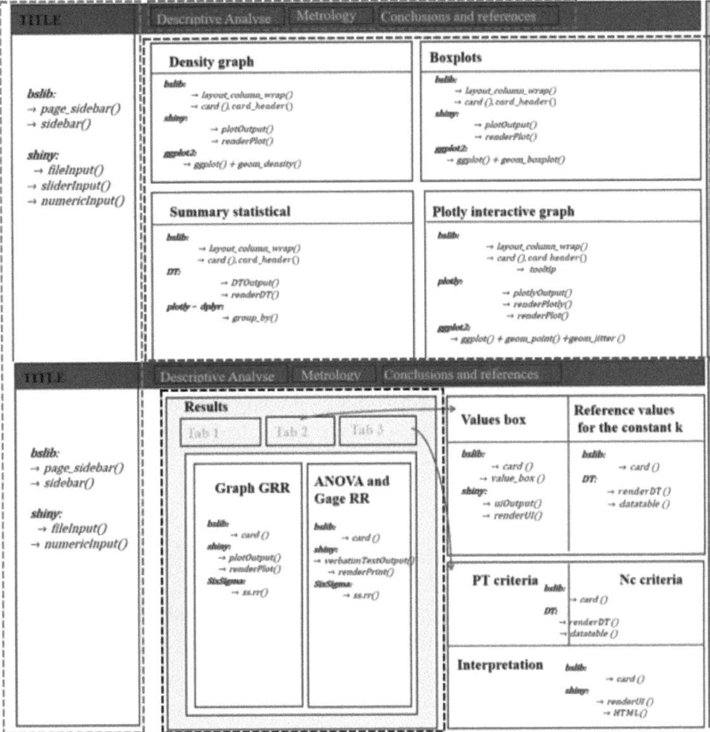

Fig. 4. Sketch of tab and navigation panel layout with associated libraries and key functions

3 Testing and Validation

To ensure the reliability and practical utility of MetR&R, a structured validation approach was implemented, combining numerical verification, scenario-based testing, and usability assessment. This process aimed to guarantee coherence between user inputs, intermediate computations, and final outputs, while confirming that the tool behaves as expected under realistic measurement conditions.

3.1 Systematic Numerical Validation

MetR&R was subjected to rigorous numerical validation by cross-checking results against benchmark calculations performed in Microsoft Excel, R routines, and Python, following well-established MSA procedures and formulas widely taught in quality control and standardization courses [14–16]. Classical and ANOVA-based GR&R metrics computed by the application consistently matched theoretical expectations and were reproducible across software platforms.

In addition, realistic measurement scenarios reflecting common industrial practices were employed to verify that variability across operators, parts, and repeated measurements was accurately captured. This multi-platform and scenario-based approach allowed identification and correction of potential discrepancies, enhancing confidence in the correctness of the implemented algorithms. Selected outputs were also compared against equivalent calculations in `Excel` and `Python` to provide a multi-platform verification of numerical accuracy; detailed comparisons are presented in Sect. 4.

3.2 Usability Assessment

Alongside numerical validation, usability testing evaluated the application's interface for intuitive navigation, responsiveness under varying dataset sizes, and robustness in handling user inputs and errors. This assessment confirmed that `MetR&R` provides a transparent, interactive, and user-friendly environment suitable for both industrial and educational applications.

By combining theoretical, numerical, and experiential validation, this dual strategy demonstrates that `MetR&R` not only reproduces measurement system variability in line with MSA principles but also facilitates exploration, learning, and informed decision-making.

4 Results

This section presents the outcomes of the metrological analysis, evaluating variability across operators and parts. Results are structured to provide an initial descriptive overview, followed by a critical assessment of measurement system performance through Classical and ANOVA-based GR&R methods. By integrating graphical exploration with inferential analysis, the section demonstrates both the operational capabilities of `MetR&R` and the robustness of its outputs in comparison with alternative computational approaches.

4.1 Systematic Numerical Validation: GR&R Outputs from the App

Descriptive Analysis. The descriptive analysis offers a preliminary overview of measurement distributions across operators and parts (Fig. 6). Density plots illustrate the general shape and overlap of operator measurements, revealing substantial agreement at lower ranges but divergence at higher values, which may reflect differences in technique or instrument handling. Boxplots per part highlight within-part variability and potential inconsistencies in median or interquartile ranges, signaling the need for statistical validation. Interactive scatter plots and summary tables provide detailed numerical and graphical insights, facilitating identification of outliers and guiding subsequent GR&R analyses. While this descriptive layer demonstrates the functionality of `MetR&R`, it also establishes a baseline for evaluating the measurement system's performance under controlled and realistic scenarios (Fig. 5).

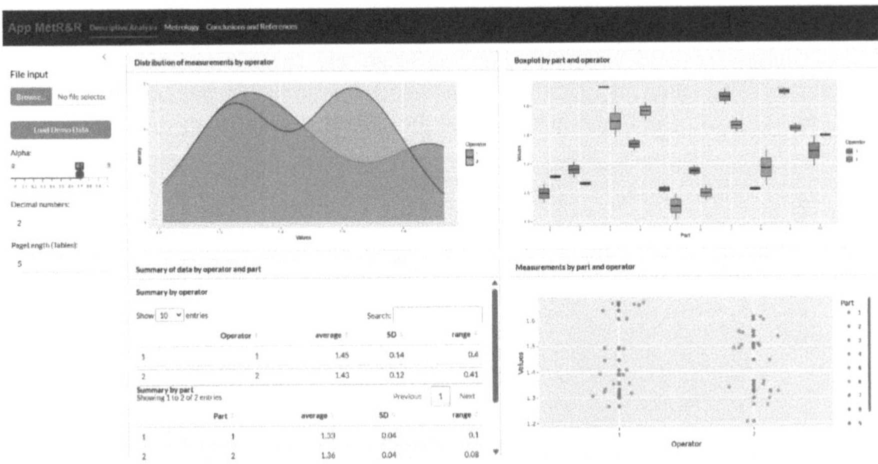

Fig. 5. Integrated results from the descriptive analysis panel.

Metrology: Classical and ANOVA-Based GR&R. The ANOVA-based GR&R analysis decomposes total variance into part, operator, and interaction components (Fig. 6). Results indicate that the part factor and part × operator interaction are statistically significant ($\alpha = 0.05$), while the operator effect is minimal. Reproducibility accounts for the largest share of total error (16.82% of 26.44%), and only two parts can be reliably distinguished, confirming limited system resolution.

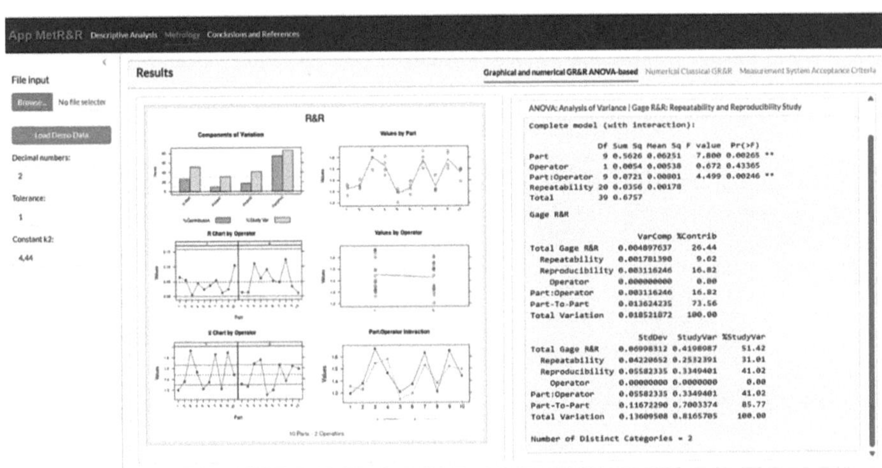

Fig. 6. Graphical and numerical GR&R ANOVA-based.

Classical GR&R metrics provide complementary insights into measurement variability. The analysis includes the average range per part, measurement error, and precision-to-tolerance (P/T) ratio (Fig. 7). These results corroborate the findings of the ANOVA-based analysis, confirming the overall reliability and repeatability of the system under study.

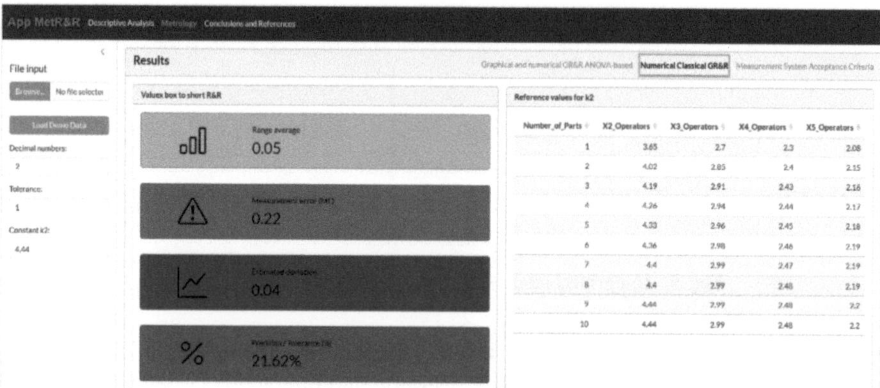

Fig. 7. Numerical Classical GR&R.

The evaluation of Measurement System Acceptance Criteria contextualizes the results by classifying the system performance. Based on both the P/T ratio and the number of distinguishable categories (Nc), the system is considered *Marginal*, suggesting the need for *review and improvement where possible* (Fig. 8). Together, these analyses offer a rigorous and comprehensive assessment of measurement variability and system performance.

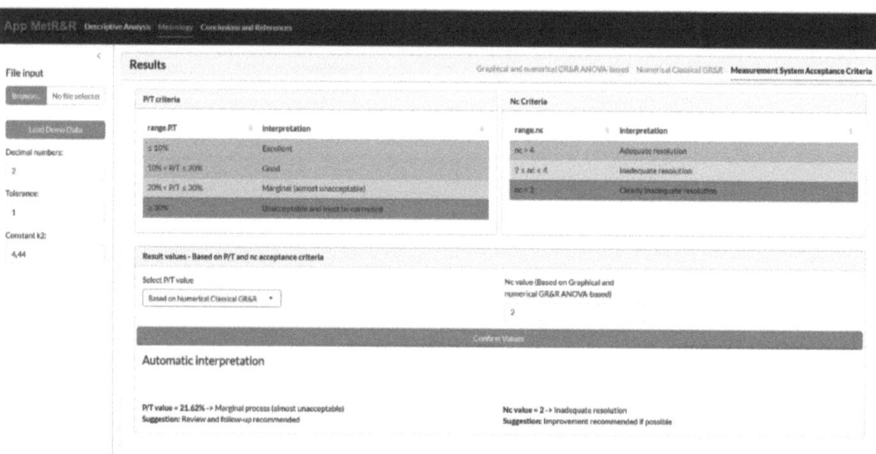

Fig. 8. Measurement System Acceptance Criteria.

Methodological Implications in Industry 4.0. In the modern industrial landscape, the methodological foundations of Measurement System Analysis (MSA) have evolved significantly due to advances in digital metrology, artificial intelligence, and data connectivity. Recent works emphasize the transition from static, periodic evaluations toward dynamic, data-driven MSA [11, 19, 20], leveraging big data analytics and machine-learning algorithms to continuously monitor measurement variability and identify contextual patterns associated with environmental, operator, or equipment-related factors [5, 11, 24]. Bayesian modeling and uncertainty propagation techniques further expand the scope of MSA by incorporating hierarchical models that capture correlations and temporal dependencies, rather than treating sources of variation as independent [50–52]. The integration of MSA within cyber-physical systems and digital twins introduces new paradigms for real-time calibration and adaptive measurement control, enabling automatic recalibration or adjustment of control limits based on predictive analytics. Consequently, methodological rigor is shifting from isolated experimentation toward embedded intelligence within Industry 4.0 ecosystems [53]. While classical MSA frameworks remain valid for reference and regulatory compliance, modern approaches extend their scope to support continuous verification of measurement performance, self-adaptive learning from historical data, and dynamic traceability, interoperability, and resilience across digital manufacturing environments [38, 53, 54].

4.2 Comparison with Python/Excel Results

A detailed comparison of `MetR&R` outputs with `Python` and `Excel` results (Table 4) provides robust validation and highlights practical implications. Average Range ($\bar{R}$) is identical across all platforms (0.0487), confirming repeatability computations. Measurement Error (ME) matches `Python` (0.2163) but is slightly higher in `Excel` (0.2325), likely due to differences in floating-point arithmetic or degrees of freedom handling; this minor discrepancy does not affect overall system evaluation. Estimated standard deviation is consistent between `MetR&R` and `Python` (0.0420), whereas `Excel` reports 0.0500, reflecting a potential overestimation of variability. Precision-to-tolerance ratios (P/T) are comparable across software, with `MetR&R` at 21.62% and `Excel` at 23.24%, classifying the system as Marginal. ANOVA metrics (part, operator, part × operator) closely match `Python` and are unavailable in `Excel`, demonstrating MetR&R's enhanced analytical depth. Graphical outputs reinforce interpretability: `MetR&R` enables interactive exploration of density, boxplots, and scatter plots, while `Python` reproduces similar static plots, and `Excel` lacks these dynamic features. Overall, these results confirm MetR&R's numerical accuracy, methodological rigor, and added value in visualization and interpretive guidance, making it a reliable tool for both industrial applications and educational settings.

5 Deployment

The deployment of an R Shiny application is a key stage, as it determines how end users can access and interact with the tool. According to Posit [55], Shiny applications can be shared either by distributing the source files requiring users to have R and the necessary packages installed or by publishing the application on a web platform that enables access from any standard browser. In this project, both methods were implemented to maximize accessibility, reproducibility, and ease of use.

Table 4. Numerical Validation of GR&R Outputs: MetR&R vs. Python and Excel

Metric - Output	MetR&R (R/Shiny)	Python	Excel/Manual
Average Range ($\bar{R}$)	0.0487	0.0487	0.0487
Measurement Error (ME)	0.2163	0.2163	0.2325
Estimated Standard Deviation	0.0420	0.0420	0.0500
P/T (%)	21.62%	21.63%	23.24%
ANOVA: Part	0.5633	0.5626	–
ANOVA: Operator	0.0054	0.0053	–
ANOVA: Part × Operator	0.0719	0.0721	–
Boxplot / Density	✓	✓	–

5.1 Shinyapps.Io + GitHub

The application was deployed on Shinyapps.io, allowing users to access it directly through a web browser at: https://4rh7ur-juliana-garcia0villada.shinyapps.io/Metrol ogy/. In parallel, a public GitHub repository was created (https://github.com/jgvill ada/Shiny-app-MetR-R), containing the essential files (`app.R`, `data.csv`, and `README.md`) so that users with R and the `shiny` package installed can run the application locally by following the provided instructions. This dual deployment strategy enhances accessibility, supports reproducibility, and accommodates both users seeking immediate online execution and those who prefer or require local deployment within their own analytical environments.

5.2 Demo Data Feature

To support users who may not have a dataset readily available, the application integrates a *Load Demo Data* button. This feature automatically loads a pre-formatted example dataset directly into the interface, enabling users to immediately explore the application's visualizations and perform GR&R calculations without uploading a personal CSV file. The demo dataset is included in the GitHub repository to ensure transparency, reproducibility, and ease of access, making it suitable for both educational use and industrial practice.

5.3 Accessibility and Educational Use

The combination of web deployment, local execution options, and the integrated demo dataset significantly enhances the accessibility of the tool. Users can test, explore, and understand the workflow without prior technical knowledge or data preparation.

These features make the application suitable not only for practical metrology tasks but also for educational and training contexts, where immediate hands-on interaction is essential. The interface design outlined in Fig. 9 supports an intuitive user experience, consistent across both the web and local execution environments.

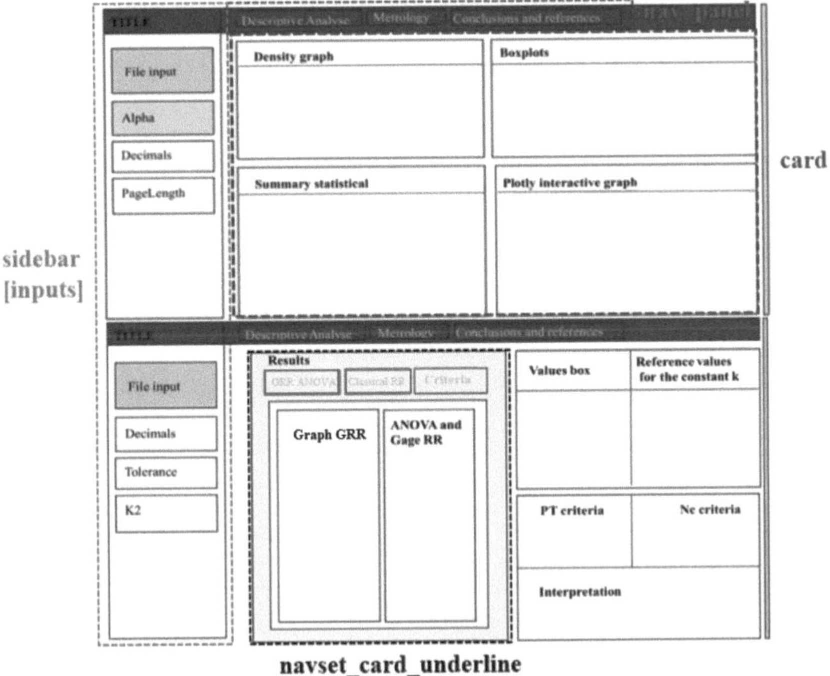

Fig. 9. Final sketch of the interface

6   Conclusions

One of the main challenges in Measurement System Analysis (MSA) is the inconsistent application and interpretation of Classical GR&R and ANOVA-based GR&R methods. These approaches often produce results that are difficult to compare, depending on process conditions, available resources, and personnel training. `MetR&R` addresses this issue by integrating both methods within a single interactive environment, allowing users to analyze operator-part variability and residual errors simultaneously. Interactive graphs, boxplots, and summary tables, combined with the ability to compare outputs with `Python` and `Excel`, facilitate clearer insights into measurement behavior, standardize evaluation practices, and improve communication across industrial and academic settings. Visualization thus becomes a critical component for enhancing the reliability, transparency, and interpretability of measurement assessments.

By combining Classical and ANOVA-based GR&R, `MetR&R` enables both rapid assessments under controlled conditions and detailed variance decomposition to identify the contributions of parts, operators, and their interactions. The Load Demo Data feature allows immediate exploration of these analyses, supporting educational purposes and hands-on training without requiring prior dataset preparation. Implemented in R Shiny, the platform offers real-time interactivity, web-based deployment, and accessible visualization of complex statistical computations. This combination of usability, methodological rigor, and open-source accessibility promotes a shared analytical framework between industry and academia, reducing the risk of misinterpretation and fostering consistent evaluation criteria.

In the context of Industry 4.0, characterized by connectivity, automation, and data integration, visualization-oriented MSA tools like MetR&R represent a key element of intelligent quality systems. By automating calculations, visualizing uncertainty, integrating criteria such as precision-to-tolerance (P/T) and the number of distinct categories (Nc), and enabling real-time comparison across software platforms, the tool supports continuous monitoring and adaptive control of measurement systems. This facilitates timely identification of recalibration needs, minimizes errors, improves compliance with quality standards, and enhances operational efficiency. For educational contexts, MetR&R provides an intuitive, interactive interface that strengthens learning in data-driven decision-making, while in industrial practice, it fosters more reliable, transparent, and standardized measurement evaluation aligned with Industry 4.0 principles.

References

1. Montgomery, D.C.: Introduction to Statistical Quality Control. Wiley (2020)
2. Senvar, O., Firat, S.U.O.: An overview of capability evaluation of measurement systems and gauge repeatability and reproducibility studies. Int. J. Metrol. Qual. Eng. **1**, 121–127 (2010)
3. Soares, W.D.O.S., Peruchi, R.S., Silva, R.A.V., Rotella, J.P.: Gage R&R studies in measurement system analysis: a systematic literature review. Qual. Eng. **34**, 382–403 (2022). https://doi.org/10.1080/08982112.2022.2069505
4. Durivage, M.A.: Practical Attribute and Variable Measurement Systems Analysis (MSA): A guide for Conducting Gage R&R Studies and test Method Validations. Quality Press (2015)
5. Sardi, A., Sorano, E., Cantino, V., Garengo, P.: Big data and performance measurement research: trends, evolution and future opportunities. Meas. Bus. Excell. (2020). https://doi.org/10.1108/MBE-06-2019-0053
6. Nembhard, H.B., Valverde-Ventura, R.: Integrating experimental design and statistical control for quality improvement. J. Qual. Technol. **35**, 406–423 (2003). https://doi.org/10.1080/00224065.2003.11980238
7. Harris, G., Pendrill, L.R.: Comparing and contrasting studies of metrology education and training in Europe and North America. NCSLI Meas. **3**, 38–45 (2008)
8. Burdick, R.K., Graybill, F.A.: Confidence intervals on variance components. CRC Press (1992)
9. Lupan, R., Bacivarof, I.C., Kobi, A., Robledo, C.: A relationship between Six Sigma and ISO 9000: 2000. Qual. Eng. **17**, 719–725 (2005)
10. Salazar-Hernández, K.-D., Hernández-Balandrán, L.-E., Purata-Sifuentes, O.-J.: Teaching GAGE repeatability and reproducibility analysis of measurement systems: choice of the best didactic cases. Meas Sens. 101313 (2024)
11. Ghernaout, D., Aichouni, M., Alghamdi, A.: Overlapping ISO/IEC 17025: 2017 into big data: a review and perspectives. Int. J. Sci. Qual. Anal. **4**, 83–92 (2018)

12. Kumar, T.V., Tharani, P., Priyanka, C.: A case study of adopting the measurement system analysis in pump manufacturing industry. In: AIP Conference Proceedings, p. 020215. AIP Publishing LLC (2023)
13. Gutiérrez, H., De La Vara, R.: Statistical Quality Control and Six Sigma. McGrawHill (2013)
14. Antony, J., Gowles, K., Roberts, P.: Gauge capability analysis: classical versus ANOVA. Qual. Assur. **6**, 173–181 (1999)
15. Burdick, R.K., Borror, C.M., Montgomery, D.C.: Design and Analysis of Gauge R&R Studies: Making Decisions with Confidence Intervals in Random and Mixed ANOVA Models. SIAM (2005)
16. Pan, J.-N.: Evaluating the gauge repeatability and reproducibility for different industries. Qual. Quant. **40**, 499–518 (2006). https://doi.org/10.1007/s11135-005-1100-y
17. Fawcett, L.: Using interactive shiny applications to facilitate research-informed learning and teaching. J. Stat. Educ. **26**, 2–16 (2018)
18. Chen, K.S., Wu, C.H., Chen, S.C.: Criteria of determining the P/T upper limits of GR&R in MSA. Qual. Quant. **42**, 23–33 (2008)
19. AIAG. AIAG Measurement System Analysis Reference Manual (2010)
20. ISO 22514-7:2021. ISO 22514-7:2021 Statistical methods in process management—Capability and performance. International Organization for Standardization and International Electrotechnical Commission (2021)
21. IATF16949:2016. IATF16949:2016 Quality management system requirements for automotive production and relevant service parts organizations. International Automotive Task Force (2016)
22. AS 9100 D/EN 9100:2018. AS 9100 D/EN 9100:2018 Quality management systems Requirements for aviation, space and defense organizations. Society of Automotive Engineers (2016)
23. Arumun, J., Ekanem, V., Ashigwuike, E.: Review of gauge repeatability and reproducibility for dimensional measurements. Sch Bull. **10**, 186–194 (2024)
24. Ojo, O.R.: Application of deep learning on gage R&R for anomaly detection. East Tennessee State University (2025)
25. Barbosa, C.R.H., Sousa, M.C., Almeida, M.F.L., Calili, R.F.: Smart manufacturing and digitalization of metrology: a systematic literature review and a research agenda. Sensors **22**, 6114 (2022). https://doi.org/10.3390/s22166114
26. Lai, G.-Y.: Integration of enhanced coordinate measuring machine systems with manufacturing engineering laboratories and curriculum at kettering university. In: 2001 Annual Conference, p. 6.625. 1–6.625. 15 (2001)
27. Strong, S., Amos, S., Callahan, R.: Developing practical skills for quality assurance and metrology applications in manufacturing. In: 2004 Annual Conference Proceedings, pp. 9.416.1–9.416.9. ASEE Conferences, Salt Lake City (2004). https://doi.org/10.18260/1-2--14069. Accessed 24 Oct 2025
28. Ibrahim, S.M., Bills, P.J., Allport, J.: Metrology education impediment in developing countries considering the case of Nigeria (2017)
29. Almeida, L.A., Frota, M.N., Frota, M.N.: Metrology education and citizenship: the Brazilian experience. In: Proceedings of the XVII IMEKO World Congress Metrology in the 3rd Millennium, Dubrovnik, Croatia, pp. 22–27 (2003)
30. da Rocha, G.M., Landim, R.P.: A pioneer metrology technical course in the Latin America. In: XIX IMEKO World Congress, Lisbon Portugal (2009)
31. Aiupov, T.A., Golovanova, I.I.: Metrological skills development of technical university students in a digital educational resource with a smart laboratory. SCOPUS16130073-2021-3057-SID85122813668 (2021)

32. Ho, S.L., Xie, M., Goh, T.N.: Adopting Six Sigma in higher education: some issues and challenges. Int. J. Six Sigma Compet. Advant. **2**, 335 (2006). https://doi.org/10.1504/IJSSCA.2006.011564

33. Chang, W., et al.: Shiny: web application framework for R (2021). https://shiny.posit.co/

34. Sievert, C.: Interactive Web-Based Data Visualization with R, Plotly, and Shiny. Chapman and Hall/CRC (2020)

35. Beeley, C., Sukhdeve, S.R.: Web Application Development with R Using Shiny: Build Stunning Graphics and Interactive Data Visualizations to Deliver Cutting-Edge Analytics. Packt Publishing Ltd. (2018)

36. RStudio. Shiny from RStudio: Web Applications Made Easy. https://shiny.posit.co/

37. Liao, M., Attali, Y., von Davier, A.A.: AQuAA: analytics for quality assurance in assessment. EDM (2021)

38. Islam, M.R., Akter, S., Islam, L., Razzak, I., Wang, X., Xu, G.: Strategies for evaluating visual analytics systems: a systematic review and new perspectives. Inf. Vis. **23**, 84–101 (2024)

39. Scrucca, L.: Qcc: an R package for quality control charting and statistical process control. R News. **4**, 11–17 (2004)

40. ISO/IEC 17025: 2017. ISO/IEC 17025: 2017. General requirements for the competence of testing and calibration laboratories. International Organization for Standardization and International Electrotechnical Commission (2017)

41. Soares, W.D.O.S., Peruchi, R.S., Silva, R.A.V., Rotella, J.P.: Gage R&R studies in measurement system analysis: a systematic literature review. Qual. Eng. **34**, 382–403 (2022)

42. R Core Team. R: A language and environment for statistical computing. R Foundation for Statistical Computing, Viena, Austria (2025)

43. Ginestet, C.: Ggplot2: Elegant Graphics for Data Analysis. Oxford University Press (2011)

44. Gómez-Rubio, V.: Ggplot2-elegant graphics for data analysis. J. Stat. Softw. **77**, 1–3 (2017)

45. Hadley, W., François, R., Henry, L., Müller, K., Vaughan, D.: dplyr: a grammar of data manipulation (2025). https://dplyr.tidyverse.org

46. Cano, E.L., Moguerza, J.M., Redchuk, A.: Six Sigma with R: Statistical Engineering for Process Improvement. Springer, New York (2012). https://doi.org/10.1007/978-1-4614-3652-2. Accessed 20 Oct 2025

47. Xie, Y., Cheng, J., Tan, X.: DT: a wrapper of the javascript library 'DataTables'. R package version 0.2 (2016)

48. Sievert, C., Cheng, J., Aden-Buie, G.: bslib: custom «Bootstrap» «Sass» themes for «shiny» and «rmarkdown» (2025). https://rstudio.github.io/bslib/

49. Wickham, H., et al.: Package 'readxl'. Version, p. 13 (2019)

50. Silver, R.M., Zhang, N.F., Barnes, B.M., Qin, J., Zhou, H., Dixson, R.: A Bayesian statistical model for hybrid metrology to improve measurement accuracy. In: Model Asp Opt Metrol III, pp. 59–69. SPIE (2011)

51. Natvig, B., Eide, H.: Bayesian estimation of system reliability. Scand. J. Stat. **14**, 319–327 (2013)

52. Yeh, T.-M., Sun, J.-J.: Using the Monte Carlo simulation methods in gauge repeatability and reproducibility of measurement system analysis. J. Appl. Res. Technol. **11**, 780–796 (2013)

53. Pongboonchai-Empl, T., Antony, J., Garza-Reyes, J.A., Komkowski, T., Tortorella, G.L.: Integration of Industry 4.0 technologies into Lean Six Sigma DMAIC: a systematic review. Prod. Plan Control 35, 1403–1428 (2024). https://doi.org/10.1080/09537287.2023.2188496

54. Hung, B.K.H., To, C.C.N., Fung, R.K.H., Chan, C.C.S.: Addressing proficiency gaps in future skills between employers and learners through data visualization. SN Comput. Sci. (2023). https://doi.org/10.1007/s42979-023-01722-3

55. Posit. Lesson 7: Share your apps. Shiny Tutor (2023). https://shiny.posit.co/r/getstarted/shiny-basics/lesson7/. Accessed 10 July 2025

Implementation and Applications of the Unit Maxwell-Boltzmann Distribution in the Gamlss Framework for Environmental Data Analysis

David Villegas Ceballos[1]([✉]) [iD], Freddy Hernandez-Barajas[2] [iD], and Olga Cecilia Usuga-Manco[1] [iD]

[1] Universidad de Antioquia, Medellín, Colombia
`david.villegas1@udea.deu.co`
[2] Universidad Nacional de Colombia, Medellín, Colombia

Abstract. In this work, we present the implementation of the Unit Maxwell-Boltzmann (UMB) distribution within the gamlss package in R, specifically included in the ZeroOneDists package, providing users with a flexible and robust tool for statistical modeling. This implementation enables the computation of probabilities, quantiles, and random numbers, as well as parameter estimation and the development of regression models within the gamlss framework. To evaluate its performance, we conducted a comprehensive simulation study comparing the UMB distribution with other commonly used distributions. The results highlight the effectiveness of the UMB distribution in capturing data variability and improving model accuracy. This contribution extends the capabilities of the gamlss package, offering R users an advanced resource for data analysis and regression modeling, especially in contexts where flexibility and precision are crucial.

Keywords: Maxwell-Boltzmann · Statistics · Pollutants

1 Introduction

In the statistical literature, a wide variety of probability models exist to describe real-world phenomena. However, it is a known fact that there is a great need for new models with bounded support Biçer et al. (2024). In response to this need, Biçer et al. (2024) recently proposed the Unit Maxwell-Boltzmann (UMB) distribution, a flexible single-parameter model derived from the Maxwell-Boltzmann distribution of statistical physics. The objective of its development was to enrich the family of unit distributions, offering an effective, single-parameter alternative to more traditional models. Thanks to its single parameter μ, the UMB is capable of modeling data that exhibit various forms of asymmetry, making it very useful in a wide range of disciplines, from engineering to healthcare.

In their original study, Biçer et al. (2024) conclusively demonstrated that the UMB is an effective model for analyzing air pollutant concentration data, outperforming competing models. This positions it as a valuable tool for researchers

B. M. Suárez et al. (Eds.): R Day 2025, CCIS 2824, pp. 61–77, 2026.
https://doi.org/10.1007/978-3-032-18455-9_4

in environmental sciences and related fields who need to model data on the unit interval.

Following this same methodological line, subsequent work such as that of Genç and Özbilen (2025) introduced variations like the Unit Inverse Maxwell-Boltzmann (UIMB) distribution, which is an exponential transformation of the inverse Maxwell-Boltzmann distribution. This development underscores that the Maxwell-Boltzmann family of distributions represents a fertile and active field for the creation of new statistical models. However, despite these theoretical proposals, their practical application remains limited due to the lack of implementations in accessible statistical software. Specifically, the UMB distribution is not available within the gamlss framework, a gap that this work seeks to fill through its implementation in the R package ZeroOneDists.

Despite being a highly useful model due to its performance and simplicity, the utility of the UMB is limited for the community because it is not available in statistical software. Specifically, the UMB is not implemented within a gamlss regression framework, which prevents its use for modeling the distribution's parameter as a function of covariates. Implementing this distribution is crucial so that the scientific community can use it and leverage its full potential.

Therefore, the main objective of this work is to implement the Unit Maxwell-Boltzmann distribution as a family within ZeroOneDists package for the gamlss framework in R, thereby providing a robust tool for modeling data on the unit interval.

This paper is structured as follows: Sect. 2 provides an overview of the UMB distribution. Section 3 describes the maximum likelihood estimation (MLE) method. In Sect. 4, we present the UMB regression model. Section 5 describes the ZeroOneDists package. Section 6 presents the simulation study and its results. Section 7 shows some applications of the studied distribution using real datasets. Finally, Sect. 8 presents the conclusions drawn from the work.

2   UMB Distribution

The Unit Maxwell-Boltzmann (UMB) distribution was proposed by Biçer et al. (2024) as a flexible, single-parameter model for data on the unit interval $(0, 1)$. In their original work, the authors demonstrated its utility for modeling air pollutant concentrations, where it outperformed other established models thanks to its ability to capture various forms of asymmetry. Expansions on the same methodological idea can be found in the related literature; for example, Genç and Özbilen (2025) developed the Unit Inverse Maxwell-Boltzmann (UIMB) distribution from the transformation of the inverse Maxwell-Boltzmann distribution. This highlights that the Maxwell-Boltzmann family of distributions is an active and relevant area of research.

The origin of the UMB lies in the Maxwell-Boltzmann distribution (MBD), which is a fundamental probability distribution in statistical physics that describes the distribution of the speed or energy of gas particles in thermal equilibrium. Although its main application is in physics and chemistry, its domain on the positive real line ($x > 0$) also makes it a viable model for analyzing positive continuous data in other fields.

Formally, if a random variable X follows a Maxwell-Boltzmann distribution with a shape parameter $\mu > 0$, its probability density function (pdf) and cumulative distribution function (cdf) are defined by Eq. (1) and Eq. (2), respectively:

$$g(x, \mu) = \sqrt{\frac{2}{\pi}} \frac{x^2}{\mu^3} \exp\left(-\frac{x^2}{2\mu^2}\right), \quad x > 0 \tag{1}$$

$$G(x, \mu) = \mathrm{erf}\left(\frac{x}{\sqrt{2}\mu}\right) - \sqrt{\frac{2}{\pi}} \frac{x}{\mu} \exp\left(-\frac{x^2}{2\mu^2}\right), x > 0 \tag{2}$$

where $\mathrm{erf}(\cdot)$ function represents the Gauss error function. The UMB distribution arises directly from the MBD through an exponential transformation. This transformation maps the domain of the variable X, which is $(0, \infty)$, to the domain of the new variable Y, which is $(0, 1)$. This procedure allows for the generation of a new probability model whose support lies on the unit interval $(0, 1)$, making it ideal for modeling proportions, rates, or fractions.

Thus, a new random variable Y is defined, according to Biçer et al. (2024), as follows. Let X be a random variable from the MBD with parameter ($\mu > 0$). Then, the pdf of the random variable $Y = e^{-X}$ is:

$$f(y, \mu) = \frac{\sqrt{\frac{2}{\pi}} \ln^2\left(\frac{1}{y}\right) e^{-\frac{\ln^2\left(\frac{1}{y}\right)}{2\mu^2}}}{\mu^3 y}, \quad 0 < y < 1 \tag{3}$$

And the corresponding cdf is:

$$F(y; \mu) = 1 - \frac{\sqrt{\ln^2\left(\frac{1}{y}\right)} \mathrm{erf}\left(\frac{\sqrt{\ln^2\left(\frac{1}{y}\right)}}{\sqrt{2}\mu}\right)}{\ln\left(\frac{1}{y}\right)} + \frac{\sqrt{\frac{2}{\pi}} \ln\left(\frac{1}{y}\right) e^{-\frac{\ln^2\left(\frac{1}{y}\right)}{2\mu^2}}}{\mu}, \quad 0 < y < 1 \tag{4}$$

In this way, the UMB inherits the influence of the parameter μ from its base distribution but adapts it to the context of bounded data. The UMB is a single-parameter model capable of capturing various forms of asymmetry, whether left-skewed or right-skewed, within the unit interval $(0, 1)$, as shown in Fig. 1.

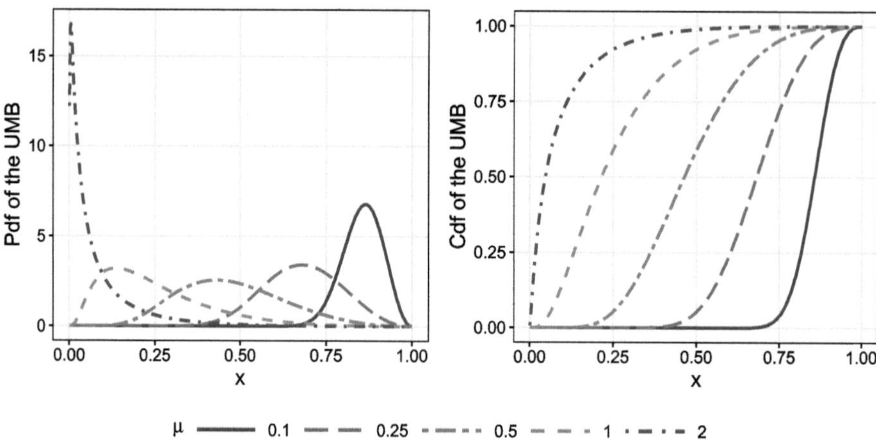

Fig. 1. Probability density function (left) and cumulative density function (right) for different values of the parameter μ.

3   Parameter Estimation

This section presents the maximum likelihood estimation method for the UMB distribution. According to Biçer et al. (2024), let $Y_1, Y_2, ..., Y_n$ be an independent and identically distributed sample from a UMB with a one-dimensional parameter μ, and let $y_1, y_2, ..., y_n$ be a realization of this sample. The likelihood function $L(\mu; y_1, y_2, ..., y_n)$ is:

$$L(\mu; y_1, y_2, \ldots, y_n) = \prod_{i=1}^{n} \frac{\sqrt{\frac{2}{\pi}} \ln^2 \left(\frac{1}{y_i} \right) e^{-\frac{\ln^2 \left(\frac{1}{y_i} \right)}{2\mu^2}}}{\mu^3 y_i} \tag{5}$$

And the log-likelihood function is:

$$\ln L(\mu; y_1, y_2, \ldots, y_n) = \frac{1}{2} \left(-\frac{\sum_{i=1}^{n} \ln^2(y_i)}{\mu^2} + 4 \sum_{i=1}^{n} \ln(\ln(y_i)) - 2 \sum_{i=1}^{n} \ln(y_i) - 6n \ln(\mu) \right)$$
$$+ \frac{1}{2} (n \ln(2) - n \ln(\pi)). \tag{6}$$

The maximum likelihood estimator (MLE) for μ can be obtained by taking the derivative of Eq. (6) with respect to μ and setting the resulting derivative to zero. The derivative of Eq. (6) with respect to μ is:

$$\frac{\delta \ln L(\mu, y_1, y_2, \ldots, y_n)}{\delta \mu} = \left(\frac{\sum_{i=1}^{n} \ln^2(y_i)}{\mu^3} - \frac{3n}{\mu} \right) = 0. \tag{7}$$

Therefore, from the solution to Eq. (7), we have the MLE of the parameter μ as:

$$\hat{\mu} = \sqrt{\frac{\sum_{i=1}^{n} \ln^2(y_i)}{3n}}. \tag{8}$$

Regarding computational performance, the UMB offers a distinct advantage over competitors like the Beta or Kumaraswamy distributions, which typically require iterative numerical optimization. The existence of a closed-form solution for the MLE, given in Eq. (8), ensures immediate convergence and superior computational efficiency. The ZeroOneDists package implements this estimator to fit the UMB distribution within the gamlss framework.

4   UMB Regression Model

This section presents the UMB regression model, which uses the MLE method to estimate its parameters. The objective of this model is to incorporate explanatory variables to estimate the single parameter of the UMB distribution. The UMB regression model assumes that the observations y_i (for $i = 1, 2, ..., n$) are independent and follow the probability density function $f(y; \mu)$ given in Eq. (3). The structure of the UMB regression model is as follows:

$$g(\mu_i) = \eta_i = \boldsymbol{X}_i \boldsymbol{\beta},$$

where g is a known monotonic function used to transform the linear predictor η_i into the appropriate domain of the parameter μ. Typically, the log function is used as the default link function. The design matrix $\boldsymbol{X}$ is known, and its columns correspond to the explanatory variables used to model μ. The vector $\boldsymbol{\beta}$ contains the fixed effects of the explanatory variables included in the model.

As in the previous section, the estimated parameters $\hat{\boldsymbol{\beta}}$ can be obtained using the UMB family implemented in the ZeroOneDists package.

5   ZeroOneDists Package

The `ZeroOneDists` package proposed by Hernandez-Barajas and Usuga-Manco (2025) implements several distributions, including the Unit Maxwell-Boltzmann. Within a gamlss framework, this package allows users to estimate parameters for the UMB distribution or estimate effects for the UMB regression model. The current version of the package is available on GitHub (https://github.com/fhernanb/ZeroOneDists). Users can download it using the following code:

```
install.packages("ZeroOneDists")
library(ZeroOneDists)
```

Regarding the UMB distribution, the `ZeroOneDists` package provides the functions `dUMB()` and `pUMB()` to obtain the pdf and cdf respectively. In these functions, the argument mu corresponds to the parameter μ defined in the previous sections. They can be used as follows:

```
dUMB(x, mu, log = FALSE)
pUMB(q, mu, lower.tail = TRUE, log.p = FALSE)
```

To obtain the reliability function $R(x)$, the user can use the pUMB() function with the argument lower.tail=FALSE. The ZeroOneDists package also provides other useful functions. For instance, rUMB() generates random samples and qUMB() obtains quantiles. To use these functions, one can follow the syntax below:

```
qUMB(p, mu, lower.tail = TRUE, log.p = FALSE)
rUMB(n, mu)
```

The function UMB() represents the family function that allows the estimation of parameters for the UMB distribution or of the effects for the UMB regression model, taking advantage of the gamlss package. The help documentation of the UMB() function provides two examples that demonstrate how any user can estimate the parameters for the UMB distribution or estimate the effects for the UMB regression model. A summary of the functions included in the package is presented in Table 1.

Table 1. Summary of functions available in the ZeroOneDists package for the UMB distribution.

Function	Description
dUMB()	Computes the probability density function (pdf)
pUMB()	Computes the cumulative distribution function (cdf)
qUMB()	Computes the quantile function (inverse cdf)
rUMB()	Generates random numbers from the distribution
UMB()	Defines the gamlss.family object for regression modeling

To demonstrate the practical utility of the package, we present two illustrative examples. The first example focuses on basic parameter estimation from a random sample, while the second demonstrates the flexibility of the package for regression modeling with covariates.

Example 1: Parameter Estimation

In this first example, we generate a random sample of size $n = 300$ from the UMB distribution with a true parameter value of $\mu = 0.5$. The purpose is to demonstrate how the gamlss framework, integrated with ZeroOneDists, can be used to recover the parameter of the distribution.

```
library(gamlss)
library(ZeroOneDists)
```

```
# Generating some random values with known mu
set.seed(12)
y <- rUMB(n=300, mu=0.5)

# Fitting the model
library(gamlss)
mod1 <- gamlss(y~1, family=UMB)
#> GAMLSS-RS iteration 1: Global Deviance = -323.7278

# Extracting the fitted values for mu
# using the inverse link function
exp(coef(mod1, what="mu"))
#> (Intercept)
#>    0.4951016
```

The fitted model yields an estimate of $\hat{\mu} \approx 0.495$, which is very close to the true parameter value ($\mu = 0.5$), illustrating the accuracy of the maximum likelihood estimation procedure implemented in the package.

Example 2: Regression Model

This second example illustrates how to fit a regression model where the parameter μ depends on explanatory variables. We simulate a dataset where the response variable Y follows a UMB distribution, and μ is modeled as a function of a covariate X using a logarithmic link function: $\log(\mu) = 0.3 + 0.2X$.

```
library(gamlss)
library(ZeroOneDists)

# Data generation function
set.seed(12)
gendat <- function(n) {
  x1 <- runif(n)
  # True model: log(mu) = 0.3 + 0.2 * x1
  # This corresponds to beta_0 = 0.3 and beta_1 = 0.2
  mu <- exp(0.3 + 0.2 * x1)
  y  <- rUMB(n = n, mu = mu)
  data.frame(y = y, x1 = x1)
}

# Generate data and fit the model
dat <- gendat(n=1000)
mod <- gamlss(y ~ x1, family = UMB, data = dat,
              control = gamlss.control(trace = FALSE))
```

```
# Extraction of estimated coefficients
beta_0_hat <- coef(mod, what = "mu")[1]
print(beta_0_hat)
#> (Intercept)
#>    0.3033388

beta_1_hat <- coef(mod, what = "mu")[2]
print(beta_1_hat)
#>         x1
#> 0.1972036
```

The estimated coefficients ($\hat{\beta}_0 \approx 0.303$ and $\hat{\beta}_1 \approx 0.197$) are remarkably close to the true values used for simulation (0.3 and 0.2, respectively), confirming that the UMB() family function correctly interfaces with the gamlss regression engine to model covariate effects. Furthermore, the package includes the datasets used in the applications section of this paper to facilitate reproducibility.

6   Simulation Study

This section presents the results of a Monte Carlo simulation study aimed at exploring parameter estimation for models with and without covariates. To evaluate the performance of the estimations of the parameter μ, the mean bias and the mean squared error (MSE) were used. These metrics are defined as follows:

$$\text{Bias} = \frac{1}{m} \sum_{i=1}^{m} \hat{\mu}_i - \mu \tag{9}$$

$$\text{MSE} = \frac{1}{m} \sum_{i=1}^{m} (\mu - \hat{\mu}_i)^2 \tag{10}$$

6.1   Results Without Covariates

In the first part of the simulation study, we replicate the same simulation study as Biçer et al. (2024). The model used in this part can be summarized as follows:

$$Y_i \sim \text{UMB}(\mu), \tag{11}$$

The parameter μ can take four values according to the following four cases:

- Case 1: $\mu = 0.25$.
- Case 2: $\mu = 0.50$.
- Case 3: $\mu = 1.0$.
- Case 4: $\mu = 2.0$.

For each value of μ and sample size $n = 50, 100, 500, 1000$, we completed the following steps to simulate the data and estimate the parameter:

1. Generate a sample of size n from the UMB(μ).
2. Obtain and store the estimates $\hat{\mu}$ using the proposed procedure, which is the MLE analytical solution given by Eq. (8).
3. Repeat steps 1 and 2 a total of m = 5000 times.

Figure 2 illustrates the performance of the estimator for μ in the four evaluated scenarios.

The left panel shows that the bias is negligible in all cases, with values oscillating very close to zero and approaching it as the sample size increases. The magnitude of the bias tends to be slightly larger for higher values of μ (as in case 4, $\mu = 2.0$), although it remains negligible. The right panel, which shows the mean squared error (MSE), reveals two clear patterns. First, the MSE shows a clear downward trend as the sample size n increases for all values of μ, which empirically illustrates the consistency of the estimator. Second, it is evident that the MSE is larger when the value of the parameter μ is larger.

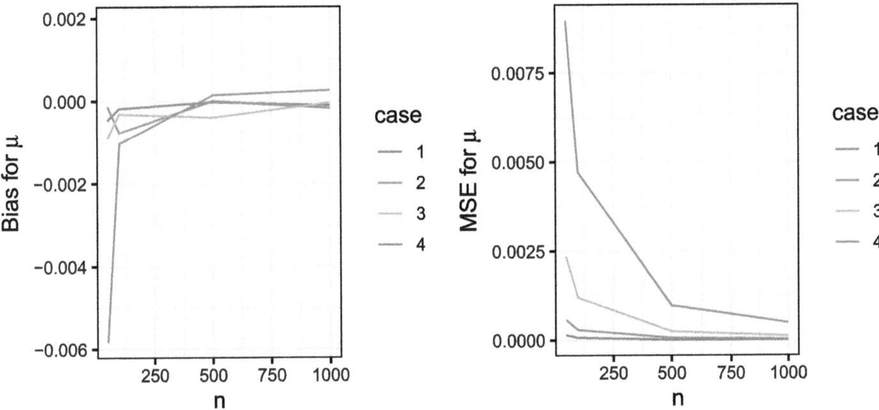

Fig. 2. Mean bias (left) and MSE (right) for $\hat{\mu}$ versus n. Each line corresponds to one of the four cases.

Furthermore, Table 2 presents the numerical results for the mean, absolute bias (AB), and mean squared error (MSE) of each estimated parameter for all cases considered in the scenario without covariates. The table shows that the estimated means converge to their true values as the sample size n increases, while both the absolute bias and the MSE progressively decrease toward zero with a larger n.

6.2   Results with Covariates

In the second part of the simulation study, we consider data generated from the following model:

Table 2. Mean, absolute bias, and MSE for the estimation of μ under each case and sample size n.

Case	n	Mean($\hat{\mu}$)	AB($\hat{\mu}$)	MSE($\hat{\mu}$)
$\mu = 0.25$	50	0.2495	0.01046	0.000143
	100	0.2498	0.00765	0.000077
	500	0.2500	0.00330	0.000014
	1000	0.2499	0.00231	0.000007
$\mu = 0.50$	50	0.4997	0.02106	0.000579
	100	0.4992	0.01493	0.000289
	500	0.5000	0.00678	0.000060
	1000	0.4998	0.00481	0.000030
$\mu = 1.00$	50	0.9991	0.04255	0.002363
	100	0.9997	0.03031	0.001192
	500	0.9996	0.01357	0.000240
	1000	1.0000	0.00948	0.000118
$\mu = 2.00$	50	1.9942	0.08265	0.008975
	100	1.9990	0.06002	0.004712
	500	2.0001	0.02711	0.000966
	1000	2.0003	0.01934	0.000483

$$Y_i \sim \text{UMB}(\mu_i),$$
$$\log(\mu_i) = \beta_0 + \beta_1 \times X_i, \tag{12}$$
$$X_i \sim \text{U}(0, 1).$$

The vector parameter was set as $\boldsymbol{\Theta} = (\beta_0 = 0.3, \beta_1 = 0.2)^\top$. These values were selected to ensure a response variable with an approximate UMB ($\mu = 1.5$) to emulate case 4 presented in the previous section. For each sample size $n = 50, 100, 500$ and 1000, we followed these steps to simulate the data and estimate the parameters:

1. Generate a random sample of n values using the `gendat` function, which implements the model given in expression (12).
2. Obtain and store the estimates $\hat{\beta}_0$ and $\hat{\beta}_1$ by fitting the regression model using the gamlss framework, following the procedure illustrated in Example 2.
3. Repeat steps 1 and 2 a total of $m = 5000$ times.

Figure 3 shows the performance of the maximum likelihood estimators for the regression model parameters.

The left panel indicates that the bias is negligible for both estimators, with values on a very small scale. Both lines clearly converge towards zero as the sample size increases, which illustrates that the estimators are asymptotically unbiased.

The right panel shows a clear downward trend in the MSE for both parameters as n increases, which empirically suggests their consistency. Furthermore, it is evident that the MSE for the slope estimator $\hat{\beta}_1$ is consistently higher than that of the intercept $\hat{\beta}_0$, a common result in regression model fitting.

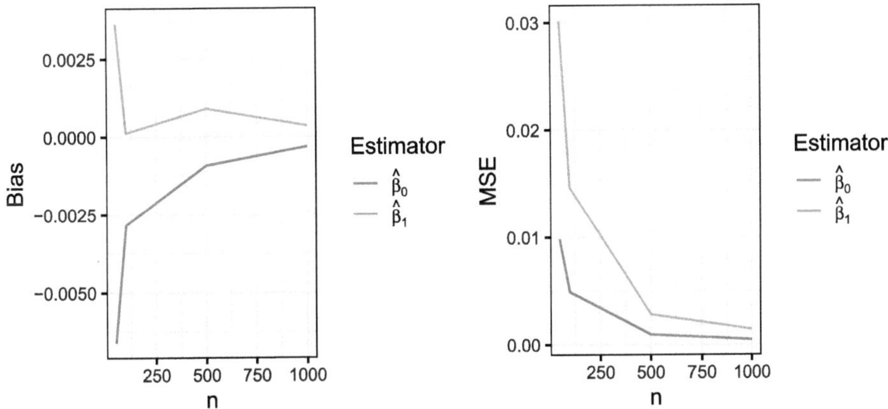

Fig. 3. Mean bias (left) and MSE (right) for the estimated parameters $\hat{\beta}_0$, $\hat{\beta}_1$ versus n.

7   Applications

In this section, the performance of the implemented UMB family is evaluated using three environmental datasets. The first two datasets correspond to those analyzed by Biçer et al. (2024). For these cases, the UMB distribution is compared against the unit Topp Leone (UTL), unit log-Lindley (ULL), unit log-weighted exponential (ULWE), and unit Kumaraswamy (UKw) distributions, as well as the Beta distribution, which serves as a standard benchmark. Additionally, a third dataset from Chambers et al. (1983) is analyzed to assess the UMB regression model with covariates, specifically comparing it against the Beta regression model. The pdfs of the competing unit distributions are presented in Table 3.

Table 3. Competing models with distribution functions.

Distribution	pdf	Domain	Reference
UTL	$2\mu x^{\mu-1}(1-x)(2-x)^{\mu-1}$	$x \in (0,1), \mu > 0$	Sangsanit and Bodhisuwan (2016)
ULL	$\frac{\mu^2}{1+\lambda\mu}(\lambda - \log x)x^{\mu-1}$	$x \in (0,1), \mu > 0, \lambda > 0$	Gómez-Déniz et al. (2014)
ULWE	$\frac{\mu\lambda}{x}(\log(x^{-1}))^{\lambda-1}\exp\left(-\mu(\log(x^{-1}))^{\lambda}\right)$	$x \in (0,1), \mu > 0, \lambda > 0$	Altun (2021)
UKw	$\lambda\mu x^{\mu-1}(1-x^{\mu})^{\lambda-1}$	$x \in (0,1), \mu > 0, \lambda > 0$	Kumaraswamy (1980)
BE	$\frac{1}{B(\mu,\lambda)}x^{\mu-1}(1-x)^{\lambda-1}$	$x \in (0,1), \mu > 0, \lambda > 0$	Johnson et al. (1995)

7.1   Dataset I

The first dataset contains the concentration of sulfate in Calgary during 31 different periods in 1995. Sulfate particles are pollutants associated with numerous health problems, such as reduced lung function.

The R code to fit the model is shown below.

```
data3 <- c(0.048, 0.013, 0.040, 0.082, 0.073, 0.732, 0.302,
           0.728, 0.305, 0.322,  0.045, 0.261, 0.192,
           0.357, 0.022, 0.143, 0.208, 0.104, 0.330, 0.453,
           0.135, 0.114, 0.049, 0.011, 0.008, 0.037, 0.034,
           0.015, 0.028, 0.069, 0.029)


# Fitting the model
library(ZeroOneDists)
library(gamlss)


mod3 <- gamlss(data3 ~ 1, family=UMB)
#> GAMLSS-RS iteration 1: Global Deviance = -46.7013


# Extracting the fitted values for mu
# using the inverse link function
exp(coef(mod3, what="mu"))
#> (Intercept)
#>    1.582003


# Extraction of the log likelihood
logLik(mod3)
#> 'log Lik.' 23.35066 (df=1)
# Extraction of the AIC
AIC(mod3)
#> [1] -44.70131
```

Table 4 shows the comparative results. The UMB model achieves the lowest AIC value, outperforming all competing distributions, including the two-parameter Beta model. This demonstrates that the UMB provides an excellent balance between simplicity and goodness-of-fit for this dataset. The results align with those presented in Tables 13 and 14 of Biçer et al. (2024).

Table 4. Estimated parameters and goodness-of-fit criteria for the models (Dataset I). Results for UTL, ULL, ULWE, and UKw distributions were taken from Biçer et al. (2024).

Distribution	$\hat{\mu}$	$\hat{\lambda}$	l	AIC
UMB	1.5820	–	23.3506	**−44.7012**
UTL	0.5424	–	21.6018	−41.2036
ULL	0.0000	0.8192	23.2327	−42.4654
ULWE	0.0000	0.8192	23.2327	−42.4654
UKw	0.7388	2.9782	23.6241	−43.2482
BE	0.1796	0.4505	23.3024	−42.6049

7.2 Dataset II

The second dataset also contains the CO concentration in Alberta, Canada, during 1995, but from the Calgary Northwest (residential) Monitoring Unit (CRMU) station.

The R code to fit the model is shown below.

```
data4 <- c(0.16, 0.19, 0.24, 0.25, 0.30, 0.41, 0.40,
           0.33, 0.23, 0.27, 0.30, 0.32, 0.26, 0.25,
           0.22, 0.22, 0.18, 0.18, 0.20, 0.23)


mod4 <- gamlss(data4 ~ 1, family=UMB)
#> GAMLSS-RS iteration 1: Global Deviance = -35.9771


# Extracting the fitted values for mu
# using the inverse link function
exp(coef(mod4, what="mu"))
#> (Intercept)
#>    0.8161202


# Extraction of the log likelihood
logLik(mod4)
#> 'log Lik.' 17.98855 (df=1)
# Extraction of the AIC
AIC(mod4)
#> [1] -33.9771
```

Table 5 summarizes the results for this dataset. Similar to the previous case, the UMB model proves to be clearly superior to all competitors, achieving the highest log-likelihood and the lowest AIC value, significantly surpassing the fit of the Beta distribution. This once again confirms the findings of Biçer et al. (2024) (Tables 16 and 17) and highlights the robustness of the implementation.

Table 5. Estimated parameters and goodness-of-fit criteria for the models (Dataset II). Results for UTL, ULL, ULWE, and UKw distributions were taken from Biçer et al. (2024).

Distribution	$\hat{\mu}$	$\hat{\lambda}$	l	AIC
UMB	0.8161	–	17.9885	**−33.9770**
UTL	1.1959	–	8.1365	−14.2730
ULL	2.2250	D7.04×10^{-33}	8.5914	−13.1828
ULWE	1.4378	2.89×10^{-8}	8.5914	−13.1828
UKw	1.2312	4.6236	12.7916	−21.5832
BE	0.2547	0.2166	14.4938	−24.9876

Furthermore, the goodness of fit is visually confirmed in Fig. 4, where the UMB density curve shows an excellent alignment with the histograms of both datasets.

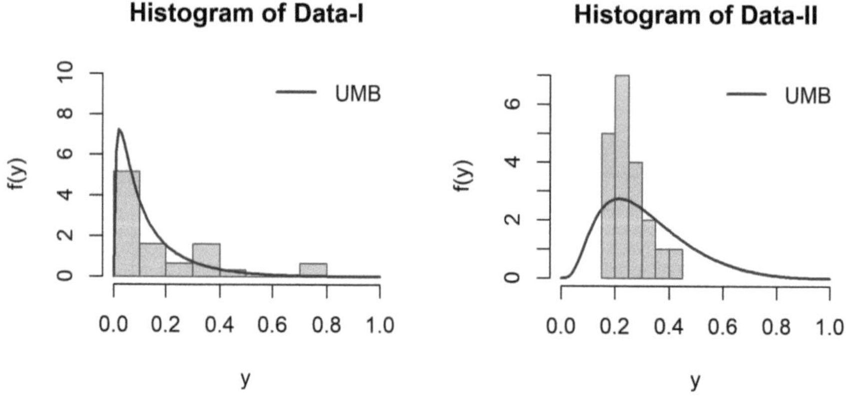

Fig. 4. Histogram of UMB for Dataset I (left) and Dataset II (right).

7.3   Dataset III

To evaluate the performance of the UMB regression model in a realistic setting, we fitted a regression model including covariates. For this analysis, the UMB model is compared specifically against the Beta regression model. The Beta distribution was selected as the sole competitor because it represents the standard benchmark for modeling data on the unit interval and, unlike other unit distributions discussed previously, possesses a standardized and robust regression implementation in R software. We utilized the airquality dataset reported by Chambers et al. (1983), which contains daily air quality measurements from the New York metropolitan area (May–September 1973). After removing missing values, the final sample size was $n = 111$. The response variable, mean Ozone concentration (in ppb), was first scaled to the closed interval $[0, 1]$ by dividing by the maximum observed value. To handle boundary values and strictly satisfy the open unit interval $(0, 1)$ requirement of unit distributions such as the UMB and Beta, we applied the transformation proposed by Smithson and Verkuilen (2006):

$$y' = \frac{y(n - 1) + 0.5}{n}, \tag{13}$$

where y is the initial scaled ozone concentration, y' is the transformed response variable used in the regression, and n is the sample size. This approach avoids numerical singularities while preserving the distributional properties of the data.

The regression model defines the location parameter μ as a function of Wind speed (mph) and Temperature (°F). Using the logarithmic link function, the model is given by:

$$\log(\mu_i) = \beta_0 + \beta_1 \times \text{Wind}_i + \beta_2 \times \text{Temp}_i \tag{14}$$

The models were fitted using the gamlss package. The following R code snippet illustrates the procedure for fitting the UMB model and the competitor Beta model:

```
library(gamlss)
library(ZeroOneDists)


# Data
data("airquality")
df_new <- na.omit(airquality)


# Normalization (Smithson & Verkuilen, 2006)
y_raw <- df_new$Ozone / max(df_new$Ozone)
n <- nrow(df_new)
df_new$y <- (y_raw * (n - 1) + 0.5) / n


# Model Fitting
fit_umb <- gamlss(y ~ Wind + Temp, family = UMB, data = df_new)
fit_beta <- gamlss(y ~ Wind + Temp, family = BE, data = df_new)
```

```
# Comparison
AIC(fit_umb, fit_beta)
```

Table 6 summarizes the parameter estimates and goodness-of-fit statistics. The results indicate that the UMB model provides a superior fit compared to the Beta model, achieving a lower AIC value. This suggests that the UMB distribution is more efficient for these data, achieving a better fit with a more parsimonious structure than the two-parameter Beta distribution.

Table 6. Parameter estimates and goodness-of-fit statistics for the New York Ozone dataset (using Smithson and Verkuilen transformation).

Model	Parameter	Estimate	Std. Error	t-value	p-value
UMB	(Intercept) β_0	2.2407	0.4653	4.82	< 0.001
	Wind β_1	0.0467	0.0139	3.36	0.001
	Temp β_2	-0.0356	0.0050	-7.08	< 0.001
	AIC	**-172.80**			
Beta	(Intercept) β_0	-4.1129	0.8053	-5.11	< 0.001
	Wind β_1	-0.1290	0.0247	-5.23	< 0.001
	Temp β_2	0.0547	0.0086	6.36	< 0.001
	AIC	-161.24			

The estimated coefficients for the UMB model are highly significant and physically consistent. The positive coefficient for Wind indicates that higher wind speeds increase the parameter μ, which shifts the distribution towards zero, leading to lower concentrations, consistent with the physical mechanism of wind acting as a ventilation factor that disperses pollutants. Conversely, the negative coefficient for Temperature indicates that higher temperatures decrease μ, thereby shifting the distribution towards higher values, thereby increasing the expected Ozone concentration. This positive association between temperature and ozone levels aligns with the exploratory analysis of this dataset by Chambers et al. (1983), reflecting the role of solar radiation and heat in ozone production.

8 Conclusions

This paper presents the implementation and validation of the Unit Maxwell-Boltzmann (UMB) distribution as a new family for the gamlss framework in R, available through the `ZeroOneDists` package. The main contribution is bridging the gap between the model's theoretical proposal and its practical application, allowing its use in regression models where the parameter μ can be modeled as a function of covariates.

The robustness of the implementation was validated through two complementary approaches. First, a Monte Carlo simulation study indicated that the maximum likelihood estimator (MLE) recovers the true parameters with very small bias and a mean squared error (MSE) that decreases as the sample size increases. Second, the analysis of three real datasets of environmental pollutants confirmed the superior fit of the UMB compared to other unit models. Notably, in the regression analysis applied to the New York ozone data (Dataset III), the UMB model outperformed the Beta regression model in terms of AIC, demonstrating its capability to efficiently capture complex environmental relationships, such as seasonality and meteorological effects, using a parsimonious single-parameter structure.

It is important to note that, as a single-parameter model, the UMB implies a functional relationship between the mean and the variance. This characteristic is advantageous for physical and environmental processes where variability naturally scales with the mean level, allowing for parsimonious modeling without the risk of overfitting. However, this also implies that the UMB is not suitable for data structures where the dispersion varies independently of the mean, or for multimodal distributions (as observed in Fig. 1), scenarios where multiparameter families would be theoretically required.

References

Altun, E.: The log-weighted exponential regression model: alternative to the beta regression model. Commun. Stat.-Theory Methods **50**(10), 2306–2321 (2021)

Biçer, C., Bakouch, H.S., Biçer, H.D., Alomair, G., Hussain, T., Almohisen, A.: Unit maxwell-Boltzmann distribution and its application to concentrations pollutant data. Axioms **13**(4), 226 (2024)

Chambers, J.M., Cleveland, W.S., Kleiner, B., Tukey, P.A.: Graphical Methods for Data Analysis. Wadsworth, Belmont (1983)

Genç, M., Özbilen, Ö.: The unit inverse maxwell-Boltzmann distribution: a novel single-parameter model for unit-interval data. Axioms **14**(8), 647 (2025)

Gómez-Déniz, E., Sordo, M.A., Calderín-Ojeda, E.: The log-lindley distribution as an alternative to the beta regression model with applications in insurance. Insurance Math. Econom. **54**, 49–57 (2014)

Hernandez-Barajas, F., Usuga-Manco, O.: ZeroOneDists: one zero statistical distributions. R package version 0.0.0.9000 (2025)

Johnson, N.L., Kotz, S., Balakrishnan, N.: Continuous Univariate Distributions, vol. 2. Wiley (1995)

Kumaraswamy, P.: A generalized probability density function for double-bounded random processes. J. Hydrol. **46**(1–2), 79–88 (1980)

Sangsanit, Y., Bodhisuwan, W.: The topp-leone generator of distributions: properties and inferences. Songklanakarin J. Sci. Technol. **38**(5) (2016)

Smithson, M., Verkuilen, J.: A better lemon squeezer? Maximum-likelihood regression with beta-distributed dependent variables. Psychol. Methods **11**(1), 54 (2006)

FoodpriceR: An R Package for Assessing the Affordability of Least-Cost Diets in Urban Contexts

Sergio A. Barona-Montoya[1]([✉]) [iD], Diego Báez Palencia[2] [iD],
José Julián Mosquera-Angulo[3] [iD], and Daniela Valdés Cárdenas[1] [iD]

[1] Department of Economics and Finance, Pontificia Universidad Javeriana Cali,
Cali, Colombia
`sergio.barona@javerianacali.edu.co`
[2] Department of Marketing and Business, Pontificia Universidad Javeriana Cali, Cali,
Colombia
[3] Universidad del Valle,
Cali, Colombia

Abstract. Economic access to sufficient and nutritious food is a necessary (but not sufficient) condition for food security. We present `FoodpriceR`, an R package for assessing the affordability of three least-cost diets: the Cost of Caloric Adequacy (CoCA), the Cost of Nutrient Adequacy (CoNA), and the Cost of a Recommended Diet (CoRD), which adheres to national FBDGs. Using linear programming, the package identifies limiting nutrients and computes shadow price elasticities, providing policy-relevant evidence. By comparing representative household diet costs with observed food expenditures, `FoodpriceR` identifies households unable to afford each diet and the affordability gap across income levels. The package contributes locally by providing preprocessed data to facilitate affordability analyses for urban households in Colombia, and globally by remaining adaptable to diverse case studies, offering a flexible, accessible, and reproducible framework in R to assess the affordability of least-cost diets.

Keywords: Food security · Healthy diets · Diet affordability · R package

1 Introduction

Economic access is a necessary, though not sufficient, condition for achieving food security. Contemporary frameworks define food security as the stable interaction of three dimensions: (i) physical availability of food, (ii) economic and physical access, and (iii) appropriate utilization to ensure nutritional well-being and the fulfillment of physiological needs [11]. A growing body of literature has focused on the second dimension —economic and physical access— by conducting empirical assessments of the affordability of least-cost diets [2,13,17,19,26]. These metrics

© The Author(s), under exclusive license to Springer Nature Switzerland AG 2026
B. M. Suárez et al. (Eds.): R Day 2025, CCIS 2824, pp. 78–96, 2026.
https://doi.org/10.1007/978-3-032-18455-9_5

quantify the minimum expenditure required to purchase locally available foods that meet dietary requirements and can sustain long-term health. Least-cost diet analysis provides a coherent framework for evaluating the economic constraints on adequate nutrition and healthy diets, revealing the limitations of poverty indicators that rely solely on energy-based standards [18,21].

Recent studies generally estimate one or more of three least-cost diet metrics: the Cost of Caloric Adequacy (CoCA) [1,14,19,22], the Cost of Nutrient Adequacy (CoNA) [2,7,9,13,23,25], and the Cost of a Recommended Diet (CoRD) [10,18,22]. The CoCA represents the minimum cost of acquiring locally available foods that provide sufficient calories to meet an individual's estimated energy requirements by age, sex, and physical activity level —an outcome that is usually dominated by a single starchy staple—. The CoNA extends this approach by requiring that minimum-cost diets also provide adequate levels of macro- and micronutrients within established lower and upper bounds. The CoRD adds adherence to national Food-Based Dietary Guidelines (FBDGs) to promote dietary diversity and long-term health.

In this paper, we introduce `FoodpriceR`, an R package specifically designed to estimate the minimum cost and affordability of urban households' diets across the three standard least-cost metrics (CoCA, CoNA, and CoRD) [16]. The package also implements complementary analytical features, including sensitivity analyses to identify limiting nutrients and computation of shadow price elasticities. To demonstrate its applicability, `FoodpriceR` includes functions to load and process publicly available data on food prices, nutrient composition, household income, and food expenditures, enabling direct implementation for Colombia's thirteen principal urban areas, as illustrated in recent applications [30]. Beyond its local relevance, `FoodpriceR` provides a flexible and reproducible analytical framework that can be adapted to other urban contexts, provided that the underlying data structures and initial conditions align with the package's functional requirements.

The remainder of this paper is organized as follows. Section 2 presents the methodological framework for estimating and assessing the affordability of least-cost diets. Section 3 describes the `FoodpriceR` package, including its architecture, main functions, and required input structures. Section 4 illustrates the package through an empirical application for Bogotá (Colombia) and via built-in routines for national replication. Section 5 concludes with key insights and directions for future work.

2   Methods

2.1   Notation

This section introduces the notation used to define the least-cost diet optimization model. Let $j \in \{1,\ldots,J\}$ index food items and $m \in \{1,\ldots,M\}$ index nutrients. All quantities are expressed on an edible-portion basis, per 100 g of food. The parameters and variables listed in Table 1 are used to formulate the LP models.

Table 1. Notation used in the least-cost diet optimization model (all per 100 g edible portion)

p_j	Retail price of food j, expressed per 100 g of the edible portion
$x_j \geq 0$	Quantity (hundreds of grams) of food j included in the least-cost diet
e_j	Energy content (kcal) per 100 g of the edible portion of food j
a_{mj}	Amount of nutrient m per 100 g of the edible portion of food j
EER	Individual's estimated energy requirement (kcal/day)
LL_m, UL_m	Lower and upper bounds for the daily intake of nutrient m

2.2 Least-Cost Diet Metrics

The CoCA represents the minimum daily cost of achieving caloric sufficiency given local food prices and energy content. The LP formulation is given by:

$$\text{CoCA} = \min_{\{x_j \geq 0\}} \sum_{j=1}^{J} p_j x_j \quad \text{s.t.} \quad \sum_{j=1}^{J} e_j x_j = \text{EER} \tag{1}$$

The optimal solution is typically determined by the least-cost starchy staple that satisfies the individual's EER [22].

Following the classical formulation of Stigler's diet problem and its subsequent extensions [24], the CoNA is formulated as the following LP problem:

$$\text{CoNA} = \min_{\{x_j \geq 0\}} \sum_{j=1}^{J} p_j x_j \quad \text{s.t.} \quad \begin{cases} \sum_{j=1}^{J} e_j x_j = \text{EER}, \\ LL_m \leq \sum_{j=1}^{J} a_{mj} x_j \leq UL_m, \quad \forall m \in \{1, \dots, M\}. \end{cases} \tag{2}$$

The CoRD represents the minimum expenditure required to adhere to national Food-Based Dietary Guidelines (FBDGs), while ensuring diversity within and between food groups. The CoRD is derived through a relatively simple procedure based on the selection of least-cost food items from each group as described in prior research [10,18]. First, assuming that food prices are reported per 100 g of the edible portion, the price per edible serving is estimated by dividing the unit price by 100 and multiplying by the serving size specified in the FBDG. Second, the number of items to be selected from each food group is defined to ensure both intra- and inter-group diversity. Third, the least-cost food items within each group are selected based on their price per edible serving. Finally, the CoRD is computed as the sum of the average cost per serving for each food group, weighted by the number of recommended daily servings.

2.3   Complementary Analysis: Limiting Nutrients and Shadow Prices

Within the LP model defined in Equation (2), *limiting nutrients* are those whose lower-bound requirements are exactly met in the optimal solution, implying that they constrain the cost-minimizing combination of foods [6]. Formally, given the optimal vector of food quantities $\tilde{x} = (\tilde{x}_1, \dots, \tilde{x}_J)$, a nutrient m is limiting if $\sum_{j=1}^{J} a_{mj}\tilde{x}_j = LL_m$.

The sensitivity of the minimum diet cost to changes in nutrient requirements is assessed through the *shadow prices* derived from the LP solution [2]. Each shadow price represents the marginal increase in the estimated CoNA associated with a one-unit rise in the corresponding nutrient requirement. To ensure comparability across nutrients expressed in different measurement units, *shadow price elasticities* (SPE) are computed following Masters et al. [23]. SPE quantify the percentage change in the estimated CoNA per one-percent change in the nutrient requirement, providing policy-relevant evidence for identifying which nutrient fortification or food-price intervention would most efficiently reduce the cost of nutrient adequacy and improve diet affordability.

2.4   Affordability Analysis

FoodpriceR adopts the standard affordability framework based on individualized diets aggregated to a representative household [3,5,14,15]. The minimum cost of each diet for the household is obtained by summing the least-cost diets estimated for all members. Affordability is then evaluated by comparing the per capita cost of each diet with the per capita food expenditure of urban households. Two indicators are computed: (i) the share of households whose per capita food expenditure is below the corresponding diet cost, and (ii) the cost-to-expenditure ratio. A ratio above one indicates that the least-cost diet exceeds typical food spending; a ratio below one represents the proportion of current expenditure required to afford it.

To capture the intensity and severity of unaffordability, two complementary indices are computed. Let Z denote the threshold given by the estimated per capita minimum diet cost, and let Y_h be the per capita food expenditure of household $h \in \{1, \dots, H\}$. Define the affordability gap as $\mathrm{Gap}_h = (Z - Y_h)\,\mathbf{1}(Y_h < Z)$, where $\mathbf{1}(\cdot)$ is the indicator function. The affordability gap index (I_1) and its squared version (I_2) are given by

$$I_1 = \frac{1}{H} \sum_{h=1}^{H} \left(\frac{\mathrm{Gap}_h}{Z} \right), \qquad I_2 = \frac{1}{H} \sum_{h=1}^{H} \left(\frac{\mathrm{Gap}_h}{Z} \right)^2.$$

Here, I_1 measures the average proportional shortfall among all households (assigning zero to households that can afford the diet), while I_2 puts greater weight on households with the largest affordability gaps.

3   Description of the FoodpriceR Package

3.1   Installation

The `FoodpriceR` package is openly available via GitHub and can be installed using `devtools` [28][1]. Installation proceeds as follows:

```
R> devtools::install_github("lea-puj/FoodpriceR")
```

Once installed, the package must be loaded into the R session:

```
R> library(FoodpriceR)
```

These commands also install the package dependencies. The code follows the tidy syntax, relying on `tidyverse` for data manipulation [27], `stringdist` for approximate matching when harmonizing food names and subgroups [20], `lpSolve` to solve the LP models underlying CoCA/CoNA [8], `rio` for format-agnostic import/export of tabular inputs and outputs [4], `janitor` for data cleaning [12], and `knitr` for vignette and example generation [29].

Table 2. Main `FoodpriceR` functions: purpose, minimal arguments, and principal return values.

Function (signature)	Purpose	Core arguments (minimal)	Key returns (top-level list)
`CoCA(data, EER)`	Minimum cost to satisfy energy requirements	`data`: food table with unit price, serving size, energy; `EER`: energy requirement by demographic group	`cost` (daily and per 1,000 kcal), `p` (price vector), `x` (optimal quantities)
`CoNA(data, EER_LL, UL, Exclude = NULL)`	Minimum cost to satisfy energy and nutrient bounds	`data`: food table including nutrient columns; `EER_LL`: lower bounds by group; `UL`: upper bounds; `Exclude`: optional vector of foods to omit	`cost`, `comp` (diet composition), `limit` (binding nutrients), `spe` (shadow prices & elasticities), `constraints`
`CoRD(data, serv, diverse)`	Minimum cost to adhere to FBDGs with intra-/inter-group diversity	`data`: group/subgroup, serving size, serving/gram prices; `serv`: recommended daily servings by subgroup and demography; `diverse`: number of distinct items per subgroup	`cost`, `comp` (servings by item/subgroup), `serv` (resolved targets)
`HCost(Data, Household, EER = NULL, EER_LL = NULL, UL = NULL, Serv = NULL, Diverse = NULL)`	Aggregate individual least-cost diets to a representative household	`Household`: members and demography; `Data`: food table; plus any subset of `EER`/`EER_LL`/`UL` (CoCA/CoNA) and `Serv`/`Diverse` (CoRD)	Nested `Model_CoCA`, `Model_CoNA`, `Model_CoRD` with per-capita and total household costs (daily, monthly, annual)
`Afford(Hexpense, Model_CoCA, Model_CoNA = NULL, Model_CoRD = NULL)`	Affordability indicators combining diet costs with budgets	`Hexpense`: income and food expenditure by household; `Model_Co*`: per-capita diet costs (from `HCost()` or constants)	`Mean_income_food` (decile summaries), `poverty_outcome` (headcount, gap, squared gap; cost-to-expenditure ratios)
`DataCol(Month, Year, City, Percentile = 0.25)`	(Colombia) Build `foodtable` from official sources	`Month`, `Year`, `City`, price `Percentile`	`foodtable` ready for CoCA/CoNA/CoRD
`IncomeCol(Month, Year, City)`	(Colombia) Build `income_file` for urban areas	`Month`, `Year`, `City`	`income_file` ready for `Afford()`.

---

[1] All source code, documentation, example datasets, and replication materials are publicly available in the FoodpriceR repository, maintained by the Applied Economics Laboratory at Pontificia Universidad Javeriana Cali: https://github.com/lea-puj/FoodpriceR.

3.2 Software Architecture and Main Functions

`FoodpriceR` follows a functional design in which each routine maps tidy data frames into S3-style list outputs. Inputs are case-sensitive and must use consistent units (prices in a single currency; quantities on an edible-portion basis; energy in kcal). Demographic keys (`Age`, `Sex`) must align across all input objects. Table 2 summarizes the exported functions, stating their purpose, the minimal argument set required for a valid call, and the main components of the returned lists. Table 3 details the corresponding input-object schemas. Figure 1 links both tables by depicting the end-to-end workflow from data ingestion to affordability indicators.

Table 3. Input object schemas: minimal required columns and key notes.

Input object	Used by	Required columns (minimal)	Optional/notes
`foodtable`	CoCA, CoNA, CoRD, HCost	Food; unit price for edible portion (`Price_100g` or `Price_g`); Serving or `Serving_g`; Energy. For CoNA: nutrient columns `Nut_1, ..., Nut_K`. For CoRD: Group, Subgroup	`Price_serving` (overrides derivation from unit price and `Serving_g`); units must be globally consistent
EER	CoCA, HCost	Age, Sex, Energy	Demographic labels must match `Household`
EER_LL	CoNA, HCost	Age, Sex, lower bounds for `Nut_1, ..., Nut_K` (optionally Energy)	Nutrient column names must match `foodtable`
UL	CoNA, HCost	Age, Sex, upper bounds for `Nut_1, ..., Nut_K`	–
serv	CoRD, HCost	Age, Sex, Subgroup, Serving (recommended daily servings)	–
diverse	CoRD, HCost	Subgroup, Number (distinct items required)	–
Household	HCost	Person, Sex, Demogroup	Demogroups align with entries in `EER/EER_LL/UL/serv`
Hexpense (income_file)	Afford	`id_hogar` (ID), `ung` (household size), income (monthly), `food_exp` (monthly)	If per-capita fields are present, `ung` is used to reconcile totals

4 Functionality via an Empirical Example

4.1 Example 1 Case Study (Bogotá, Colombia): Least-Cost Diets and Affordability

To illustrate the main functionalities, inputs, and outputs of the `FoodpriceR` package, we adapt the empirical framework proposed by Yoshioka et al. [30], originally applied to estimate the minimum cost and affordability of three diet-quality levels for urban households in Cali, Colombia, using data from September 2022. Here, based on the same reference period, we conduct a comparable analysis for Bogotá, showing the package's applicability across different urban contexts.

Data. `FoodpriceR` operates on a structured input dataset —called `foodtable` in this example— that lists the foods available in local markets. In this example, `foodtable` comprises 97 records, each corresponding to a single food item. Key variables include: `Food` (item name), `Serving` (reference portion, typically 100 g), `Price_100g` (retail price per 100 g of edible portion), `Group` and `Subgroup`

Data Inputs
Functions: DataCol(), IncomeCol().
Food & nutrient data: retail prices (per 100 g, edible portion), nutrient composition, serving sizes, food groups/subgroups (for CoRD / GABA).
Demography and requirements: EER and nutrient bounds LL_m, UL_m by sex/age/physiological status.
Household data: income and per-capita food expenditure (urban households; main cities as in application).

Individual Diet Metrics
CoCA(): min cost s.t. $\sum e_j x_j = $ EER; returns cost (daily, per 1,000 kcal), price vector p, optimal quantities x.
CoNA(): min cost with energy and nutrient bounds $LL_m \leq \sum a_{mj} x_j \leq UL_m$; returns cost, composition (comp), limiting nutrients (limit), shadow prices / elasticities (spe).
CoRD(): min cost to adhere to FBDGs (GABA): price per edible serving, intra/inter-group diversity; returns cost, servings by item/subgroup (comp).

Household Aggregation
Function: HCost().
Input: representative household (members & demography) + outputs from CoCA/CoNA/CoRD.
Output: per-capita and total household costs (daily / monthly / annual) for each metric; nested Model_CoCA, Model_CoNA, Model_CoRD.

Affordability Analysis
Function: Afford().
Input: household food expenditure (Hexpense) + per-capita costs from HCost().
Indicators: headcount (share with $Y_h < Z$), cost-to-expenditure ratio, affordability gaps I_1 and I_2.
Outputs: indicators by diet type (CoCA, CoNA, CoRD) for the representative household.

Fig. 1. Workflow of FoodpriceR from data inputs to affordability indicators, showing functions and required inputs at each stage.

(food classifications), Serving_g (edible grams per serving), and Price_serving (serving-level price derived from Price_100g and Serving_g). The remaining columns report the energy and nutrient composition (macro- and micronutrients) of each food item. (The code below displays the first rows of the dataset).

```
R> print(head(foodtable, 6))
```

```
  Cod_TCAC            Food Serving Price_100g Serving_g Price_serving
1     A010  Arroz de primera     100   446.1377  24.92918      111.2185
2     A012 Avena en hojuelas     100   909.3720  21.41119      194.7074
```

3	A027	Galletas saladas	100	1423.5926	20.75472	295.4626
4	A029	Avena molida	100	909.3720	21.25604	193.2965
5	A053	Chócolo mazorca	100	500.5285	57.89474	289.7797
6	A072	Pastas alimenticias	100	761.4441	23.71968	180.6121

	Group	Subgroup	Energy	Protein	Carbohydrates	Lipids	Calcium
1	Cereales y raíces	Cereales	353	6.7	80.1	0.4	9
2	Cereales y raíces	Cereales	411	16.9	64.1	7.5	54
3	Cereales y raíces	Cereales	424	9.6	72.9	9.7	38
4	Cereales y raíces	Cereales	414	14.7	66.2	8.6	54
5	Cereales y raíces	Cereales	152	3.4	30.5	1.2	5
6	Cereales y raíces	Cereales	371	13.0	74.7	1.5	25

FoodpriceR also requires three parameter datasets that define energy and nutrient constraints by demographic group (age and sex). The EER dataset specifies individual estimated energy requirements (kcal/day), and the remaining parameter tables specify the corresponding nutrient requirements. Together with foodtable, these inputs enable the implementation of LP models that minimize diet cost subject to nutritional constraints.

```
R> print(head(EER, 6))
```

```
# A tibble: 6 × 3
  Age            Sex  Energy
  <chr>        <dbl>   <dbl>
1 1 a 3 años       0   1111.
2 4 a 8 años       0   1436.
3 9 a 13 años      0   2007.
4 14 a 18 años     0   2747.
5 19 a 30 años     0   2820.
6 31 a 50 años     0   2671.
```

The EER_LL dataset provides the lower nutrient limits for each macro- and micronutrient, using variable names consistent with foodtable[2].

```
R> print(head(EER_LL, 6))
```

```
# A tibble: 6 × 9
  Age            Sex  Energy Protein Lipids Carbohydrates VitaminC Folate VitaminA
  <chr>        <dbl>   <dbl>   <dbl>  <dbl>         <dbl>    <dbl>  <dbl>    <dbl>
1 1 a 3 años       0   1111.    27.8   37.0          139.       13    120      210
2 4 a 8 años       0   1436.    35.9   39.9          180.       22    160      275
3 9 a 13 años      0   2007.    50.2   55.7          251.       39    250      445
4 14 a 18 años     0   2747.    68.7   76.3          343.       63    330      630
5 19 a 30 años     0   2820.    98.7   62.7          352.       75    320      625
6 31 a 50 años     0   2671.    93.5   59.4          334.       75    320      625
```

Similarly, the UL dataset defines the upper bounds for each nutrient, using the same variable naming structure as foodtable.

---

[2] Energy may be included as a reference, but it is optional if it has already been specified in EER.

```
R> print(head(UL, 6))
```

```
# A tibble: 6 × 8
   Age          Sex Protein Lipids Carbohydrates VitaminC Folate VitaminA
   <chr>      <dbl>   <dbl>  <dbl>         <dbl>    <dbl>  <dbl>    <dbl>
1 1 a 3 años      0    55.5   49.4          180.      400    300      600
2 4 a 8 años      0    71.8   55.9          233.      650    400      900
3 9 a 13 años     0   100.    78.0          326.     1200    600     1700
4 14 a 18 años    0   137.   107.           446.     1800    800     2800
5 19 a 30 años    0   141.   110.           458.     2000   1000     3000
6 31 a 50 años    0   134.   104.           434.     2000   1000     3000
```

For the CoRD metric, two additional datasets are required. The `serv` table records the daily servings needed by demographic profile and food subgroup, with columns `Age`, `Sex`, `Subgroup`, and `Serving` (servings/day). These fields indicate how many portions from each subgroup must be included in the optimized diet.

```
R> print(head(serv, 6))
```

```
# A tibble: 6 × 4
   Age          Serving Subgroup     Sex
   <chr>          <dbl> <chr>      <dbl>
1 1 a 3 años       1.41 Cereales       1
2 4 a 8 años       1.32 Cereales       1
3 9 a 13 años      1.82 Cereales       1
4 14 a 18 años     2.19 Cereales       1
5 19 a 30 años     2.11 Cereales       1
6 31 a 50 años     1.99 Cereales       1
```

The `diverse` table then specifies inter- and intra-group diversity requirements, indicating for each `Subgroup` the number of distinct items (`Number`) that must be selected to ensure dietary variety.

```
R> print(head(diverse, 4))
```

```
# A tibble: 10 × 2
   Subgroup   Number
   <chr>       <dbl>
1 Lácteos         1
2 Grasas          1
3 Azúcares        1
4 Verduras        2
```

Estimation of CoCA, CoNA, and CoRD Metrics. Using the datasets `foodtable` and `EER`, the Cost of Caloric Adequacy (CoCA) is estimated through the `CoCA()` function:

```
R> Model_CoCA <- CoCA(data = foodtable, EER = EER)
```

```
CoCA: Average daily cost per 1000 kilocalories is   1263.846
```

```
R> Model_CoCA$cost
```

```
                  Food quantity      Demo_Group Sex           Group cost_day Cost_1000kcal
1   Arroz de primera 288.6984        1 a 3 años   1 Cereales y raíces 1287.993      1263.846
2   Arroz de primera 372.9665        4 a 8 años   1 Cereales y raíces 1663.944      1263.846
3   Arroz de primera 513.0898       9 a 13 años   1 Cereales y raíces 2289.087      1263.846
4   Arroz de primera 598.5667      14 a 18 años   1 Cereales y raíces 2670.432      1263.846
5   Arroz de primera 575.5889      19 a 30 años   1 Cereales y raíces 2567.919      1263.846
6   Arroz de primera 542.9140      31 a 50 años   1 Cereales y raíces 2422.144      1263.846
7   Arroz de primera 493.2539      51 a 70 años   1 Cereales y raíces 2200.591      1263.846
8   Arroz de primera 463.7615          >70 años   1 Cereales y raíces 2069.015      1263.846
```

The output reports the estimated daily CoCA (in COP) for each demographic group, as well as the mean CoCA per 1,000 kcal (1,263.85 COP)[3]. Using the nutrient bounds provided by EER_LL and UL, the Cost of Nutrient Adequacy (CoNA) is obtained via:

```
R> Model_CoNA <- CoNA(data = foodtable, EER_LL = EER_LL, UL = UL)
```

```
[1] The CoNA for individuals of sex 0 and the age group >70 years was estimated to fulfill
97% of the minimum required intake of Magnesium.
No feasible solution was found to meet all nutrient requirements simultaneously.

The nutrients to use in the model are: Protein, Carbohydrates, Lipids, Calcium, Zinc, Iron,
Magnesium, Phosphorus, VitaminC, Thiamine, Riboflavin, Niacin, Folate, VitaminB12,
VitaminA, Sodium

CoNA: Average daily cost per 1000 kilocalories is:   2561.523
```

```
R> Model_CoNA$cost
```

```
    Demo_Group Sex cost_day Cost_1000kcal
1     >70 años   1 4529.171      2766.619
2    1 a 3 años   1 2762.490      2710.701
3  14 a 18 años   1 5300.573      2508.624
4  19 a 30 años   1 4618.040      2272.849
5  31 a 50 años   1 4485.705      2340.588
6    4 a 8 años   1 3591.385      2727.830
7  51 a 70 años   1 4630.655      2659.483
8   9 a 13 años   1 4795.816      2647.856
```

---

[3] The cost per 1,000 kcal is constant across groups because the CoCA diet typically consists of a single least-cost starchy staple.

The output returns the daily CoNA and the mean CoNA per 1,000 kcal (2,561.52 COP). The model also identifies limiting nutrients —those nutrients for which the constraint is binding in the optimal solution—. The column `Diff` reports the difference between the required amount (`Rest`) and the optimized intake (`Opt`); when `Diff = 0`, the nutrient is limiting:

```
R> head(Model_CoNA$limit)
```

```
    Nutrients        Opt       Rest   Diff Limiting        Age Sex
1     Protein   76.73786   70.89926   8.24        0 >70 años   0
2 Carbohydrates 313.52813 253.21165  23.82        0 >70 años   0
3      Lipids   45.01541   45.01541   0.00        1 >70 años   0
4     Calcium  933.00000  933.00000   0.00        1 >70 años   0
5        Zinc   11.19600   11.19600   0.00        1 >70 años   0
6        Iron   15.29978    8.39700  82.21        0 >70 años   0
7   Magnesium  326.55000  326.55000   0.00        1 >70 años   0
8  Phosphorus 1194.34910  541.14000 120.71        0 >70 años   0
```

To analyze the sensitivity of diet cost to changes in nutrient requirements, SPE are computed as:

```
R> head(Model_CoNA$spe)
```

```
       Age Sex     Nutrients         SP         SPE constraint
1 >70 años   0       Protein   0.000000  0.00000000        Min
2 >70 años   0 Carbohydrates   0.000000  0.00000000        Min
3 >70 años   0        Lipids   3.371589  0.03067214        Min
4 >70 años   0       Calcium   1.328057  0.25040706        Min
5 >70 años   0          Zinc  47.747203  0.10803368        Min
6 >70 años   0          Iron   0.000000  0.00000000        Min
7 >70 años   0     Magnesium   0.000000  0.00000000        Min
8 >70 años   0    Phosphorus   0.000000  0.00000000        Min
```

The highest SPE values are observed for iron among pregnant adolescents younger than 18 years (1.18) and pregnant women aged 31–50 years (1.12). Calcium displays a similarly high influence among women aged 70 years and older (0.50). These results indicate that even modest increases in these nutrient requirements substantially increase the minimum cost of adequacy.

Finally, the Cost of a Recommended Diet (CoRD) is estimated using the `serv` and `diverse` datasets, which specify recommended servings and dietary diversity parameters:

```
R> Model_CoRD <- CoRD(data = foodtable, serv = serv, diverse = diverse)
```

```
CoRD: Average daily cost per 1000 kilocalories is 4106.867
```

```
R> Model_CoRD$cost
```

```
  Demo_Group cost_day Cost_1000kcal Sex
1    > 70 años 6612.165      3985.318   1
2    1 a 3 años 4654.673      4134.708   1
3 14 a 18 años 8670.220      4067.672   1
4 19 a 30 años 8337.387      4067.672   1
5 31 a 50 años 7864.092      4067.672   1
6    4 a 8 años 5506.928      4134.477   1
7 51 a 70 años 7144.767      4067.672   1
8    9 a 13 años 7366.724      4037.198   1
```

The average CoRD per 1,000 kcal is 4,106.87 COP, and the function returns the corresponding daily CoRD for each demographic group. These estimates, together with the CoCA and CoNA values, are visualized in Fig. 2, which displays the minimum daily cost of each diet across demographic groups in Bogotá.

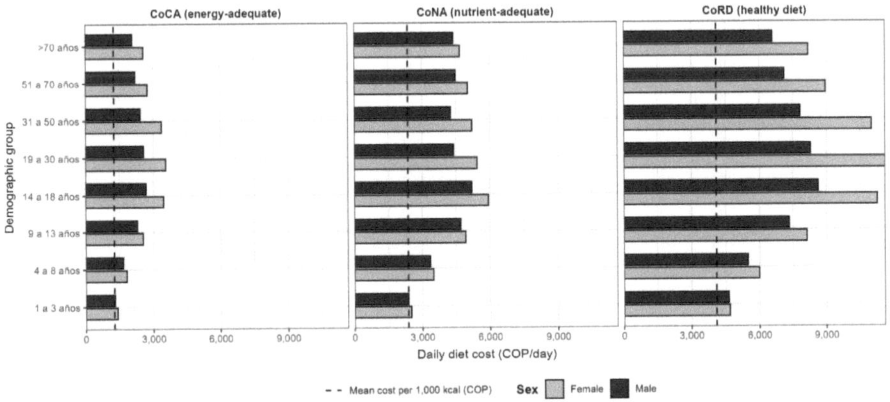

Fig. 2. Minimum daily cost of nutrient-adequate (CoNA), energy-sufficient (CoCA), and healthy (CoRD) diets by demographic group—Bogotá, Colombia, September 2022.

Representative Household. `FoodpriceR` assesses affordability by aggregating individualized diets to the household level. A representative household is defined using the `Household` dataset, which specifies members' sex and demographic group (`Person, Sex, Demogroup`). In this example, the representative household comprises three members: a male aged 31–50 years, a female aged 31–50 years, and a girl aged 9–13 years. The `HCost()` function aggregates these individual least-cost diets to obtain daily per-capita and total household costs for each diet metric:

```
R> Diet_HC <- HCost(Data = foodtable, Household = Household,
   EER = EER, EER_LL = EER_LL, UL = UL,
   Serv = serv, Diverse = diverse)
```

```
Running CoCA model...
Running CoNA model...
Running CoRD model...
Process completed.
```

```
R> print(Diet_HC$Model_CoCA)
```

```
    Demo_Group Sex Person cost_day total_household per_capita
1 31 a 50 años   0      1 2939.617        7042.361   2347.454
2 31 a 50 años   1      2 2109.308        7042.361   2347.454
3  9 a 13 años   1      3 1993.436        7042.361   2347.454
  per_capita_year per_capita_month
1        856820.6         70423.61
2        856820.6         70423.61
3        856820.6         70423.61
```

```
R> print(Diet_HC$Model_CoNA)
```

```
    Demo_Group Sex Person cost_day total_household per_capita
1 31 a 50 años   0      1 5090.451       13885.03   4628.343
2 31 a 50 años   1      2 4156.898       13885.03   4628.343
3  9 a 13 años   1      3 4637.679       13885.03   4628.343
  per_capita_year per_capita_month
1         1689345         138850.3
2         1689345         138850.3
3         1689345         138850.3
```

```
R> print(Diet_HC$Model_CoRD)
```

```
    Demo_Group Sex Person  cost_day total_household per_capita
1 31 a 50 años   0      1 11036.813        26213.02   8737.673
2 31 a 50 años   1      2  7835.569        26213.02   8737.673
3  9 a 13 años   1      3  7340.637        26213.02   8737.673
  per_capita_year per_capita_month
1         3189251         262130.2
2         3189251         262130.2
3         3189251         262130.2
```

The outputs include `cost_day` (daily cost per person), `total_household` (total daily cost for the representative household), and their monthly and annual equivalents (e.g., `per_capita_month`, `per_capita_year`).

```
R> print(head(income_file, 6))
```

```
  deciles household_id ung  income per_capita_income      share  food_exp
1 Decil 3     7187760-1   3 1909500            636500.0 0.3138554  599307.0
2 Decil 2     7187761-1   2  800000            400000.0 0.3466354  277308.3
3 Decil 6     7187762-1   4 4600833           1150208.3 0.3149129 1448862.0
4 Decil 3     7187764-1   3 2035000            678333.3 0.3138554  638695.8
5 Decil 8     7187765-1   2 4500000           2250000.0 0.2302423 1036090.3
6 Decil 8     7187767-1   4 9329833           2332458.3 0.2302423 2148122.2
  food_exp_per_capita food_exp_per_capita_year
1            199769.0                  2397228
2            138654.1                  1663850
3            362215.5                  4346586
4            212898.6                  2554783
5            518045.2                  6216542
6            537030.6                  6444367
```

Affordability indicators are computed with the `Afford()` function:

```
R> Aff <- Afford(Hexpense = income_file, Model_CoCA = Diet_HC$Model_CoCA,
  Model_CoNA = Diet_HC$Model_CoNA, Model_CoRD = Diet_HC$Model_CoRD)

Running affordability for Model_CoCA...
Running affordability for Model_CoNA...
Running affordability for Model_CoRD...
Done.
```

```
R> print(head(Aff$Mean_income_food, 6))
```

```
  decile_groups food_per_capita_avg threshold_1 threshold_2 threshold_3
1       Decil 1            82473.96    70423.61    138850.3    262130.2
2       Decil 2           150371.76    70423.61    138850.3    262130.2
3       Decil 3           196885.25    70423.61    138850.3    262130.2
4       Decil 4           271517.02    70423.61    138850.3    262130.2
5       Decil 5           277930.70    70423.61    138850.3    262130.2
6       Decil 6           401490.70    70423.61    138850.3    262130.2
    ratio_1   ratio_2   ratio_3
1 0.8538890 1.6835650 3.1783386
2 0.4683300 0.9233800 1.7432142
3 0.3576886 0.7052345 1.3313856
4 0.2593709 0.5113870 0.9654282
5 0.2533855 0.4995860 0.9431494
6 0.1754053 0.3458368 0.6528923
```

```
R> print(head(Aff$poverty_outcome, 6))
```

```
  deciles     rate       gap  severity model
1 Decil 1 33.86863 0.1518935 0.1002924  CoCA
2 Decil 2  0.00000 0.0000000 0.0000000  CoCA
3 Decil 3  0.00000 0.0000000 0.0000000  CoCA
4 Decil 4  0.00000 0.0000000 0.0000000  CoCA
5 Decil 5  0.00000 0.0000000 0.0000000  CoCA
6 Decil 6  0.00000 0.0000000 0.0000000  CoCA
```

The output summarizes per-capita food expenditures across income deciles and the corresponding "poverty lines" defined by each diet's per-capita cost. For Bogotá, results show that 33.9% of households in the lowest income decile cannot afford even the least-cost caloric diet, which corresponds to roughly 3.39% of all urban households. The `poverty_outcome` table reports affordability indicators— headcount rate, affordability gap, and squared gap—consistent with the measures defined in Sect. 2.4 (Fig. 3).

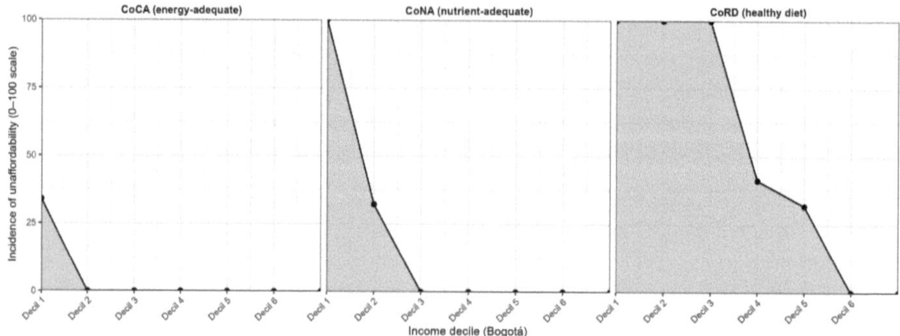

Fig. 3. Proportion (%) of urban households in Bogotá, by income decile, unable to afford each type of diet—Bogotá, September 2022.

4.2 Example 2 Built-in Defaults for Colombia (`DataCol`, `IncomeCol`)

Beyond user-supplied data, `FoodpriceR` provides built-in functions that automatically retrieve and preprocess official data for Colombian case studies. These functions reproduce the inputs required for estimating least-cost diets and affordability analyses using standardized national sources.

The `DataCol()` function constructs the `foodtable` dataset by integrating monthly food price and availability data from the Information System for Prices and Supply of the Agricultural Sector (SIPSA) of the national statistics office (DANE). For instance, the dataset used for Bogotá in September 2022 can be generated as follows:

```
R> foodtable <- DataCol(
  Month = 9, Year = 2022, City = "Bogota"
)
```

Similarly, the `IncomeCol()` function processes microdata from Colombia's GIHS (Great Integrated Household Survey) and QLS (Quality-of-Life Survey) to generate household income and food expenditure indicators for the 13 principal urban areas.

```
R> income_file <- IncomeCol(
  Month = 9, Year = 2022, City = "Bogota"
)
```

These functions allow for straightforward replication of national and subnational affordability analyses without requiring manual data preprocessing, ensuring comparability and reproducibility across applications

5 Conclusions and Future Work

`FoodpriceR` contributes to the methodological and computational framework for estimating least-cost diets and their affordability using open, structured data in R. By formalizing the three metrics of diet quality —Cost of Caloric Adequacy (CoCA), Cost of Nutrient Adequacy (CoNA), and Cost of a Recommended Diet (CoRD)— within a reproducible linear-programming framework, it enables replication and comparability across demographic groups, cities, and countries, allowing researchers and policymakers to assess the cost of these diet types and household affordability across Colombian urban areas. The empirical implementation for Bogotá, Colombia, demonstrates the practical relevance of the package: CoCA represents the lowest threshold for caloric sufficiency, often composed of a single food item, usually the least-cost starchy staple; CoNA nearly doubles that amount once nutrient adequacy constraints are introduced, identifying iron and calcium as binding nutrients with the highest shadow price elasticities; and CoRD is the highest, reflecting the additional requirements of dietary diversity and adherence to the national Food-Based Dietary Guidelines (GABA). These results highlight the economic gradient between caloric sufficiency, nutrient adequacy, and guideline-consistent diets, underscoring the challenges households face in achieving healthy diets.

Future developments of `FoodpriceR` will expand the temporal coverage of the analysis and introduce methodological updates to the package, including revised energy requirements and adaptations of the CoRD metric to the latest national Food-Based Dietary Guidelines. Looking ahead, the package aims to develop an interactive web platform for visualizing national results. This platform will also allow uploading datasets from other countries to compute the three diet metrics and their affordability indicators directly within `FoodpriceR`, thereby strengthening its role as an open and extensible tool for analyzing food and nutrition security in diverse urban contexts.

Acknowledgments. This research was financially supported by the Alliance Bioversity International – CIAT through the project "Construcción y documentación en R de la metodología para la estimación del costo mínimo a una dieta suficiente en energía, a una dieta adecuada en nutrientes y a una dieta saludable y la asequibilidad para hogares urbanos", and by Project No. 5.271, "Análisis sobre la asequibilidad de los hogares urbanos a dietas de costo mínimo para tres ciudades principales de Colombia durante 2019–2023", funded by FPIT – Banco de la República.

We further acknowledge the Centro de Innovación y Emprendimiento at Pontificia Universidad Javeriana Cali for its financial support, which contributed to the scaling and continued development of this project.

Disclosure of Interests. The authors have no competing interests to declare that are relevant to the content of this article.

References

1. Bai, Y., Alemu, R., Block, S.A., Headey, D., Masters, W.A.: Cost and affordability of nutritious diets at retail prices: evidence from 177 countries. Food Policy **99**, 101983 (2021). https://doi.org/10.1016/j.foodpol.2020.101983
2. Bai, Y., Herforth, A., Masters, W.A.: Global variation in the cost of a nutrient-adequate diet by population group: an observational study. Lancet Planetary Health **6**(1), e19–e28 (2022). https://doi.org/10.1016/S2542-5196(21)00285-0
3. Baldi, G., Martini, E., Catharina, M., et al.: Cost of the diet (CoD) tool: first results from Indonesia and applications for policy discussion on food and nutrition security. Food Nutr. Bull. **34**, S35–S42 (2013). https://doi.org/10.1177/15648265130342S105
4. Becker, J., Chan, C.H., Schoch, D., et al.: RIO: a Swiss-Army Knife for Data I/O (2025). https://doi.org/10.32614/CRAN.package.rio, https://CRAN.R-project.org/package=rio. R package version 1.2.4
5. Biehl, E., Klemm, R.D., Manohar, S., et al.: What does it cost to improve household diets in Nepal? Using the cost of the diet method to model lowest cost dietary changes. Food Nutr. Bull. **37**(3), 247–260 (2016)
6. Briend, A., Darmon, N., Ferguson, E., Erhardt, J.: Linear programming: a mathematical tool for analyzing and optimizing children's diets during the complementary feeding period. J. Pediatr. Gastroenterol. Nutr. **36**(1), 12–22 (2003)
7. Chungchunlam, S., Moughan, P.J., Garrick, D.P., Drewnowski, A.: Animal-sourced foods are required for minimum-cost nutritionally adequate food patterns for the united states. Nat. Food **16**(1), 376–381 (2020). https://doi.org/10.1038/s43016-020-0096-8
8. Csárdi, G., Berkelaar, M.: lpSolve: interface to lp_solve vol 5.5 to solve linear/integer programs (2024). https://doi.org/10.32614/CRAN.package.lpSolve, https://CRAN.R-project.org/package=lpSolve. R package version 5.6-23
9. Deptford, A., Allieri, T., Childs, R., Damu, C., Ferguson, E., et al.: Cost of the diet: a method and software to calculate the lowest cost of meeting recommended intakes of energy and nutrients from local foods. BMC Nutr. **3**(26) (2017). https://doi.org/10.1186/s40795-017-0136-4
10. Dizon, F., Herforth, A., Wang, Z.: The cost of a nutritious diet in Afghanistan, Bangladesh, Pakistan, and Sri Lanka. Global Food Security **21**, 38–51 (2019). https://doi.org/10.1016/j.gfs.2019.07.003
11. FAO, IFAD, UNICEF, WFP, WHO: The state of food security and nutrition in the world 2023 (2023)
12. Firke, S., Denney, B., Haid, C., Knight, R., Grosser, M., Zadra, J.: janitor: simple tools for examining and cleaning dirty data (2025). https://CRAN.R-project.org/package=janitor. R package version 2.2.1

13. Garille, S.G., Gass, S.I.: Stigler's diet problem revisited. Oper. Res. **49**(1), 1–13 (2001). https://doi.org/10.1287/opre.49.1.1.11187
14. Geniez, P., Mathiassen, A., de Pee, S., Grede, N., Rose, D.: Integrating food poverty and minimum cost diet methods into a single framework: a case study using a Nepalese household expenditure survey. Food Nutr. Bull. **35**(2), 151–159 (2014). https://doi.org/10.1177/156482651403500201
15. Giacobone, G., Tiscornia, M.V., Guarnieri, L., Castronuovo, L., Mackay, S., Allemandi, L.: Measuring cost and affordability of current vs. healthy diets in Argentina: an application of linear programming and the informas protocol. BMC Public Health **21**(1), 891 (2021). https://doi.org/10.1186/s12889-021-10914-6
16. González, D., Barona-Montoya, S.A., Ordoñez, J.C.: Foodprice: Un paquete en R para la estimación de dietas de costo mínimo. Working and discussion paper, Alliance Bioversity International and CIAT (2024)
17. Herforth, A., Bai, Y., Venkat, A., Mahrt, K., Ebel, A., Masters, W.A.: Cost and affordability of healthy diets across and within countries. FAO Agricultural Development Economics Technical Study No. 9, Food and Agriculture Organization of the United Nations, Rome (2020). https://doi.org/10.4060/cb2431en
18. Herforth, A., Venkat, A., Bai, Y., Costlow, L., Holleman, C., Masters, W.A.: Methods and options to monitor the cost and affordability of a healthy diet globally. FAO Agricultural Development Economics Working Paper 22–03, Food and Agriculture Organization of the United Nations, Rome (2022). https://doi.org/10.4060/cc1169en
19. Kachwaha, S., Nguyen, P., DeFreese, M., et al.: Assessing the economic feasibility of assuring nutritionally adequate diets for vulnerable populations in Uttar Pradesh, India: findings from a "cost of the diet" analysis. Curr. Dev. Nutr. **4**, 1–9 (2020). https://doi.org/10.1093/cdn/nzaa169
20. van der Loo, M., van der Laan, J., Logan, N., Muir, C., Gruber, J., Ripley, B.: stringdist: approximate string matching, fuzzy text search, and string distance functions (2025). https://CRAN.R-project.org/package=stringdist. R package version 0.9.15
21. Mahrt, K., Herforth, A.W., Robinson, S., et al.: Nutrition as a basic need: a new method for utility-consistent and nutritionally adequate food poverty lines. IFPRI Discussion Paper 02120, International Food Policy Research Institute (IFPRI), Washington, D.C. (2022)
22. Mahrt, K., Mather, D., Herforth, A., Headey, D.: Household dietary patterns and the cost of a nutritious diet in Myanmar. IFPRI Discussion Paper 1854, International Food Policy Research Institute (IFPRI) (2019)
23. Masters, W.A., Bai, Y., Herforth, A., Sarpong, D., Mishili, F., et al.: Measuring the affordability of nutritious diets in Africa: price indexes for diet diversity and the cost of nutrient adequacy. Am. J. Agr. Econ. **100**(5), 1285–1301 (2018)
24. Paris, Q.: The diet problem revisited. In: An Economic Interpretation of Linear Programming, pp. 241–253. Palgrave Macmillan, New York (2016)
25. de Pee, S., Hardinsyah, R., Jalal, F., et al.: Balancing a sustained pursuit of nutrition, health, affordability and climate goals: exploring the case of Indonesia. Am. J. Clin. Nutr. **114**(5), 1686–1697 (2021). https://doi.org/10.1093/ajcn/nqab258
26. Schneider, K.R., Christiaensen, L., Webb, P., Masters, W.A.: Assessing the affordability of nutrient-adequate diets. Am. J. Agr. Econ. **105**(2), 503–524 (2022). https://doi.org/10.1111/ajae.12334
27. Wickham, H., Averick, M., Bryan, J., et al.: Welcome to the `tidyverse`. J. Open Sour. Softw. **4**(43), 1686 (2019). https://doi.org/10.21105/joss.01686

28. Wickham, H., Hester, J., Chang, W., Bryan, J., Posit Software, P.: devtools: tools to make developing R packages easier (2025). https://doi.org/10.32614/CRAN. paquete.devtools, https://CRAN.R-project.org/package=devtools. R package version 2.4.6
29. Xie, Y.: knitr: a general-purpose package for dynamic report generation in R (2025). https://doi.org/10.32614/CRAN.package.knitr, https://yihui.org/knitr/. R package version 1.50
30. Yoshioka, A.M., et al.: Cost and affordability of three levels of diet quality for urban households in Colombia. Public Health Nutr. 1–34 (2025). https://doi.org/ 10.1017/S1368980025000564

ShinyCFA as a Tool for Understanding the Wonderful World of Categorical Data

Luis Gabriel Osorno-Muñoz$^{(\boxtimes)}$ [ID], Yamid Urrego-Galvis [ID],
and Raúl Alberto Pérez-Agamez [ID]

National University of Colombia, Medellín Campus, Medellín, Colombia
`{losornom,yurregog,raperez1}@unal.edu.co`

Abstract. Categorical data analysis is highly relevant and in demand, as it allows for the identification of patterns, relationships, and associations between qualitative variables. The integration of Shiny applications into statistical analysis has proven to be a powerful tool, facilitating learning and practice. Therefore, an app has been developed using Shiny in R software, addressing statistical topics such as simple and multiple correspondence factor analysis. This tool presents an accessible, dynamic, intuitive, and versatile interface that allows interaction with the information through various panels, buttons, and selectors. The application offers the ability to upload individual data, allowing users to perform their own analyses. Each section includes theoretical elements to contextualize the user; automated components that generate tables, graphs, and valuable results, facilitating the exploration of categorical data and the extraction of relevant knowledge for various studies.

Keywords: Shiny · Learning · Interaction · Correspondence factor analysis · Categorical data

1 Introduction

Due to recent advances in information technology, there has been an increase in the development and use of tools aimed at facilitating teaching and learning processes in different areas of knowledge. Today, there are a multitude of computer aids for teaching in certain specific areas, ranging from simple communication blogs or chat groups to sophisticated graphical user interfaces that aid the teaching process and specialized software, see. [9,11,19,20], [21,25,37,42].

When teaching statistical topics that involve the use and analysis of databases, from which some type of statistical modeling can be implemented and the subsequent analysis of results through the visualization of information using appropriate tables and graphs, it is important to have technological tools that help facilitate the understanding of the different concepts covered in these teaching processes. The R software [42] has several powerful packages for developing applications or graphical user interfaces that aid in this teaching process. Among these packages is Shiny, see [6,47]. Shiny is a package that allows us

B. M. Suárez et al. (Eds.): R Day 2025, CCIS 2824, pp. 97–114, 2026.
https://doi.org/10.1007/978-3-032-18455-9_6

to develop Shiny applications in different types of presentation formats. These applications in graphical user interface format are very useful because they allow direct integration with the user, who can use their own databases that may be available in different formats and whose origin may come from different sources that need to be analyzed using appropriate statistical processes. A large list of examples of shiny applications of different kinds can be found in the shiny example gallery, see. [6, 13, 35, 38, 40].

Current technologies have radically transformed the way we access knowledge, how learning takes place, and academic research. They have gone from being simple complementary tools to indispensable tools in education and scientific research. Some of the reasons that have led us to this point are: access to an overwhelming amount of information, collective work at a global level, which has facilitated the exchange of ideas between researchers and developers, innovative teaching methodologies, the personalization of learning, the automation of calculations and processes, among others. There is a growing desire to use more tools to answer research questions, but the aforementioned drawbacks of the complexity of the necessary statistical procedures and the proper use of statistical software are some of the limitations encountered on a daily basis. This is where simulation applications or environments come in, allowing the creation of automated calculation and visualization spaces that change based on interaction with user data. The best thing about these tools is that the user can find a user-friendly interface with which they can interact and understand different theoretical concepts experimentally; it is like playing with the data. All this without losing the ethical component with which the data must be analyzed, avoiding manipulations that consciously or unconsciously lead to erroneous decisions.

Multivariate statistical analysis is an area of statistics that deals with the study of databases for which several variables, which may be given on different measurement scales, are to be analyzed simultaneously. It is a very important area due to the large number of multivariate statistical procedures that exist for data analysis, which is directly linked to the area of Big Data, which every day involves more procedures to tackle different types of practical problems, see. [17, 18, 22], [28, 41]. In the area of Multivariate Statistical Analysis, there are some shiny applications that focus mainly on the global presentation of descriptive summaries using appropriate tables and graphs, other applications focus on data modeling using specific statistical processes, and others focus only on the digital presentation of data without any statistical analysis, see gallery of shiny examples: [2, 6, 44]. Within Multivariate Data Analysis, there is a sub-area known as Correspondence Analysis (CA), which is a useful statistical technique that works with categorical data, for example, data obtained from social surveys. The method is particularly effective for analyzing contingency tables with numerical frequency data, as it provides an elegant and simple graphical representation that allows for quick interpretation and understanding of the data, see. [16].

Currently, different applications have been developed, generally to interact with separate concepts, which contain a single window where the user can modify values and generate a graph or table based on those parameters, see. [5,8,34]. Implementations of this type of environment exist for probability distributions, linear regression, correlations, hypothesis testing, sample sizes, among others, see [12,29,48]. However, there are no Shiny applications in the literature that are specifically geared toward the CA teaching process and that facilitate direct interaction between the application user and their own data.

This article presents a Shiny application that facilitates CA teaching. The proposed application is very user-friendly and facilitates direct use and interaction with the user. The application in Spanish can be accessed directly from the following link: https://raulperezagamez.shinyapps.io/afc_/. The application is divided into several modules ranging from Simple Correspondence Analysis (SCA) to Multiple Correspondence Analysis (MCA). The first part of each panel defines the data that will be used during application execution, followed by the different results of the applied CA, which are briefly described in the article. Each panel of the application initially presents a descriptive summary of the data, followed by the different CA options. The web application that has been created allows all these elements to be compiled, such as graphs, tables, and basic theory, so that the academic community and researchers have everything they need at their fingertips to perform analyses that enrich their projects and knowledge.

This article is structured as follows. Section 2 presents the basic theoretical foundations necessary for understanding CA, which are subsequently applied in shinyApps. In Sect. 3, provide information about simple correspondence analysis. Section 4 presents the basic theoretical foundations necessary for understanding MCA, which are subsequently applied in shinyApps. Section 5 provides an overview of how shiny applications work. Section 6 provides a detailed description of how to use the proposed application, with a graphic representation of most of its parts or panels. Section 7 provides some conclusions and recommendations and finally presents the respective bibliographic references.

2   Correspondence Factor Analysis

In many surveys, it is common to see categorical variables predominate over continuous ones, since it is easier for a person or observer to describe a phenomenon or frame it in a fixed category than to explain it openly. For example, consider a survey that contains the following two questions: What is your hair color among three or four options? and What is your height? Note that both questions are easy to answer, with answers that anyone could provide, but isn't it easier to answer the question about hair color? Answering about height is simple, but it requires a measuring instrument to have a certain degree of accuracy since it is a continuous variable. Many studies on marketing, politics, education, security, among others, involve questions whose answers have to do with an opinion, with the degree of satisfaction, with the quality of a product or a particular characteristic, among others. The information obtained through the application of a survey can be analyzed using different statistical methodologies that can lead us to improve the development of some industrial procedures, improve public policies, improve customer service, advertising, the description of a phenomenon, among others. There are many statistical techniques for analyzing information obtained from the measurement of continuous or numerical variables, from the univariate perspective to the multivariate case [1,23,45] whereas the opposite is true for information obtained from the measurement of categorical variables, where existing statistical methodologies for this type of information may be fewer.

Another multivariate statistical methodology is principal component analysis (PCA) [23], also known as a dimensionality reduction technique that allows the graphical representation in low dimensions of tables of information obtained by measuring continuous or numerical variables, which allows us to discover possible relationships between the variables considered in the analysis in a practical way. On the other hand, PCA is also considered a multivariate analysis technique used especially for the exploration and identification of relationships or associations between two or more variables, but of a nominal or ordinal categorical type [16]. Similar to PCA, it also allows the visualization of large contingency tables in low-dimensional spaces. In this sense, it is a potential tool in market research, characterization of consumer preferences, characterization of individuals according to their choice trends in given categories, customer segmentation, ecological studies, Text analysis, Demography, Public health, etc. Unlike other analyses, both PCA and CA have no distributional assumptions since they are exploratory techniques, see [22].

3   Simple Correspondence Analysis (SCA)

As mentioned above, Correspondence Analysis is a graphical procedure for representing the associations in a frequency or count table. In the case of a two-way frequency table or Contingency Table (CT), the CA methodology is called Simple Correspondence Analysis (SCA) and is used to analyze the frequencies formed by the two categorical variables involved in that table. SCA provides scores (or coordinates) for the row points and column points of the contingency table. These coordinates are used to graphically visualize the association between row elements and column elements in the table. Similarly, SCA can also be understood as a geometric approach to visualizing the rows and columns of a CT as points in a low-dimensional space, so that the positions of the row points and column points are consistent with their associations in the original table. One of the objectives of SCA is to provide an overview of the data based on its graphical representation, which is useful for interpretation, see. [16,22]. Part of the fundamental theory of SCA is described below.

Contingency Table. The contingency table consists of I categories of the row variable and J categories of the column variable.

Table 1 shows the structure of the contingency table, where:

Table 1. Contingency Table for SCA

Row I / Col J	J_1 J_2 ... J_J	Marginal Row
I_1	$n_{11}\, n_{12} \ldots n_{1J}$	$n_{1.}$
I_2	$n_{21}\, n_{22} \ldots n_{2J}$	$n_{2.}$
$\vdots$	$\vdots \quad \vdots \quad \ddots \quad \vdots$	$\vdots$
I_I	$n_{I1}\, n_{I2} \ldots n_{IJ}$	$n_{I.}$
Marginal Column	$n_{.1}\, n_{.2} \ldots n_{.J}$	N

- Each cell $\mathbf{n_{ij}}$ corresponds to the frequency of the simultaneous occurrence of category i and category j.
- $\mathbf{n_{i.}}$ and $\mathbf{n_{.j}}$ are the row and column margins, respectively.
- $\mathbf{N}$ is the sum of all frequencies n_{ij}.

However, this analysis does not work directly with the contingency table, but rather processes it to obtain the relative frequency table (called $\mathbf{P}$) where the cells are given by: $p_{ij} = \frac{n_{ij}}{N}$

Row and Column Profiles. Represent the relative percentage distribution [15] and are defined as:

$$\mathbf{R} = \mathbf{D}_r^{-1}\mathbf{P} \quad \text{y} \quad \mathbf{C} = \mathbf{D}_c^{-1}\mathbf{P}^T \tag{1}$$

where $\mathbf{D}_r$ and $\mathbf{D}_c$ are the diagonal matrices of the relative marginal frequencies (masses) of the rows and columns, respectively.

Chi-Square Distance. It is a metric used to measure dissimilarity between profiles, giving greater weight to deviations in profiles in categories with low marginal frequencies, which is important for detecting interesting associations.

Singular Value Decomposition (SVD). It is one of the mathematical bases of factorial correspondence analysis [43]. Based on the marginal frequencies per row and column, the observed matrix P is centered and normalized, yielding the standardized residual matrix, given by the equation:

$$\mathbf{S} = \mathbf{D}_r^{-1/2}(\mathbf{P} - \mathbf{rc}^T)\mathbf{D}_c^{-1/2} \tag{2}$$

where $\mathbf{r}$ and $\mathbf{c}$ are the mass vectors of the rows and columns. The SVD of $\mathbf{S}$ is:

$$\mathbf{S} = \mathbf{U\Lambda V}^T \tag{3}$$

- $\mathbf{U}$ and $\mathbf{V}$: Left and right singular vector matrices (related to factorial coordinates).
- $\mathbf{\Lambda}$: Diagonal matrix of singular values λ_k (square roots of eigenvalues α_k).

Inertia. Total inertia is the weighted sum of the χ^2 distances between the profiles and the average profile. It is a measure of the total dispersion or variability of the data around its center of gravity.

One of the objectives of PCA is to project the profiles onto a lower-dimensional space (factor planes) in such a way as to maximize the portion of total inertia captured by the new axes (factors).

Factorial or Principal Coordinates. They allow the relationships between categories to be represented on a low-dimensional plane (usually two-dimensional). The coordinates of the points on the factorial map are:

$$\mathbf{F} = \mathbf{D}_r^{-1/2}\mathbf{U}\boldsymbol{\Lambda} \quad \text{y} \quad \mathbf{G} = \mathbf{D}_c^{-1/2}\mathbf{V}\boldsymbol{\Lambda} \tag{4}$$

Where F represents the coordinates in the plane for the rows and G the coordinates for the columns.

The graph produced by the ACS contains two sets of points: 1) a set of I points corresponding to the rows, and 2) a set of J points corresponding to the columns. The positions of these points will reflect the possible associations between rows, the possible associations between columns, and the possible associations between rows and columns of the CT. The row points that are closest indicate similar rows (or row profiles) (i.e., similar conditional distributions) across columns, while the column points that are closest indicate similar columns (or column profiles) (i.e., similar conditional distributions) across rows. Finally, row points that are close to column points represent combinations that occur more frequently than would be expected in an independence model, i.e., a model in which row categories are unrelated to column categories. The usual result of an ACS includes the "best" two-dimensional representation of the data, along with the coordinates (or scores) of the row/column points in the graph and a measure of the amount of information (called inertia) retained in each dimension [30].

4   Multiple Correspondence Analysis (MCA)

Simple correspondence analysis deals with contingency tables or two-dimensional absolute frequency tables. MCA can be extended to tables with three or more entries, in which case the relationship with the principal component analysis (PCA) method is more apparent. The rows of these tables are considered to be the objects or individuals, and the columns are considered to be the modalities of the categorical variables considered in the study. This type of information is very common in surveys, where the rows represent individuals, human groups, institutions, objects, etc., and the columns represent modalities of categorical variables or responses to the questions asked in the questionnaire or data collection instrument. One of the approaches of Multiple Correspondence Analysis (MCA) can be considered as a simple correspondence analysis (SCA) applied to a table called a Complete Disjunctive Table (CDT) in the sense that a categorical variable assigns each individual in a population one and only one modality of that variable, and, consequently, partitions (in a disjunctive and exhaustive manner) the individuals in the study population. It is denoted by X and encoded in binary form with zeros (0) and ones (1), where one (1) indicates that the individual belongs to that category. The structure of the CDT is shown in the Table 2.

Table 2. Matrix Representation of the Complete Disjoint Table ($\mathbf{X}$)

	$\mathbf{X_1}$			$\mathbf{X_2}$		$\mathbf{X_l}$					$\mathbf{X_k}$		$\mathbf{x_{i\cdot}}$
	1	2	p_1	1	p_2	1	$\dots$	m	$\dots$	p_l	1	p_k	
1	0	1	0	1	0	0	$\dots$	1	$\dots$	0	1	0	k
2	1	0	0	1	0	0	$\dots$	0	$\dots$	1	1	0	k
$\vdots$	$\vdots$	$\vdots$	$\vdots$	$\vdots$	$\vdots$	$\vdots$	$\ddots$	$\vdots$	$\ddots$	$\vdots$	$\vdots$	$\vdots$	$\vdots$
i	0	0	1	0	1	1	$\dots$	$\mathbf{x_{im}}$	$\dots$	0	1	0	$x_{i\cdot} = k$
$\vdots$	$\vdots$	$\vdots$	$\vdots$	$\vdots$	$\vdots$	$\vdots$	$\ddots$	$\vdots$	$\ddots$	$\vdots$	$\vdots$	$\vdots$	$\vdots$
n	0	1	0	0	1	0	$\dots$	1	$\dots$	0	1	0	k
$\mathbf{x_{\cdot j}}$	$x_{\cdot 1}$	$x_{\cdot 2}$	$x_{\cdot p_1}$	$x_{\cdot 1}$	$x_{\cdot p_2}$	$x_{\cdot 1}$	$\dots$	$\mathbf{x_{\cdot m}}$	$\dots$	$x_{\cdot p_l}$	$x_{\cdot 1}$	$x_{\cdot p_k}$	nk

The matrix is divided into blocks by variable ($\mathbf{X_1}, \mathbf{X_2}, \dots, \mathbf{X_k}$). Each block contains as many columns as categories that variable has (p_l). The right column ($X_{i\cdot}$) indicates that the sum of the values in each row is equal to the total number of variables (k), illustrating the Complete Disjointness property [22]. The last row represents the absolute marginal frequencies for each category ($x_{\cdot j}$). This is simply the number of individuals who belong to that category. Finally, the element in the lower right corner (nk) is the sum of all cells in the matrix (the number of individuals n multiplied by the number of variables k).

Burt Matrix or Table. It is a $p \times p$ matrix where p is the total number of categories of the k variables considered, which contains all possible contingency tables based on the crossings between all variables. It is symmetrical and composed of blocks. The diagonal blocks are diagonal matrices with the marginal frequencies of each variable, and the blocks outside the diagonal are the aforementioned contingency tables. This table is given by the following equation:

$$\underset{p \times p}{\mathbf{B}} = \underset{p \times n}{\mathbf{X}^T} \underset{n \times p}{\mathbf{X}} \tag{5}$$

For this analysis, the concept of decomposition into singular values, row and column profiles, inertia, and factorial coordinates is used in a similar way to SCA. In this case, these are obtained for both categories and individuals, in addition to other metrics such as representation quality, contribution, and grouping of individuals.

Despite its similarities to SCA, MCA has some peculiarities due to the very nature of the CTT, see [4,16,22,27].

The objectives of MCA include the following: 1) To provide a typology of individuals (or rows), that is, to study the similarities between individuals from a multidimensional perspective, 2) to evaluate the relationships between the variables considered and study the associations between the categories or modalities of the variables considered, and 3) to unify the study of individuals and variables in order to characterize individuals using the variables.

MCA compares observations or individuals through the modalities or categories of the variables considered in the study. A group of individuals is similar if they assume approximately the same modalities of the variables considered. The association between variables (or between modalities of variables) occurs because it is almost the same individuals who assume the same modalities of different variables. MCA finds associations between the categorical variables considered through the respective categories or modalities of those variables. As in the SCA, the MCA considers the n-point row cloud and the p-point column cloud. Again, through the use of dimensionality reduction techniques, rows or individuals and variables or variable modalities are graphically represented in appropriate low-dimensional spaces, which makes it easier to study the information contained in the data table.

5   What Is a Shiny Application and How Is It Created?

Shiny [6] is an R package [36] designed to create interactive web applications in a simple and effective way. Its main objective is to allow users, even those without web development experience, to build applications that facilitate interaction with data and statistical analysis. This package was developed by RStudio, now known as Posit [33], and has gained popularity in the data science community due to its ability to integrate real-time visualizations and analysis. The package is currently implemented in Python. Shiny is used to create dashboards, data visualizations, pilot applications, among others, without the need to write code in HTML, JavaScript, or CSS, which can be complex for novice programmers. Major industries use Shiny, including the University of Oxford, Stack Overflow, Harvard Medical School, Pfizer, Johnson & Johnson, Janssen, and AstraZeneca, among others. These are very important industries that make significant contributions to science, especially medicine.

Shiny allows users to create web applications that can interact with R code. Through a graphical interface, users can use controls such as sliders, drop-down menus, and buttons to manipulate data and visualize results instantly. Reactive programming is one of the fundamental concepts of Shiny, where the behavior of the application automatically adapts to user input. This means that any change in an interface control can trigger updates in other elements of the application without the need to reload the page [46].

To create a basic shiny application, two files are mainly required: the `ui.R` file containing the user interface and the `server.R` file containing all the operations performed by the application, or two functions *ui* and *server* in the same file. Figure 1 shows a simple application that has a slider for the user to choose the number of intervals that the histogram built in the application should have.

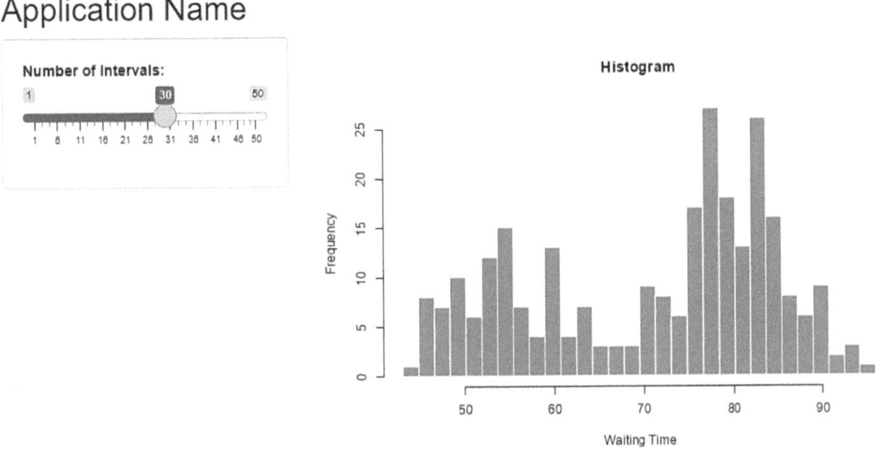

Fig. 1. Basic Shiny application

The .R file containing the two functions that generate the app is shown in Fig. 1 below.

To develop a shiny application, you need the R or Python programming language to write the code and the shiny package installed.

One of the most notable features of shiny is its ability to provide interactivity. Users can modify parameters using sliders or drop-down menus and see how the visualizations change in real time. This is especially useful in educational or analytical contexts where the user wants to explore different scenarios [5, 12, 29].

Shiny applications can be easily deployed on local servers or cloud platforms such as shinyapps.io. This allows applications to be shared with other users without additional technical complications. The community around shiny is active and offers numerous resources, from tutorials to forums where developers can share their experiences and resolve questions. There are books and scientific articles that address both technical aspects and practical applications of using Shiny. The official Shiny website, https://shiny.posit.co/, contains demonstrations of applications created with R and Python, offers a quick start guide,

and has an extensive gallery with many examples that showcase the potential of
Shiny.

```
library(shiny)                          server <- function(input, output,
ui <- fluidPage(                          session) {
  titlePanel("Application name"),         output$distPlot <- renderPlot({
  sidebarLayout(                            x <- faithful[, 2]
    sidebarPanel(                           bins <- seq(min(x), max(x),
      sliderInput("bins",                   length.out=input$bins+1)
        "Number of intervals:",             hist(x, breaks=bins,
        min=1,                                   col='darkgray',
        max=50,                                  border='white',
        value=30)                                xlab='Waiting time',
    ),                                           main='Histogram')
    mainPanel(                             })
      plotOutput("distPlot")             }
    )
  )                                       shinyApp(ui = ui, server = server)
)
```

A Shiny application has three fundamental elements for its execution: the user
interface, the server function, and the "shinyApp" function. The user interface
controls the appearance of the application: position, size, shape, colors, among
other qualities that the user will see. The server is the engine of the application,
where all the calculations required to be presented to the user are performed. The
shinyApp is the function that creates Shiny elements, which are a combination
of the server and UI.

6   Using the Shiny Application on CA

Today, there are several powerful packages directly or indirectly related to Correspondence Factor Analysis that can be used to obtain all the numerical and
graphical results required in different studies. Some of these are:

Package FactoMineR. A package dedicated to exploratory analysis of multivariate data, with a wide variety of visualization options and the ability to
handle partitions of individual variables and hierarchies [26].

Package Factoextra. Provides easy-to-use functions for extracting and visualizing the results of multivariate analyses with the versatility of ggplot2 [7].

Package FactoClass. It contains a set of tools for combining factorial methods
with cluster analysis, with the aim of exploring and understanding the structure
of a data table [31].

Package Ade4. It is one of the classic packages that has several tools for multivariate data analysis such as PCA, CFA, and multiple table analysis with great efficiency for large data sets [10].

Many more packages with different specific functionalities depending on academic or research needs can be found on the website.

However, using these elements requires certain knowledge and proficiency with R software. Thus, the proposed Shiny application integrates each of the tools to perform a complete factorial correspondence analysis. This eliminates code barriers, broadens the scope to non-technical users, allows for standardization and error prevention, generates a consistent workflow, facilitates the creation of reports with all results at hand, avoids repetitive tasks, and allows researchers to focus on interpreting results. In this way, a complex code-based statistical process becomes a robust research and learning tool with a centralized and user-friendly environment. [5,32,48].

This section presents the application on CA created with the shiny package, which facilitates the teaching and use of Correspondence Analysis, as it allows users to interact easily with a variety of widgets and outputs using their own data. The application is available in Spanish at the following link: https:// raulperezagamez.shinyapps.io/afc_/ and can be used freely by any user.

The application has three modules or panels, and each module has tabs that allow different tasks to be performed. Below is a list of the three modules: Introduction, Simple Correspondence Analysis (SCA), and Multiple Correspondence Analysis (MCA), and the different tabs (P_i) associated with each module with the tasks that can be performed.

1. Introduction
   P1. Authors.
   P2. About the APP.
   P3. Data Specifications.
   P4. Bibliography.
2. Simple Correspondence Analysis (SCA)
   P1. Definitions.
   P2. Data.
   P3. Chi-Square Test.
   P4. Relative Frequencies.
   P5. Profiles.
   P6. Ternary Graphs.
   P7. Eigenvalues and Inertia.
   P8. Row Points and Column Points.
   P9. Biplot.
   P10. Contributions.
   P11. Correlations with Axes.
3. Multiple Correspondence Analysis (MCA)
   P1. Definitions.
   P2. Data.
   P3. Frequencies by Variables.
   P4. Eigenvalues.
   P5. Variables, Individuals, and Modalities.
   P6. Biplot.
   P7. Modality Contributions.
   P8. Individual Contributions.
   P9. Correlations with Axes.
   P10. Grouping of Individuals.

The first introductory module contains general information about the authors, the type of data to be used, and some important references about CA. The other two modules each have a Data tab that allows the user to choose which data they want to work with, either using their own data or the simulated data that appears by default. This last option allows users to interact and learn about all the benefits of the application without having to worry about uploading their own data. Once users have chosen the data to use, they can continue with the various statistical procedures available in the respective module.

Below is a brief description of the results that a user can obtain when using the ACS module and some short general interpretations. The application provides a descriptive summary of the variables and different visualizations. Additionally, each section has a blue button with the word "theory," which, when clicked, displays basic theory on the specific topic. In the case of ACS, there is a section where the chi-square test is developed in order to verify the association between the variables.

Figure 2 shows part of the application menu at the top and displays the table of relative frequencies of the data in a graph called a balloon plot, where the size of the circle indicates how large the relative frequency is. Some interpretations that can be given in this regard are: The cell $(\mathbf{F_4}, \mathbf{C_4})$ has the highest frequency in the table (0.062). This indicates that row category F_4 is very strongly associated with or occurs simultaneously with column category C_4. The cell $(\mathbf{F_3}, \mathbf{C_1})$ has the lowest relative frequency (0.010). This suggests that category F_3 and category C_1 almost never occur together(Figs.2,3,4,5 and 6).

Análisis Factorial de Correspondencias

	C1	C2	C3	C4	C5	C6	
F1	0.049	0.014	0.048	0.031	0.019	0.044	0.21
F2	0.025	0.013	0.037	0.061	0.020	0.035	0.19
F3	0.010	0.044	0.020	0.039	0.039	0.022	0.17
F4	0.053	0.037	0.017	0.062	0.043	0.036	0.25
F5	0.040	0.016	0.032	0.052	0.024	0.021	0.19
	0.18	0.12	0.15	0.24	0.14	0.16	1

Fig. 2. Relative Frequency Table and Balloonplot

Figure 3 shows the row profile table together with its graph (the app also has this for column profiles). This is the conditional distribution of the columns (C_1 to C_6) within each row category (F_1 to F_5). The sum of the values in each row is 1, and the row labeled GF or marg is the marginal profile of the columns, representing the expected distribution or average profile. C_4 is the most common column category (**0.25** or 25%). Rows $\mathbf{F_3}$ and $\mathbf{F_1}$ will be the ones with the most distant points on the factorial map, as they have the most heterogeneous profiles and therefore contribute most to the total inertia (dispersion). whereas row $\mathbf{F_4}$, which is closer to the average profile (GF), will probably be drawn closer to the origin (center of gravity) of the factorial map.

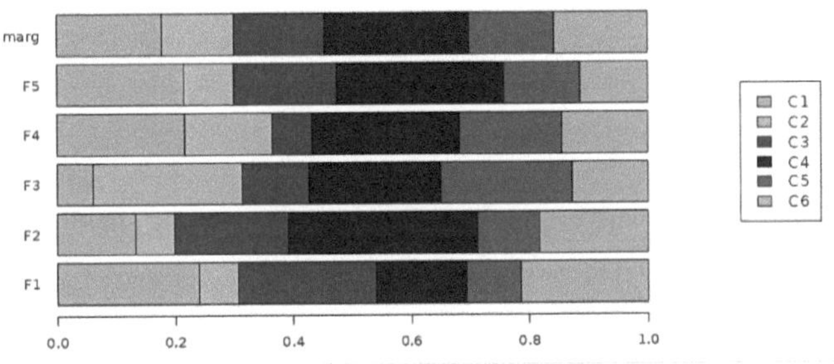

Fig. 3. Row profile table and its graph

Figure 4 shows the so-called ternary graphs, which are used to view the structure of the profiles when only three categories are taken in a row or column. Ternary graphs allow us to isolate and visually confirm the most significant affinities and oppositions detected in the general analysis. For example, for the ternary graph of column profiles, we have the distribution of each column profile (C_1 to C_5) with respect to the three selected rows ($\mathbf{F_1}$, $\mathbf{F_2}$, and $\mathbf{F_3}$). In this case, the GC point represents the center of gravity or average profile. In the left triangle, we can see that C_2 is the column most strongly attracted to F_3 (highest position). Its profile has a high proportion of F_3.

In addition to these outputs, the user will find tabs with the analysis of eigenvalues and eigenvectors involved in the process of obtaining the best low-dimensional subspace to represent the row profiles and column profiles of the SCA, another tab with the analysis of the row profiles and column profiles, which are presented in tabular form and in low-dimensional factorial planes, a

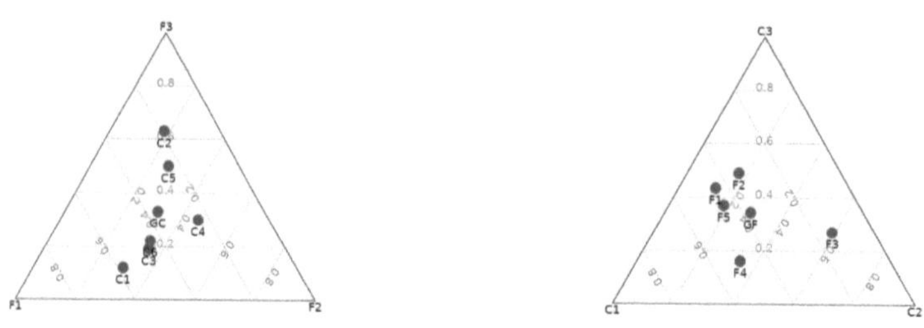

Fig. 4. Ternary graphs for row and column profiles

tab with the respective Biplot associated with the SCA results, a tab with the tables and graphs associated with the contributions of both row profiles and column profiles and their respective interpretation from a statistical point of view.

Figure 5 shows the results related to this last part, which has to do with the different types of contributions. These indicate which rows and columns are responsible for the inertia (variance) explained by the first factorial axis (the most important axis). Points with high contributions are those that define the direction and meaning of the axis. The rows with the highest contribution are those furthest from the origin along Axis 1, indicating that their profiles deviate most from the average profile in that dimension. In the case of rows, the highest contributions are given by rows F1 and F2, while for columns, the highest contributions are given by C2, C3, and C5.

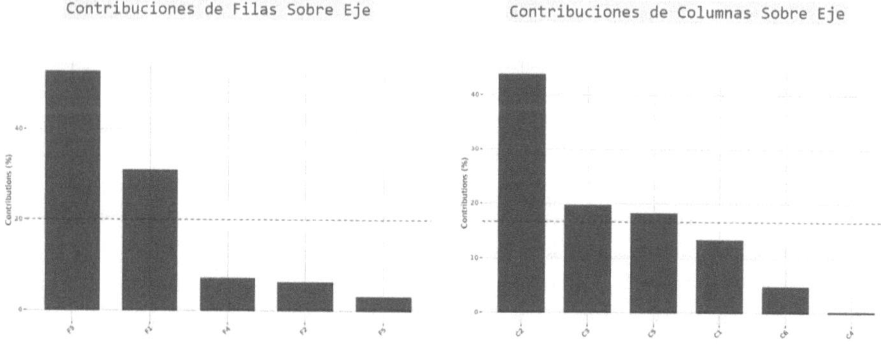

Fig. 5. Graph of row and column profile contributions on the first factorial axis

In the case of the MCA panel, there are tabs similar to those described above for SCA. Figure 6 shows the biplot with the MCA results, which is vital for providing a complete description of the statistical interpretation of these results.

This biplot contains both the points of the individuals (circles) and the categories (triangles). There, you can see a contrast or opposition (negative association) between characteristics F and A with X and Z with respect to dimension 1 (vertical axis). Similarly, there is a contrast between characteristics A and B with respect to dimension 2 (horizontal axis). Categories that are close to each other are strongly associated, for example, categories A and X. On the other hand, individuals (circles) are located close to the categories they possess, which allows for segmentation. For example, individuals 27 and 12 possess characteristics X and A, while individuals close to the origin are those who are not strongly characterized by any of the main dimensions; their profiles are closer to the overall average profile of the population.

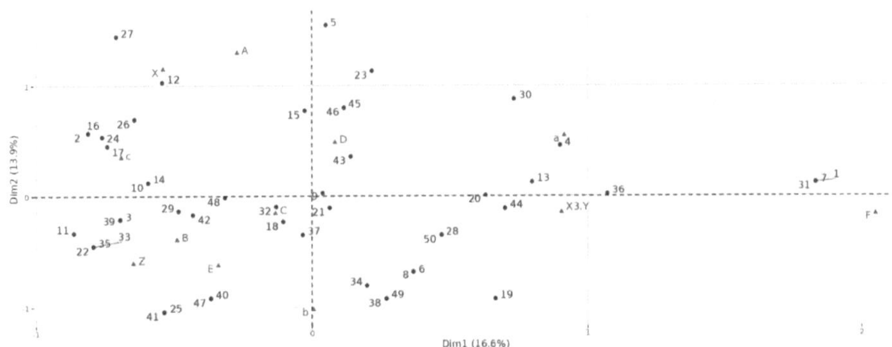

Fig. 6. Biplot of ACM Results

In summary, the application has a complete structure that allows for comprehensive analysis of this theory, but it is always advisable to immerse oneself in the existing examples and literary theory on the subject, some references for which are provided for the reader's consultation at the end of this article. [3, 14–16, 24, 39].

7   Conclusions

The application is primarily intended for teachers who teach classes on multivariate data analysis techniques, specifically correspondence analysis (CA), as it facilitates the illustration of examples using their own data. The application is also useful for students who have the basic knowledge necessary to understand the theory behind CA. The application allows users to easily interact with the concepts of Simple Correspondence Analysis (SCA) and Multiple Correspondence Analysis (MCA) in an intuitive way using their own data. Since the application can be used freely, teachers can use it in the classroom to illustrate CA topics through examples. The various links to references on basic theory at the beginning of the application and throughout help users easily understand the different results of each CA procedure performed.

The application, therefore, establishes itself as an effective pedagogical bridge between the complexity of multivariate theory and its practical application, especially in the field of social and market research, which predominantly handles categorical data.

Future work includes the implementation of an application that covers several topics considered in a basic course on Multivariate Data Analysis taught to university students in fields related to statistics.

Disclosure of Interests. The authors have no competing interests to declare that are relevant to the content of this article.

References

1. A., R., Wichern, Johnson, W.: Applied multivariate statistical analysis. pearson education, Filadelfia, PA, Estados Unidos de América, 6 edn. (2018)
2. Antonelli, T.M., Olivieri, A.C.: Developing and implementing an r shiny application to introduce multivariate calibration to advanced undergraduate students (2020)
3. Benzécri, J.P.: El análisis de correspondencias. Les cahiers de l'analyse des données **2**(2), 125–142 (1977)
4. Berk, K.N.: Multivariate descriptive statistical analysis-correspondence analysis and related techniques for large matrices (ludovic lebart, alain morineau, and kenneth m. warwick). SIAM Rev. **27**(3), 462 (1985)
5. Berkmann, E., Fisseler, B., Schützler, L., Christ, O.: Shiny apps in distance education: do psychology students benefit from interactive statistics applications? J. Stat. Data Sci. Educ., 1–13 (2025)
6. Chang, W., et al.: Shiny: web application framework for R (2024). https://CRAN.R-project.org/package=shiny, r package version 1.8.1.1
7. (CRAN), C.R.A.N.: Extract and visualize the results of multivariate data analyses [R package factoextra version 1.0.7] (2020). https://cran.r-project.org/web/packages/factoextra/index.html
8. Cruz, T., Jimenez, F.G., Bravo, A.R.Q., Ander, E.: Dataxplorefines: generalized data for informed decision, making, an interactive shiny application for data analysis and visualization. arXiv preprint arXiv:2307.11056 (2023)
9. Dougiamas, M., Taylor, P.: Moodle: Using learning communities to create an open source course management system. In: EdMedia+ innovate learning, pp. 171–178. Association for the Advancement of Computing in Education (AACE) (2003)
10. Dray, S., Dufour, A.B., Chessel, D.: The ade4 package - II: two-table and k-table methods. R News **7**(2), 47–52 (2007). https://cran.r-project.org/doc/Rnews/
11. Du, H.S., Wagner, C.: Learning with weblogs: enhancing cognitive and social knowledge construction. IEEE Trans. Prof. Commun. **50**(1), 1–16 (2007)
12. Fabacher, T., et al.: Medical biostatistics with gmrc shiny stats-learning by doing. In: Annales Pharmaceutiques Francaises. vol. 78, pp. 499–506 (2020)

13. Fawcett, L.: Using interactive shiny applications to facilitate research-informed learning and teaching. J. Stat. Educ. **26**(1), 2–16 (2018)
14. Fernández, E.A.: El análisis factorial de correspondencias aplicado al marketing. Boletín de Estudios Económicos **41**, 575 (1986)
15. Flores, M.J.S.: Capítulo 5 análisis de correspondencia simple (2021). https://bookdown.org/jsalinas/tecnicas_multivariadas/intro.html
16. Greenacre, M.: La práctica del análisis de correspondencias. Fundacion BBVA, La Villa y Corte de Madrid, España (2008)
17. Härdle, W.K., Hlávka, Z.: Multivariate statistics: exercises and solutions. Springer (2015)
18. Härdle, W.K., Simar, L., Härdle, W.K., Simar, L.: Multivariate distributions. Applied Multivariate Statistical Analysis, pp. 117–181 (2015)
19. Higham, D.J., Higham, N.J.: MATLAB guide. SIAM (2016)
20. Hohenwarter, M., Preiner, J.: Dynamic mathematics with geogebra. J. online Math. Appl. **7**(1), 2–12 (2007)
21. Hrastinski, S.: Asynchronous and synchronous e-learning. Educause quarterly **31**(4), 51–55 (2008)
22. Johnson, R.A., Wichern, D.W.: Applied multivariate statistical analysis. 6th. New Jersey, US: Pearson Prentice Hall (2007)
23. Jolliffe: principal component analysis. Springer, Nueva York, NY, Estados Unidos de América (1986)
24. Fernández-Avilés y José-María Montero, G.: Capítulo 35 análisis de correspondencias. https://cdr-book.github.io/correspondencias.html
25. Kluyver, T., et al.: Jupyter notebooks-a publishing format for reproducible computational workflows. In: Positioning and power in academic publishing: Players, agents and agendas, pp. 87–90. IOS press (2016)
26. Lê, S., Josse, J., Husson, F.: FactoMineR: A package for multivariate analysis. J. Stat. Softw. **25**(1), 1–18 (2008). https://doi.org/10.18637/jss.v025.i01
27. Lebart, L., Morineau, A., Fénelon, J.P.: Tratamiento estadístico de datos: Métodos y programas. Marcombo (1985)
28. Lesser, L., Melgoza, L.: Simple numbers: anova example of facilitating student learning in statistics. Teach. Stat. **29**, 102–105 (2007)
29. Martinková, P., Drabinová, A.: Shinyitemanalysis for teaching psychometrics and to enforce routine analysis of educational tests. R Journal **10**(2) (2018)
30. Nishisato, S.: Michael j. greenacre theory and applications of correspondence analysis. london: Academic press, 1984. 364+ xi pp. $56.00. Psychometrika**50**(3), 376–377 (1985)
31. Pardo, C.E., Del Campo, P.C.: Combinaci'on de m'etodos factoriales y de an'alisis de conglomerados en R: el paquete FactoClass. Revista Colombiana de Estadística **30**(2), 231–245 (2007)
32. Posit: shiny. https://shiny.posit.co/r/reference/shiny/
33. Posit Team, R., et al.: Rstudio: Integrated development environment for r. Posit Software, PBC, Boston, MA (2023)
34. Postma, M., Goedhart, J.: Plotsofdata–a web app for visualizing data together with their summaries. PLoS Biol. **17**(3), e3000202 (2019)
35. Potter, G., et al.: Web application teaching tools for statistics using r and shiny. Tech. Innov. Stat. Educ. **9**(1) (2016)
36. R Development Core Team: R: A language and environment for statistical computing. R Foundation for Statistical Computing, Vienna, Austria (2007). http://www.R-project.org, ISBN 3-900051-07-0

37. Resnick, M., et al.: Scratch: programming for all. Commun. ACM **52**(11), 60–67 (2009)
38. Rosli, I.N.I.B.M., Yusop, N.M.: Developing text mining web applications for teaching and learning using shiny apps. Int. J. Acad. Res. Progressive Educ. Dev. **12**(3) (2023)
39. Sibaja, A.E., Rojas, O.R.: Análisis factorial de correspondencias: estudio en creencias y estilos de enseñanza en docentes de matemática. Revista Educación, pp. 598–628 (2018)
40. Sievert, C.: Interactive web-based data visualization with R, plotly, and shiny. Chapman and Hall/CRC (2020)
41. Srivastava, M.S.: Methods of multivariate statistics, vol. 419. John Wiley & Sons (2002)
42. Team, R.C., et al.: R: A language and environment for statistical computing. r foundation for statistical computing, vienna, austria (2016). http://www.R-project.org/
43. Teil, H.: Correspondence factor analysis: an outline of its method. J. Int. Assoc. Math. Geol. **7**(1), 3–12 (1975). https://doi.org/10.1007/bf02080630
44. Toffalini, E., et al.: Clusters that are not there: An r tutorial and a shiny app to quantify a priori inferential risks when using clustering methods. Int. J. Psychol. (2024)
45. Wakerly, D.D.: Estadistica Matematica Con Aplicaciones, 6th edn. Cengage Learning Editores S.A. de C.V, Valle de México (2002)
46. Wickham, H.: Mastering shiny. O'reilly media (2018)
47. Wickham, H., et al.: Welcome to the tidyverse. J. Open Source Softw. **4**(43), 1686 (2019). https://doi.org/10.21105/joss.01686
48. Zhao, F., Schützler, L., Christ, O., Gaschler, R.: Learning statistics with interactive pictures using r shiny: generally preferred, but not generally advantageous. Teach. Stat. **45**(2), 106–124 (2023)

Change-Point Detection for Time Series Using the GA-Coen Algorithm: An Implementation in the Tidychange-Point Library in R

Biviana Marcela Suárez-Sierra[1], Arrigo Coen[2], and Carlos A. Taimal[1(✉)]

[1] Computing and Analytics Area. School of Applied Sciences and Engineering,
EAFIT University, Medellín, Colombia
`{bmsuarezs,cataimaly}@eafit.edu.co`

[2] Department of Mathematics, Faculty of Sciences, National Autonomous University
of México, Mexico City, Mexico

Abstract. The detection of change-points, as a recurrent problem in the time series and signal processing literature, has given rise to a wide variety of methodologies over the years. [2, 27] have taken the first steps toward standardizing this problem by defining it in terms of components that are common across almost every solution. However, at a software level, there are still a gap in R [21] that implements this conceptual framework in a practical manner.

In this context, `tidychange-point` [6] is introduced as a unifying actor for the implementation of diverse change-point detection methods. Using this package, change-points were detected across four simulated datasets, revealing consistent strengths across all applied methods and reinforcing the proposals of [15, 24, 25] regarding the use of approximate and meta-heuristic methodologies when uncertainty exists about the underlying nature of the data.

Keywords: Multiple Change-points Detection · tidyverse · Time Series Analysis · PELT · Binary Segmentation · Genetic Algorithms

1   Introduction

With around 7 million of works published on the subject as per Google Scholar [1], the change-point detection is still a current topic in the statistics, machine learning and signal processing literature. More specifically, an efficient and in some cases, early and timely detection in changes in a dynamical system can help to make decision making that have repercussions at an economic, security or ecological level. For example, the accumulation of such little abrupt changes can result in catastrophic events such as a radar navigation system failure [5].

While there is not consensus on how to approach this subject given the different types of changes that we wish to monitoring that variate according to each

B. M. Suárez et al. (Eds.): R Day 2025, CCIS 2824, pp. 115–138, 2026.
https://doi.org/10.1007/978-3-032-18455-9_7

context, there have been efforts to standardize the problem [27], furthermore, to implement it in programming tools, particulary Python [26]. Nevertheless this type of development is absent in R [21] at a general level with specific libraries for each type of approach. Thus, through the use of `tidychangepoint` we look to remedy this absencese using the principles of the tidyverse applied to the change-point detection context. As such, an implementation of three methods included in the library `tidychangepoint` will be made for four probabilistic functions analyzing on the one hand the number of detected change-points and on the other, the position associated to the true points of break. Next the problem will be introduced using the formulation by [27]. In the following section the three methods to study will be described under this approach so that a simulation exercise will be described for the experiments, next. Lastly the performance for each model will be assessed so that a conclusion will be made for the presented work and its consequences for the users of the change-points theory and the available software in R [21].

2   General Framework and Setting

2.1   Change-Points Detection

Change-point Detection (CPD from now on) in a time series or signal is a recurring topic in the statistical, signal processing and machine learning literature, present since around the XX century, with Page's work [19] being the first widely acknowledged reference on the subject [2,27]. Broadly speaking, the problem consists in determining the instant, or set of instants, at which the time series undergoes variations in its parameters and in the statistics derived from them. These may include variations in statistics of central tendency and dispersion: mean, variance, and standard deviation, autocorrelation coefficients, and parameters from an assumed generating process (ARMA and ARIMA models under the Box-Jenkins method), among others.

Although there exists a variety of methods—including control charts, non-parametric hypothesis tests, methods based on classical probabilistic distributions (likelihood-based methods), and piecewise regression methods, to name a few, these can be standardized under a framework of three components:

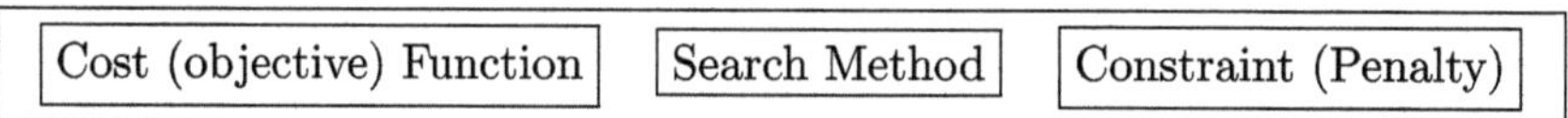

Fig. 1. Change-point Detection Framework as per [27]

which we define in detail below.

Change-Points. So what is it exactly a change-point?. Let's suppose a time series $y_t = \{y_t\}_{t=1}^{T}$, or a set of independent and identically distributed (iid) measurements in the discrete interval $(0, T)$: $y_t = \{y_1, y_2, \ldots, y_T\}$. We then define a set of indices denoted by $\mathcal{T} := \{\tau_1, \tau_2, \ldots\} \subset \{1, 2, \ldots, T\}$.

y_t is piecewise stationary, that is, some characteristics of the series change abruptly at specific instants $\{\tau_1, \tau_2, \ldots, \tau_J\}$, while remaining constant between each interval. The purpose is, therefore, to estimate the values of $\{\tau_j\}_{j=1}^{J}$, which we shall hereafter refer to as change-points.

Note that one change-point divides the series into two subseries, called regimes; two change-points into three, and so on. Hence, J change-points subdivide the series into $J + 1$ regimes. The goal of change-point estimation is to find the $J + 1$ regimes that optimize the value of a loss or cost function, which depends both on $\tau_j, \forall j = 1, \ldots, J$ and on y_t. In this way, we are dealing with an optimization problem with constraints given by a cost and a penalty.

Cost Function. The cost function can be summarized as a quantitative criterion that provides the best partition of y_t. This criterion will be denoted by $V(\mathcal{T}, y)$ and consists of the sum of the costs associated with each regime $(\tau_{j-1}, \tau_j), j = 1, \ldots, J + 1$, where $\tau_0 = 1$ and $\tau_{j+1} = T$ are, respectively, the first and last instants of the series. For each regime, the associated cost is $c(y_{\tau_j} : y_{\tau_{j+1}})$, and the overall cost function will be:

$$V(\mathcal{T}, y) := \sum_{j=0}^{J} c(y_{\tau_j} : y_{\tau_{j+1}}) \tag{1}$$

Thus, the first of the three components has been defined. Next, we require a penalty or constraint on the objective space, as well as a method to help estimate the values of the parameters.

Penalty and Search Method. As stated in the previous paragraph, the penalty represents a constraint on the set of change-points $\tau_1, \ldots, \tau_J$, seeking to satisfy the principle of parsimony. This ensures that not all instants of the series are defined as change-points, or conversely, that only significant instants or none at all are detected. In this sense, the penalty can be regarded as a measure of the segmentation's complexity. For this study, following the notation of [27], we denote the penalty as $\text{pen}(\mathcal{T})$.

Usually, $\text{pen}(\mathcal{T})$ is chosen so that, on the one hand, $\tau_{j-1} < \tau_j \,\forall\, j$, and on the other, the cardinality or number of change-points, $|\mathcal{T}|$, satisfies $|\mathcal{T}| < T$. Adding $\text{pen}(\mathcal{T})$ to (1) yields

$$V(\mathcal{T}, y) + \text{pen}(\mathcal{T}) \tag{2}$$

Finally, by inducing the search for the optimal values of τ, the following minimization problem is established:

$$\min_{\mathcal{T}} V(\mathcal{T}, y) + \text{pen}(\mathcal{T}) \tag{3}$$

The search method is thus an optimization procedure, either exact—based on dynamic programming—such as the PELT (Pruned Exact Linear Time) method [12], approximate, such as Binary Segmentation [3,8], Window Sliding [11,29], or Bottom-up Segmentation [7,10] or, to a lesser extent, metaheuristics such as Genetic Algorithms [15,24,25].

In general, this three-component framework is suitable for standardizing change-point detection procedures, allowing, on the one hand, comparison among methods of this type, and on the other, the development of implementations in specialized libraries, a gap that still exists within R. Consequently, the following sections will describe in detail three of the CPD techniques mentioned earlier: PELT, Binary Segmentation, and Genetic Algorithms, under the definition of the three components: cost function, penalty, and search method. Next, the `tidychangepoint` library is introduced, explaining how it defines the problem in R using tidy principles [30], its utilities, and its internal implementations. Subsequently, a set of experiments will be carried out using the aforementioned algorithms on four simulated datasets with known partition times.

2.2   The Pruned Exact Linear Time (PELT) Algorithm

Description. Broadly speaking, the *Pruned Exact Linear Time* method—PELT—proposes using pruning to eliminate elements of the process that may be considered unnecessary and that contribute to an increase in both computational complexity and the value of the cost function. In the context of CPD, pruning consists of removing change-point values $\tau \in \mathcal{T}$ that do not minimize (1). The condition is established such that a bound K—which coincides with pen($\mathcal{T}$) over (1)—is imposed, so that if this bound is exceeded when introducing a value $\tau_j \in \mathcal{T}, j = 1, \ldots, J+1$, the change-point is considered non-optimal and can be discarded.

A detailed definition of each component, according to the previously introduced three-part framework, is given below.

Cost Function. In the paper where PELT was introduced, the cost function is defined as the negative log-likelihood of the observations, used to satisfy the computational assumptions of linear cost over the data partitions. Thus, it is necessary to assume an underlying stochastic model parameterized by a set of values $\boldsymbol{\Theta} \in \mathbf{R}^p$. For example, if a distribution $\mathbf{Beta}(\alpha, \beta)$ is assumed for y_t, then $p = 2$. Hence, a limitation of PELT is the need to know—or at least approximate—the distributional and parametric form of the data.

Assuming a distribution $f(y_t|\boldsymbol{\Theta})$ in each regime, the cost function for PELT is

$$\mathcal{C}(y_{\tau_j} : y_{\tau_{j+1}}) = -\max_{\boldsymbol{\Theta}} \sum f(y_t|\boldsymbol{\Theta}) \tag{4}$$

Penalty. In the case of penalties, the authors suggest using values that do not depend directly on the number or location of change-points. However, PELT and its implementations are expressed directly as a function of the number of change-points, J. The most immediate penalization option for PELT is βJ, that is, a linear function that depends directly on the number of change-points. Other options include information criteria such as Akaike's (AIC): $\beta = 2p$, or the Bayesian (BIC): $\beta = p\log(T)$, where p is the number of parameters to be estimated, including one additional change-point introduced into the series.

Search Method. To define the search method, we rewrite (3) as follows:

$$\min_{\tau \in T}\left\{\sum_{i=1}^{J+1}[\mathcal{C}(y_{\tau_{i-1}}, y_{\tau_i}) + \beta]\right\} \tag{5}$$

We then define the set of change-points as $T_s := \{\tau : 0 = \tau_0 < \tau_1 < \ldots < \tau_J < \tau_{J+1} = s\}$. The summation in (5) can now be divided as follows:

$$\min_{t}\left\{\min_{\tau \in T_j}\sum_{i=1}^{J}[\mathcal{C}(y_{\tau_{i-1}}, y_{\tau_i}) + \beta] + \mathcal{C}(y_{i+1}, y_T)\right\} \tag{6}$$

Defining the first summand as the minimization over the j-th regime, denoted by $F(j)$, we then have

$$F(s) = \min_{t}\left\{F(j) + \mathcal{C}(y_{t+1} : y_T) + \beta\right\} \tag{7}$$

Now, a constant K is defined such that for an instant s, $t < s < T$,

$$\mathcal{C}(y_{t+1} : y_s) + \mathcal{C}(y_{s+1} : y_T) + K \leq \mathcal{C}(y_{t+1} : y_T) \tag{8}$$

Then, if the following condition holds,

$$F(t) + \mathcal{C}(y_{t+1} : y_s) + K \geq F(s) \tag{9}$$

for a future instant $\mathcal{N} > s$, t can never be the last optimal change-point prior to $\mathcal{N}$.

While this method is efficient and can overcome barriers related to exact methods and segmentations, still has a logarithmic complexity in time of $\mathcal{O}(T^2)$ for time series where there are not change-points and on the other side, the optimal pruning will only work in specific cases for the cost function. For example, if there are changes in the slope of the data it is not possible to segment the data without losing optimality [22]. Now, we present the structure of the PELT algorithm in pseudocode form.

R Implementations. For this method, the main implementations come in the form of the `change-point` [13] and `change-point.np` [9] libraries, with the former being a direct dependency of the latter. In both cases, the default S3 object is named `cpt`, which stores both the parameters passed to PELT as well as the time series and the results of the procedure. The basic instruction for a PELT segmentation is the following one,

---

Algorithm 1. PELT

Input:
A set of data of the form, $y_1, y_2, \ldots, y_T$ such that $y_t \in \mathrm{R}$.
A measure $\mathcal{C}(\cdot)$ depending on the data.
A penalty constant β which does not depend on the number or location of change-points.
A constant K that satisfies (8).
Initialize Let $T =$ length of data and set $F(0) = -\beta$, $cp(0) = NULL$, $R_1 = \{0\}$.
for $\tau^* = 1$ to T **do**
$\quad$ Calculate $F(\tau^*) = \min\limits_{\tau \in R_\tau^*} [F(\tau) + \mathcal{C}(y_{\tau+1} : y_{\tau^*})]$
$\quad$ Let $\tau^1 = \arg[\min\limits_{\tau \in R_\tau^*} F(\tau) + \mathcal{C}(y_{\tau+1} : y_{\tau^*}) + \beta]$
$\quad$ Set $cp(\tau^*) = [cp(\tau^1), \tau^1]$
$\quad$ Set $\mathcal{R}_{\tau^*+1} = \{\tau \in \mathcal{R}_{\tau^*} \cup \{\tau^*\} : F(\tau) + \mathcal{C}(y_{\tau+1} : y_{\tau^*}) + K \leq F(\tau^*)\}$
end for
Output: the change-points recorded in $cp(T)$.

```
cpt.mean(time_series_data, penalty="penalty_type", method = "PELT")
```

`penalty_type` can be changed as per the user's preferences and using any of the previous mentioned penalties.

2.3 CPD Through Binary Segmentation.

Description. The binary segmentation algorithm is based on the intensive division of the time series into sets of two. Starting from the defined cost function, an initial change-point is found that divides the series into two parts. Once these two parts are obtained, if another change-point is found that produces a decrease in the cost for one of the halves, the corresponding subseries is divided, and the process continues in this way until there is no significant reduction in $V(\mathcal{T}, y)$. Its computational complexity is approximately log-linear, depending directly on the cost of identifying a single change-point c^* and on the total number of observations, T. That is, the algorithm's complexity is: $\mathcal{O}(c^*T log(T))$ [27].

The version implemented in `tidychangepoint` [6] corresponds to Wild Binary Segmentation (WBS), as introduced in [8]. In that work, the authors define e subseries of the time series y_t, and compute a CUSUM statistic for each subset. Each CUSUM is maximized, and the cutoff point associated with the maximum value among the e CUSUMs is taken as a change-point, after being compared against a threshold. This process is repeated for the subseries to the right and left of the change-point until the comparison criterion relative to the threshold is no longer satisfied, or until no further improvement is observed with respect to the so-called *strengthened Schwarz information criterion*.

Cost Function. We begin by defining s and e as the two subseries or subvectors obtained after partitioning the interval $(1, T)$, which contains all time points of the original series. s and e are elements of $\mathbb{Z}^+$ such that $1 \leq s < e \leq T$. The set of subseries is denoted by $(y_s, \ldots, y_e)$, and from this, an inner product is computed with respect to a vector of "contrasts" defined as follows:

$$\tilde{y}_{s,e}^b = \sqrt{\frac{e-b}{n(b+s-1)}} \sum_{t=s}^{b} y_t - \sqrt{\frac{b-s+1}{n(e-b)}} \sum_{t=b+1}^{e} y_t \tag{10}$$

where $s \leq b < e$ and $n = e - s + 1$.

The statistic (10) is evaluated against a specific criterion, and if it is deemed significant, the instant associated with it is then determined to be a change-point. After this, the series is recursively partitioned into the resulting halves to the right and left of the detected change-point. A limitation, however, lies in the assumption about the structure of y_t, such that

$$y_t = f_t + \varepsilon_t, \ 1, 2, \ldots, T \tag{11}$$

where f_t in (11) is a piecewise constant deterministic function or one with change-points, and $\varepsilon_t \ \forall \ t, \ t = 1, 2, \ldots, T$ is assumed to follow a Gaussian or normal distribution—more specifically, $\varepsilon \sim \mathcal{N}(0, 1)$. This is a strong assumption, implying that the series should approximately follow a normal distribution in order for the contrast to be valid. For the purpose of introducing the penalty term in this method, the value ζ_T must be defined as the threshold that the objective function must exceed for an instant to be considered a change-point.

Penalty. As mentioned at the beginning of this section, the method introduces a penalization that the authors call the Strengthened Schwarz Information Criterion or sSIC.

Let us suppose a candidate model for f_t, denoted by $\mathcal{C}_k$, associated with k candidate change-points. The residual variance is defined as the difference between the model fitted via Maximum Likelihood for f_t, denoted as $\hat{f}_t^k$:

$$\hat{\sigma}_k^2 = \frac{1}{T} \sum_{t=1}^{T} (f_t - \hat{f}_t^k)^2 \tag{12}$$

Then, the penalty will be,

$$\text{sSIC}(k) = \frac{T}{2} \log \sigma_k^2 + k \log^\alpha T \tag{13}$$

α in (13) defines the strength of the penalization, such that if it equals 1, it corresponds to the traditional SIC, whereas if $\alpha > 1$, it imposes a stricter penalty on the number of change-points to be detected.

Search Method. Finally, the heuristic through which the change-points are identified is defined in the following pseudocode.

Algorithm 2. Wild Binary Segmentation

---

if $e - s < 1$ **then**
    STOP
else
    $\mathcal{M}_{f,1} :=$ set of those indices m for which $[s_m, e_m] \in F_T^M$ is such that $[s_m, e_m] \subseteq [s, e]$
    (Optional: argument $\mathcal{M}_{s,e} := \mathcal{M}_{s,e} \cup \{0\}$ where $[s_0, e_0] = [s, e]$)
    $(m_0, b_0) := \underset{m \in M_{s,e}\, b \in \{s_m, \ldots, e_{m-1}\}}{\operatorname{argmax}} |\tilde{y}^b_{s_m, e_m}|$
    if $|\tilde{y}^{b_0}_{s_{m_0}, e_{m_0}}| > \zeta_T$ **then**
        add b_0 to the set of estimated change-points
    else
        STOP
    end if
end if

It can thus be seen that not only is it necessary to meet distributional assumptions about the errors, but also to implement a set of mathematical functions with a high level of abstraction, making this procedure rather unintuitive.

On the other side, as [16] points out, additional limitations besides the distribution wise ones are the overestimation of the number of change-points when observations with IID errors are used and, furthermore, when the errors are correlated the method behaves dysfunctionally.

R Implementations. At an implementation level in R, the authors propose the `wbs` package [4]. This package provides two S3 objects for performing CPD: `sbs` and `wbs`, where the former corresponds to the standard binary segmentation case, and the latter implements the method discussed throughout this section. Likewise, it includes a `changepoints` method that acts as a wrapper around the `sbs` or `wbs` objects, applying the time series partitioning procedure as well as the penalization according to the user's chosen criterion. Other implementations can be found in `changepoint` [13] using the function `cpt.mean` with the method set as"binseg":

```
cpt.mean(time_series_data, penalty="penalty_type", method = "binseg")
```

On the other hand, using the `wbs` library [4] the basic instruction comes as:

```
wbs(x, ...)
```

with `x` our time series of interest. Additional implementation details and parameter specifications can be found in the paper associated with the package [4].

2.4   Genetic Algorithms for CPD

Description. The name Genetic Algorithm for CPD can be considered too general, since only one specific class within this metaheuristic family will be used for the CPD problem. Moreover, both the objective function and the penalization represent particular cases within a broader set of methodologies. The algorithm to be presented is GA-Coen, which is based on Bayesian inference paradigms for stochastic count processes, specifically Non-Homogeneous Poisson Processes (NHPP). Parameter estimation for the model is carried out in a fully Bayesian framework using Markov Chain Monte Carlo (MCMC) methods, which are combined with the genetic algorithm so that the set of candidate change-points is selected and subsequently fed into the overall objective function.

This method, like other metaheuristics, can be considered computationally expensive when high precision is required, as it demands a greater number of evaluations and iterations to reach values close to the true solution. However, if a smaller number of iterations is considered, the resulting approximation is sufficiently optimal for ad-hoc decision-making and remains interpretable within a purely probabilistic framework—in contrast to the frequentist paradigm. The specific details of this method will be discussed below, once again using the three-component framework for change-point detection, as was done in the other two cases.

Cost Function. Consider a threshold on the observations, y_t, denoted by ζ. This threshold may or may not coincide with the mean of the observations, or it can be linked to a specific contextual criterion; for instance, environmental policy standards ([24]).

From ζ, we identify the times t for which $y_t > \zeta$, resulting in a derived series d_t of event times. Based on this series, it becomes possible to approximate a Non-Homogeneous Poisson Process (NHPP), under the assumption that within d_t there exist partitions or change-points indicating shifts in the rate of occurrence of such events.

The rate associated with the occurrences can then be approximated through probability distributions that model the time until the event occurs, for example, the Weibull, Goel-Okumoto, Musa-Okumoto, or Generalized Goel-Okumoto distributions, among others. For simplicity, we will assume a Weibull distribution with parameters $(\alpha_j, \beta_j) \; \forall \; j \in 1, 2, \ldots, J$, meaning that there will be an intensity function for each regime. Additionally, under the Bayesian paradigm, these parameters will have an associated probability density function, where each α is assumed to follow:

$$\alpha \sim Gamma(\phi_{11}, \phi_{12})$$

and for each β it will be assumed,

$$\beta \sim Gamma(\phi_{21}, \phi_{22})$$

Finally, it will also be assumed that each change-point is uniformly distributed over the interval $(1, T)$, thus,

$$\tau_j \sim Unif(1, T)$$

that is, we assume the same probability of being selected as a change-point for each time instant.

Bringing these elements together, the cost function is defined as follows:

$$V(\mathcal{T}, y) = \sum_{j=1}^{J+1} \left(\frac{\tau_{j-1}^{\alpha_j} - \tau_j^{\alpha_j}}{\beta_j^{\alpha_j}} + (N_{\tau_j} - N_{\tau_j - 1}) \right.$$

$$\left. + (ln(\alpha_j) - \alpha_j ln(\beta_j)) + (\alpha_j - 1) \sum_{i=N_{\tau_j-1}+1}^{N_{\tau_j}} ln(d_i) \right)$$

$$+ \sum_{j=1}^{J+1} ((\phi_{12} - 1)ln(\alpha_j) - \phi_{11}\alpha_j + (\phi_{21} - 1)ln(\beta_j) - \phi_{21}\beta_j) - J\, ln(T-1)$$

$$\tag{14}$$

The first part of (14) is associated with the likelihood of the exceedance times, while the second corresponds to the prior distributions of the parameters. N_{τ_j} in (14) refers to the number of exceedances before the change-point τ_j, $j = 0, 1, 2, \ldots, J+1$, with $\tau_0 = 0$, $\tau_{J+1} = T$, and $N_0 = 0$.

Penalty. The penalty of the method is associated with the Minimum Description Length (MDL) principle, which seeks an optimal representation of an object—in this case, the piecewise time series model in terms of bits. For the mathematical details of this principle, the reader is referred to ([15,24]). The constraints for representing the time series model with the presence of change-points are thus defined as follows:

$$\text{pen}(\tau) = 2 \sum_{i=1}^{J+1} \sum_{i=1}^{J+1} \frac{ln(\tau_j - \tau_{i-1})}{2} + ln(J) + \sum_{j=2}^{J} ln(\tau_i) \tag{15}$$

Adding up (14) and (15), we have as the function to be optimized:

$$V(\mathcal{T}, y) + \text{pen}(\tau) = 2 \sum_{i=1}^{J+1} \sum_{i=1}^{J+1} \frac{ln(\tau_j - \tau_{i-1})}{2} + ln(J) + \sum_{j=2}^{J} ln(\tau_i)$$

$$- \sum_{j=1}^{J+1} \left(\frac{\tau_{j-1}^{\alpha_j} - \tau_j^{\alpha_j}}{\beta_j^{\alpha_j}} + (N_{\tau_j} - N_{\tau_j - 1}) \right.$$

$$\left. + (ln(\alpha_j) - \alpha_j ln(\beta_j)) + (\alpha_j - 1) \sum_{i=N_{\tau_j-1}+1}^{N_{\tau_j}} ln(d_i) \right)$$

$$+ \sum_{j=1}^{J+1} ((\phi_{12} - 1)ln(\alpha_j) - \phi_{11}\alpha_j + (\phi_{21} - 1)ln(\beta_j) - \phi_{21}\beta_j)$$

$$- J\, ln(T-1) \tag{16}$$

Search Method. For the search method, a simple genetic algorithm is used, with a rank-based selection for solutions, the initial population generation is based on a binomial distribution, the creation of new solution is based on the crossover between parent solutions, and a random mutation with equal probabilities of mutation or change for each solution is used.

It is emphasized that this type of method is not unique and this version represents only one of many possible implementations found in the literature. For a detailed study of the various components of the genetic algorithm, the reader is referred to ([25]). Next, through the pseudocode, we aim to explain and illustrate the steps mentioned above.

Algorithm 3. GA-Coen

1: Initialize a matrix $M_{best} \in \mathrm{R}^{r \times T}$ with zeroes.

2: Generate a sample of zeroes and ones of size T using a $Binomial(n = T - 1, p = 0.06)$. The one indicates the time is a candidate for change-point, the zero the contrary.

3: The previous step is repeated k times.

4: Then for each solution the cost is calculated using (16).

5: **for** i in $1 : r$ **do**

6:      The values for the cost are sorted inversely. The rank order is denoted as S_j such that if $S_j = k$ is the solution with the best fit and $S_j = 1$ denotes the solution with the least well fit.

7:      Using the ranks S_j, a "roulette" is created for the solutions with slots with area as follows,

$$S_j / \sum_{i=1}^{k} S_i$$

     Then, the roulette is spun and a solution is chosen. This is kept and with the rest of solutions a second one is chosen.

8:      The two solutions in the previous steps are denoted as $(\delta_1, \delta_2, \ldots, \delta_m)$ and $(\epsilon_1, \epsilon_2, \ldots, \epsilon_q)$ where m and q denote the number of change-points chosen by the algorithm, the two strings are concatenated to create a new solution, $(\tau_1, \tau_2, \ldots, \tau_{m+q})$.

9:      After the new solution is created, the duplicates are deleted and each element in the new solution is kept or deleted with probability 0.5. This new solution is denoted as $(\tau_1, \tau_2, \ldots, \tau_B)$.

10:      Then, a dice with faces (-1, 0, 1) and probabilities (0.3, 0.4, 0.3) is thrown B times. The B results are added to the solution $(\tau_1, \tau_2, \ldots, \tau_B)$. If -1 comes up, the change-point is reduced by one time, if 0 comes up the change-point is kept and otherwise, the change-points is increased in one time. This new solution is denoted $(\tau_1, \tau_2, \ldots \tau_J)$ after duplicates are removed.

11:      The best solution is stored in M_{best}

12: **end for**

The r iterations are chosen according to the user preference or until the best solution is not able to achieve an improvement in (16).

While there are not distributional assumptions while using this method, there is still exists the need to calibrate the genetic algorithm parameters and, on the other side, the stochastic nature of it doesn't provide a global optimum, only an ad-hoc solution that can help a fast decision making. Finally, the complexity in time of the method depends directly on the number of iterations or generations that could extends otherwise a stopping criterion is not defined and an optimal solution is not reached according to the user standards.

R Implementations. Although there are implementations of genetic algorithms in R, there is no specific package for this one in particular. `tidychangepoint` [6] attempts to fill this gap while incorporating the previous ones, i.e., PELT, WBS among others.

The principal S3 object that generalizes all other methods is `tidycpt`, that has three sub-objects:

- `segmenter`: stores information about the change-points detected by each type of model.
- `model`: a subclass representing a time series based on a given set of segmentations, or a specific model that produces such segmentations.
- `elapsed_time`: the processor time required to find the optimal segmentation according to the method used.

On the other hand, objects of type `tidycpt` include methods for generic functions such as `as.model()`, `as.segmenter()`, `as.ts()`, `change-points()`, `diagnose()`, `fitness()`, `model_name()`, `regions()`, `plot()`, and `summary()`. Finally, the strength of the package lies in the `segment()` wrapper, which provides the ability to generalize the previously mentioned implementations, such as:

- PELT from the `change-point` package [12]
- Wild Binary Segmentation from the `wbs` package [4]
- Genetic Algorithms [14,23,24]

As well as piecewise regression-based segmentations [18].

As observed for other methods, each library has a different syntax that makes difficult to compare between results in an immediate manner, even further, the resulting objects can be very different between each other such that a list is returned, a vector with the positions of the change-points or S3 objects. This gap allows to formulate a new type of object, `segment` that generalizes all other previously mentioned. The general structure for this type of object is:

```
segment(time_series_data, method = "segmentation_method",
penalty = "penalty_type", ...)
```

With `method` being the type of algorithm to use, `penalty` the type of penalty used by such algorithm and additionally, other options particular to each methods can be defined. The package documentation delves into its technical capabilities, including the incorporation of customized methods under a tidy paradigm [30].

3   Methods

3.1   Data

For the different implementations, four datasets are considered. Although all of them are realizations of the exponential family of distributions, they differ in their central tendency characteristics. Following the examples proposed in [24,25], data from a lognormal distribution are first used and then transformed to obtain normally distributed data. Additionally, realizations from a Pareto distribution—characterized by its heavy, unbounded tails—are included [20]. Finally, data from a Beta distribution are simulated to be highly positively skewed (left-skewed) (see Fig. 2 and Fig. 3); the first two distributions were chosen to allow similar results to those of [24,25], while the Beta and Pareto were chosen looking for data as different as possible to the previous two; for the Beta we looked for a more restricted support: $(0, 1)$ and on the other hand, we looked for data with heavy tails: Pareto.

The parameters for the normal distribution is given by μ and σ that work as centrality and dispersion parameters or of mean and standard deviation, respectively. The same ones apply for the lognormal distribution scaling through logarithm, such that under this distribution if we apply this type of transformation we get $Normal(\mu, \sigma)$ data.

The probability density function (pdf) for a normal distribution is given by [20],

$$f(x|\mu,\sigma) = \frac{1}{\sqrt{2\pi\sigma^2}}\exp\left\{-\frac{(x-\mu)^2}{2\sigma^2}\right\}, \ \mu \in \mathbb{R}, \ \sigma > 0, \text{and } x \in \mathbb{R}. \tag{17}$$

While the pdf for the lognormal distribution is [20],

$$f(x|\mu,\sigma) = \frac{1}{x\sigma\sqrt{2\pi}}\exp\left\{-\frac{(\ln(x)-\mu)^2}{2\sigma^2}\right\}, \ \mu \in \mathbb{R}, \ \sigma > 0, \text{and } x \in \mathbb{R} \tag{18}$$

The beta distribution is parametrized by the values α and β where both values are positive and control the shape of the distribution. Its pdf is given by [20],

$$f(x|\alpha,\beta) = \frac{\Gamma(\alpha+\beta)}{\Gamma(\alpha)\Gamma(\beta)}x^{\alpha-1}(1-x)^{\beta-1}, \ x \in (0,1) \tag{19}$$

and $\Gamma(\cdot)$ is the *Gamma* function defined for a non-negative value z as,

$$\Gamma(z) = \int_0^\infty t^{z-1}e^{-t}dt.$$

Finally, the Pareto distribution is parametrized by the values θ and η such that the first value controls the shape of the distribution (shape parameter) and the second one, its centrality (location parameter).Then, its pdf its given by [17],

$$f(x|\eta,\theta) = \frac{\theta\eta^\theta}{x^{\theta+1}}, \theta,\eta > 0, \ x \geq \eta. \tag{20}$$

For the lognormal and normal datasets, the change-points correspond to $\mathcal{T} = \{548, 823, 973\}$, whereas for the Pareto and Beta datasets, the series are simulated with three change-points located at $\mathcal{T} = \{274, 550, 825\}$. 1096 records were simulated such that the total coincides with that of [24,25].

All parameters used for the simulation exercise are given in the Table (3.1).

Distribution	Parameter	Regime 1 Parameters	Regime 2 Parameters	Regime 3 Parameters	Regime 4 Parameters
Normal	μ	3.5	4	4.5	5
	σ	0.32	0.32	0.32	0.32
Lognormal	μ	3.5	4	4.5	5
	σ	0.32	0.32	0.32	0.32
Beta	α	0.4	5	0.1	3.5
	β	5	500	100	4
Pareto	η	0.4	5	0.1	3.5
	θ	100	100	100	100

Using these datasets, experiments were conducted for the three algorithms described: PELT, WBS, and GA-Coen. For PELT, the penalty term was varied according to the following criteria:

- AIC
- BIC
- MBIC

For WBS, the penalty was also varied—analogously to PELT—and, in addition, the parameter Q was modified. This parameter represents an initial expert-informed estimate of the expected number of change-points in the series. Finally, tests were performed for the Genetic Algorithm on each dataset, keeping the algorithm's diversity parameters fixed and setting the number of generations to 500. For these experiments two types of results were evaluated: the number of detected change-points and how close the individual change-points were to the real ones. Other metrics taken from classification problems could be adapted such as the Rand Index, or the Recall [27] as well as the Hamming distance assuming as 0 the case of no change-points in the series and 1 as the contrary [25]. The results are presented and discussed in the next sections.

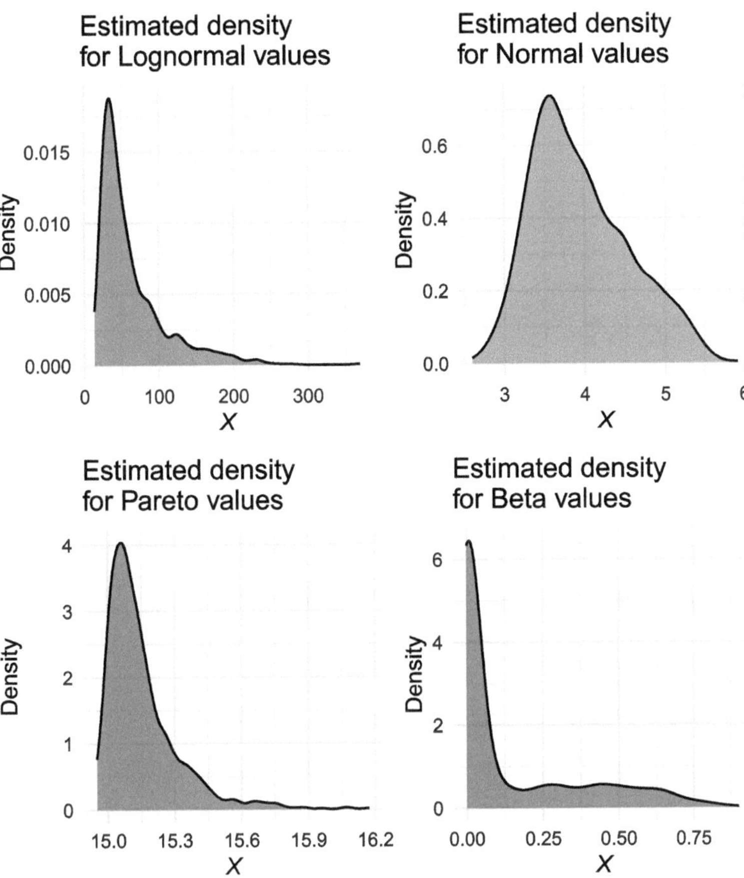

Fig. 2. *Simulated data densities.* Only the normal data is notable for its symmetry around the mean, while the Lognormal, Pareto, and Beta are positive skewed. Yet the dispersion is small in all datasets and in the case of the normal and lognormal values, this are concentrated around $\mu \approx 4$.

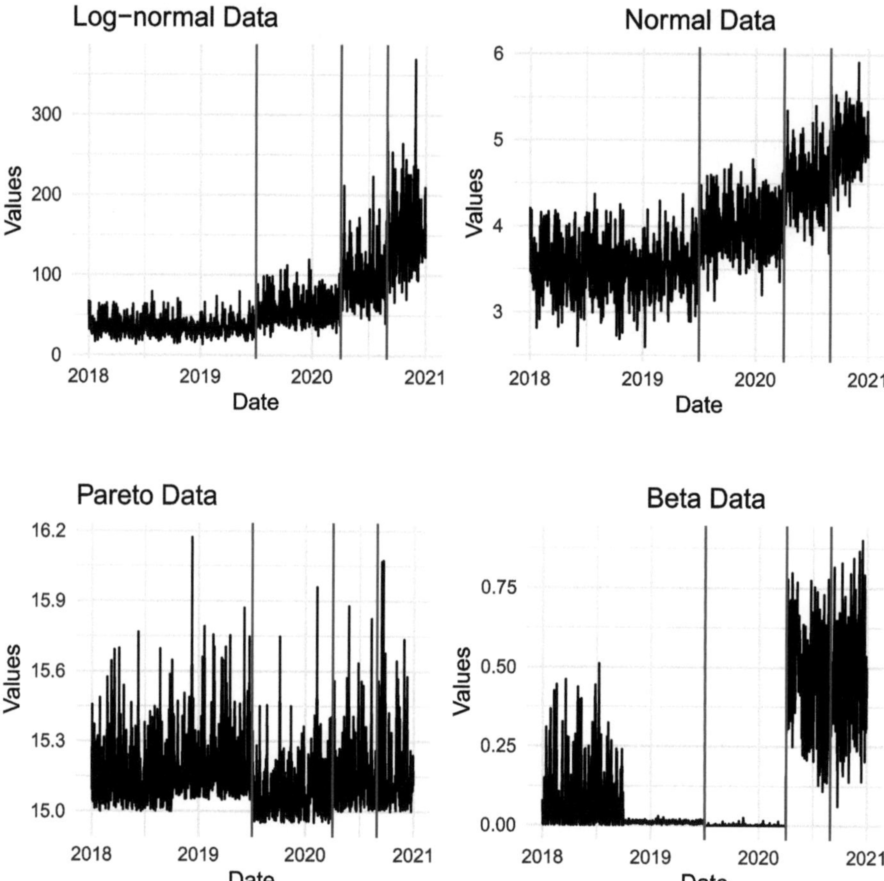

Fig. 3. *Time Series Data.* For the normal and lognormal data is constant in the variance while the mean is increased monotonically. For the Pareto data these follow the same behavior as the normal and lognormal datasets in terms of variation of the mean and constant variance, nevertheless, it should be taken into consideration the inherent heavy tails and asymmetry related to this type of distribution. Finally, for the Beta dataset the changes are abrupt both for the mean and the variance.

3.2 PELT Results

For the PELT method, the `tidychange-point` instruction was the following one:

```
segment(data_set_name, method = "pelt", penalty = "penalty_name")
```

where "`penalty_name`" was either `AIC`, `BIC` or `MBIC`.

For this method and its results reported in (Table 1) we found that the component that most strongly influences the quality of the results is the penalty term. For instance, when using AIC, all four tested distributions yielded a number of change-points greater than the true value, regardless of whether the data were easily approximated to a normal distribution or not.

In the second case, when BIC was used as the penalty criterion, some differences emerged. Although both AIC and BIC are derived from likelihood-based principles, AIC penalizes only the number of estimated parameters, while BIC also penalizes according to the number of observations. Theoretical discussions have explored the relative effectiveness of these penalties in different contexts, concluding that BIC performs better when the true model is included in the evaluated set, whereas AIC performs better when the true model is not [28].

Finally, the models obtained using the Modified BIC (MBIC)—an extension of the standard BIC—show largely similar results across the four datasets. However, improvements are observed for the Pareto dataset, since MBIC does not rely directly on likelihood values but instead on an approximation to the Bayes factor [31].

In terms of estimation accuracy, the first three change-points detected for the Log-Normal and Normal datasets closely approximate the true values. For the Log-Normal data, both BIC and MBIC locate the first change-point exactly ($\tau_1 = 547$), the second slightly below by six units ($\hat{\tau}_2 = 817$ vs. $\tau_2 = 823$), and the third three units below the true value ($\hat{\tau}_3 = 970$ vs. $\tau_3 = 973$).

For the Normal dataset, the first change-point is underestimated by one unit ($\hat{\tau}_1 = 547$ vs. $\tau_1 = 548$), the second by two units ($\hat{\tau}_2 = 821$ vs. $\tau_2 = 823$), and the last by four units ($\hat{\tau}_3 = 969$ vs. $\tau_3 = 973$). In all estimations, an additional change-point appears at $\tau_4 = 1096$, corresponding to the length of the time series—this results from how PELT is designed, always considering the final observation as a boundary solution.

For the Pareto dataset, where the true change-points are $\mathcal{T} = \{274, 550, 825\}$, the BIC criterion does not perfectly recover the exact locations but produces results within a reasonable neighborhood (less than ten units away). With more repeated experiments, a probabilistic rather than pointwise estimate could likely improve accuracy. The MBIC, however, struggles to identify close approximations, skipping directly to the second change-point (around 550), which is consistent across both BIC and MBIC, while the third change-point (825) is not well captured in either case.

Finally, for the Beta dataset, none of the three penalty criteria (AIC, BIC, MBIC) yield change-point estimates close to the true number. Nonetheless, a gradual improvement is observed as the penalty becomes stricter —from AIC to BIC to MBIC—suggesting better regularization. However, when examining the specific change-point locations, neither BIC nor MBIC manages to approximate the true $\mathcal{T}$ values, often failing to identify valid candidates or estimating them with large deviations. This is consistent with what [12] reported on how the method fails approximating changes in the slope of the data.

We now turn to discussing the results obtained from the Binary Segmentation.

Table 1. PELT experiments results

Penalty	Dataset	Number of Detected Change-points	Detected Change-points
AIC	Log-Normal	203	N/A
	Normal	180	N/A
	Pareto	278	N/A
	Beta	294	N/A
BIC	Log-Normal	4	(548 , 817, 970, 1096)
	Normal	4	(547, 821, 969, 1096)
	Pareto	10	(258, 260, 544, 761, 987, 1000, 1022, 1034, 1075, 1096)
	Beta	65	N/A
MBIC	Log-Normal	4	(548, 817, 970, 1096)
	Normal	4	(547, 821, 969, 1096)
	Pareto	3	(544, 761, 1096)
	Beta	31	N/A

3.3   WBS Results

The results for this method are reported in (Table 2). The first set of experiments were carried using the instruction:

```
segment(data_set_name, method = "BinSeg", penalty = "penalty_name",
Q = number_of_cpt)
```

from the `tidychangepoint` library.

Penalty Wise Experiments

AIC

Although all four datasets showed an overestimation in the number of change-points under the three penalization scenarios, it is noteworthy that in this case, the overestimation did not exceed the range of tens, unlike what was observed with the PELT method. Indeed, the worst scenario occurred with the AIC, where all datasets were estimated with six change-points.

For the Log-Normal and Normal cases, the method estimated an initial change-point at time 5 of the series, which does not correspond to any true change introduced during data generation, and is therefore considered a spurious detection. Excluding this value, the next estimated point, $\hat{\tau}_1$, lies slightly below the true value $\tau_1 = 548$, with $\hat{\tau}_1 = 548$ for the Log-Normal data and $\hat{\tau}_1 = 547$ for the Normal data.

The second instant, $\tau_2 = 823$, was estimated with a six-unit lag below the true value, giving $\hat{\tau}_2 = 817$ for both distributions. Finally, the last change-point, $\tau_3 = 973$, was estimated with a three-unit lag for the Log-Normal data ($\hat{\tau}_3 = 970$) and a four-unit lag for the Normal data ($\hat{\tau}_3 = 969$).

Additionally, two extra points were estimated—one at time 1073 and another at the end of the series. These can be interpreted as false positives, likely caused by the method's sensitivity to small local changes in mean or variance near the series' end.

Regarding the Pareto and Beta datasets under the AIC penalty, a spurious early change-point was again detected, not associated with any true change. The second change-point, $\tau_2 = 550$, was estimated close to the true value, with $\hat{\tau}_2 = 544$ for the Pareto data and $\hat{\tau}_2 = 548$ for the Beta data. Finally, the last change-point, $\tau_3 = 825$, was accurately estimated only for the Beta distribution, with $\hat{\tau}_3 = 822$, while for the Pareto distribution none of the remaining estimated points were near the true value.

BIC and *MBIC*

When using the BIC and MBIC as penalty criteria, the performance of the WBS method improves significantly. In fact, for WBS, there are no observable differences between using one or the other penalty—both produce identical segmentations.

For the Log-Normal and Normal distributions, the first true change-point occurs at $\tau_1 = 548$. The algorithm successfully detects this point at $\hat{\tau}_1 = 548$ for the Log-Normal data and $\hat{\tau}_1 = 547$ for the Normal data. The second true change-point, $\tau_2 = 823$, is detected at $\hat{\tau}_2 = 817$ in both cases, corresponding to a six-unit underestimation. Finally, the last change-point, $\tau_3 = 973$, is estimated as $\hat{\tau}_3 = 970$ and $\hat{\tau}_3 = 969$ for the Log-Normal and Normal datasets, respectively— that is, underestimations of three and four units.

Turning to the Pareto distribution, the first true change-point $\tau_1 = 274$ is not correctly detected by either the BIC or MBIC, neither exactly nor within a reasonable neighborhood. The second change-point, $\tau_2 = 550$, is estimated at $\hat{\tau}_2 = 544$ with the BIC and at $\hat{\tau}_2 = 547$ with the MBIC. The final change-point, $\tau_3 = 825$, is not appropriately estimated under either penalization scheme.

In the case of the Beta distribution, the algorithm performs better: the first change-point is accurately estimated at $\hat{\tau}_1 = 273$ in both cases. The second change-point, $\tau_2 = 550$, is closely approximated with $\hat{\tau}_2 = 548$, and the last one, $\tau_3 = 825$, is accurately detected as $\hat{\tau}_3 = 822$. However, in addition to the correctly identified change-points, the algorithm also produces four false positives, likely due to local fluctuations in the data that the method interprets as structural shifts.

Table 2. WBS experiments results

Penalty	Dataset	Number of Detected Change-points	Detected Change-points
AIC	Log-Normal	6	(5, 548, 817, 970, 1073, 1096)
	Normal	6	(5 , 547, 817, 969, 1073, 1096)
	Pareto	6	(265, 544, 761, 977, 1075, 1096)
	Beta	6	(273, 548, 559, 797, 822, 1096)
BIC	Log-Normal	4	(548, 817, 970, 1096)
	Normal	4	(547, 817, 969, 1096)
	Pareto	5	(265, 544, 761, 1075, 1096)
	Beta	6	(273 , 548, 559, 797, 822, 1096)
MBIC	Log-Normal	4	(548, 817, 970, 1096)
	Normal	4	(547, 817, 969, 1096)
	Pareto	3	(544, 761, 1096)
	Beta	6	(273, 548, 559, 797, 822, 1096)

3.4   Experiments on Q

Additionally, experiments were conducted on the Q parameter of the WBS method (see Fig. 4), revealing an approximately constant relationship between the number of detected change-points and the value of Q. For the Log-normal, Normal, and Pareto distributions, there is a rapid convergence toward an asymptote located around 4 to 5 detected change-points, whereas the Beta distribution exhibits a slower asymptotic approach, reaching approximately 20 detected change-points when Q approaches 50. Thus, the parameter Q influences the number of detected change-points in an approximately linear fashion up to values around 20-50, after which the number stabilizes. This convergence occurs more rapidly when the underlying distribution is less asymmetric, as observed for the Log-normal, Normal, and Pareto datasets.

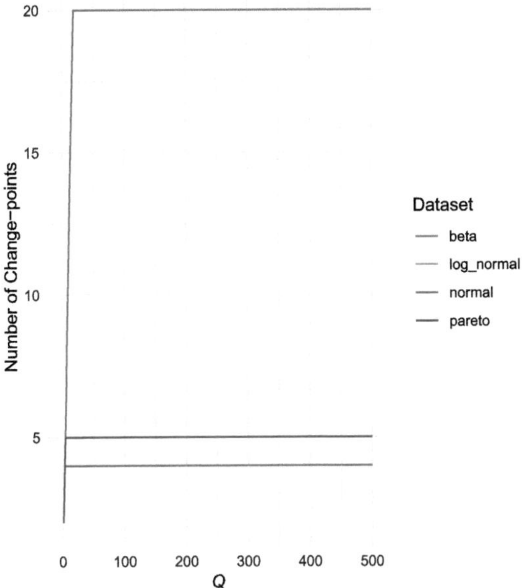

Fig. 4. Results for the Q parameter experiments

3.5   GA-Coen Results

Finally, regarding the GA-Coen method, in contrast to the other two approaches, it was not necessary to define a cost function that varied according to the dataset used. Moreover, no additional parameter required calibration to serve as an initial estimate for the number of change-points, nor was it necessary to assume any specific distributional form for the datasets. For its implementation the following instruction was used,

```
segment(data_set_name, method = "ga-coen", maxiter = 500)
```

Thus, the results for the estimated number and locations of change-points for each dataset were obtained using the genetic algorithm described in (3), with $r = 500$ generations or iterations and are recoreded in Table (3).

Table 3. GA-Coen experiments results

Dataset	Number of Detected Change-points	Detected Change-points
Log-Normal	2	(554, 816)
Normal	2	(544, 804)
Pareto	1	(616)
Beta	5	(171, 276, 630, 810, 823)

For the Log-normal and normal datasets the first change-point, $\tau_1 = 548$ is estimated approximately with $\hat{\tau}_1 = 554$, that is, an excess estimation of six units. On the other hand, $\tau_2 = 823$ was estimated with $\hat{\tau}_2 = 816$ and $\hat{\tau}_2 = 804$, respectively, that is, in both cases an underestimation of 8 and 19 units respectively. Finally for these two datasets the algorithm failed to detect the last change-point, a behavior previously documented in [24] where change-points close to the end are less likely to be easily detected.

For the Pareto dataset, only one change-point was detected with position $\hat{\tau}_1 = 616$ which is closer to the second change-point defined for the simulation data. Finally for the Beta distribution observations the first change-point $\tau_1 = 276$ was approximated by $\hat{\tau}_1 = 276$, $\tau_2 = 550$ by $\hat{\tau}_2 = 630$ and $\tau_3 = 825$ by $\hat{\tau}_3 = 823$.

Taking into account that this algorithm was based on a probabilistic iterative process and did not assume any prior hypothesis about the data—beyond the fact that exceedances could be modeled through NHPP—the results can be considered satisfactory. Moreover, they could serve as a starting point for post-hoc analyses and hypothesis formulation by the researcher, particularly in cases where the underlying nature of the data is unknown.

The results obtained here are highly subject to improvement, referring once again to the exploration of other operators associated with genetic algorithms, as discussed in [25]. Additionally, further tuning could be achieved by considering as an alternative, an informative prior distribution for the change-points, in contrast to the uniformity assumption adopted in this work.

4   Conclusions and Further Developments

A unified framework for implementing different change-point detection methods was employed through the `tidychange-point` library.

This library integrates approaches of various types, encompassing exact, approximate, and metaheuristic methods, and also provides the possibility of extending its functionality through user-defined procedures.

Using this tool, experiments were carried out on four datasets and three segmentation methods available within the library. The results show that the PELT method is highly sensitive when the AIC is used as a penalty, while Bayesian information criteria provide better results. However, this method does not perform well when dealing with observations drawn from a Pareto distribution.

In the case of the WBS method, more accurate results were obtained—indeed, the best among the three evaluated methods. A possible limitation lies in the need to define an initial value for the number of change-points, although the method exhibits low sensitivity to this parameter once its value exceeds the order of tens, at which point stability in the number of detected change-points is observed.

Finally, an alternative to the previous methods is the use of metaheuristics, which allow obtaining adequate approximations without requiring strict assumptions about the general behavior of the observations.

While the last method incurs in a computational load that could be considered significant, this is balanced by the easiness in obtaining results with only the data. An approach could be then, using the genetic algorithm and then the number of detected change-points as an input for the WBS thus facilitating the implementation while eliminating any subjective judgement.

References

1. Change-point detection papers on google scholar. https://tinyurl.com/2x9b76j4, Accessed 10 Dec 2025
2. Aminikhanghahi, S., Cook, D.J.: A survey of methods for time series change point detection. Knowl. Inf. Syst. **51**(2), 339–367 (2016). https://doi.org/10.1007/s10115-016-0987-z
3. Bai, J.: Estimating multiple breaks one at a time. Economet. Theor. **13**(3), 315–352 (1997)
4. Baranowski, R., Fryzlewicz, P.: wbs: wild binary segmentation for multiple change-point detection (2014). https://doi.org/10.32614/cran.package.wbs
5. Basseville, M., Nikiforov, I.: Detection of abrupt changes: theory and application. Prentice Hall, Prentice-Hall information and system sciences series (1993)
6. Baumer, B.S., Suárez Sierra, B.M., Coen, A., Taimal, C.A.: tidychangepoint: a tidy framework for changepoint detection analysis (2025). https://doi.org/10.32614/CRAN.package.tidychangepoint, https://CRAN.R-project.org/package=tidychangepoint, r package version 1.0.1
7. Fryzlewicz, P.: Unbalanced haar technique for nonparametric function estimation. J. Am. Stat. Assoc. **102**(480), 1318–1327 (2007). https://doi.org/10.1198/016214507000000860, http://dx.doi.org/10.1198/016214507000000860
8. Fryzlewicz, P.: Wild binary segmentation for multiple change-point detection. Ann. Stat. **42**(6), 2243–2281 (2014)
9. Haynes, K., Killick, R.: changepoint.np: methods for nonparametric changepoint detection (2016). https://doi.org/10.32614/cran.package.changepoint.np
10. Keogh, E., Chu, S., Hart, D., Pazzani, M.: An online algorithm for segmenting time series. In: Proceedings 2001 IEEE International Conference on Data Mining, pp. 289–296 (2001). https://doi.org/10.1109/ICDM.2001.989531
11. Keogh, E., Lin, J.: Clustering of time-series subsequences is meaningless: implications for previous and future research. Knowl. Inf. Syst. **8**(2), 154–177 (2005)
12. Killick, R., Fearnhead, P., Eckley, I.A.: Optimal detection of changepoints with a linear computational cost. J. Am. Stat. Assoc. **107**(500), 1590–1598 (2012). https://doi.org/10.1080/01621459.2012.737745
13. Killick, R.: changepoint: methods for changepoint detection (2011). https://doi.org/10.32614/cran.package.changepoint
14. Li, M., Lu, Q.: changepointga: changepoint detection via modified genetic algorithm (2025). https://doi.org/10.32614/cran.package.changepointga, http://dx.doi.org/10.32614/CRAN.package.changepointGA
15. Li, S., Lund, R.: Multiple changepoint detection via genetic algorithms. J. Clim. **25**(2), 674 – 686 (2012). https://doi.org/10.1175/2011JCLI4055.1
16. Lund, R.B., Beaulieu, C., Killick, R., Lu, Q., Shi, X.: Good practices and common pitfalls in climate time series changepoint techniques: A review. J. Clim. **36**(23), 8041–8057 (2023). https://doi.org/10.1175/jcli-d-22-0954.1

17. Millard, S.P.: EnvStats: an R package for environmental statistics. Springer, New York (2013). https://www.springer.com

18. Muggeo, V.M.R.: segmented: regression models with break-points / change-points estimation (with possibly random effects) (2003). https://doi.org/10.32614/cran.package.segmented, http://dx.doi.org/10.32614/CRAN.package.segmented

19. Page, E.: Continuous inspection schemes. Biometrika **41**(1–2), 100–115 (1954). https://doi.org/10.1093/biomet/41.1-2.100

20. Pal, N., Jin, C., Lim, W.: Handbook of exponential and related distributions for engineers and scientists. CRC Press (2005)

21. R Core Team: R: A language and environment for statistical computing. R Foundation for Statistical Computing, Vienna, Austria (2025). https://www.R-project.org/

22. Romano, G.: Math337: changepoint detection (2025)

23. Scrucca, L.: Ga: a package for genetic algorithms inr. J. Stat. Softw. **53**(4) (2013). https://doi.org/10.18637/jss.v053.i04, http://dx.doi.org/10.18637/jss.v053.i04

24. Suárez-Sierra, B.M., Coen, A., Taimal, C.A.: Genetic algorithm with a bayesian approach for multiple change-point detection in time series of counting exceedances for specific thresholds. J. Korean Stat. Soc. **52**(4), 982–1024 (2023). https://doi.org/10.1007/s42952-023-00227-2, http://dx.doi.org/10.1007/s42952-023-00227-2

25. Taimal, C.A., Suárez-Sierra, B.M., Rivera, J.C.: An exploration of genetic algorithms operators for the detection of multiple change-points of exceedances using non-homogeneous poisson processes and bayesian methods, pp. 230–258. Springer Nature Switzerland (2023). https://doi.org/10.1007/978-3-031-47372-2_20

26. Truong, C., Oudre, L., Vayatis, N.: ruptures: change point detection in python (2018). https://arxiv.org/abs/1801.00826

27. Truong, C., Oudre, L., Vayatis, N.: Selective review of offline change point detection methods. Signal Process. **167**, 107299 (2020). https://doi.org/10.1016/j.sigpro.2019.107299, http://dx.doi.org/10.1016/j.sigpro.2019.107299

28. Vrieze, S.I.: Model selection and psychological theory: a discussion of the differences between the akaike information criterion (aic) and the bayesian information criterion (bic). Psychol. Methods **17**(2), 228–243 (2012). https://doi.org/10.1037/a0027127

29. Wei, L., Keogh, E.: Semi-supervised time series classification. In: Proceedings of the 12th ACM SIGKDD International Conference on Knowledge Discovery and Data Mining, pp. 748–753. KDD '06, Association for Computing Machinery, New York, NY, USA (2006). https://doi.org/10.1145/1150402.1150498

30. Wickham, H.: Tidy data. J. Stat. Softw. **59**(10) (2014). https://doi.org/10.18637/jss.v059.i10, http://dx.doi.org/10.18637/jss.v059.i10

31. Zhang, N.R., Siegmund, D.O.: A modified bayes information criterion with applications to the analysis of comparative genomic hybridization data. Biometrics **63**(1), 22–32 (2007). https://doi.org/10.1111/j.1541-0420.2006.00662.x

Interactive Visualization and Analysis of Colombia's GEIH Data: A Shiny Application for Reproducible Demographic and Labor Market Research

Iván Cruz(✉)(iD), Daniel Molina(iD), and Alic Barandica(iD)

Economics, Universidad del Magdalena, Santa Marta, Colombia
`icruz@unimagdalena.edu.co`

Abstract. We present an interactive Shiny application designed to democratize access to data from the Comprehensive Household Survey, or GEIH, with special emphasis on Venezuelan migration patterns and labor market dynamics. Our application addresses the growing need for accessible and reproducible demographic research tools. Through a modular architecture encompassing demographics, education, labor markets, housing, and health indicators, we have created a platform that serves academics, policymakers, NGOs, and public officials. The application leverages advanced R packages including data. table for efficient large-scale data processing, plotly for interactive visualizations, and shinydashboard for intuitive user interfaces. Our implementation demonstrates how complex statistical analysis can be made accessible without requiring extensive programming knowledge from end users. We discuss the technical challenges encountered in processing multi-module GEIH data and performance optimization strategies for large datasets. The application facilitates reproducible research by enabling data export in multiple formats and providing public access to source code under the MIT license.

Keywords: Shiny Applications · GEIH Data · Interactive Visualization · Venezuelan Migration · Reproducible Research · Labor Market Analysis

1 Introduction

The massive exodus of Venezuelans represents the largest migration crisis in recent history of the Western Hemisphere, with Colombia receiving more than 1.8 million displaced individuals since 2015 [12]. This demographic shock, increasing Colombia's population by nearly 4%, has created unprecedented challenges and opportunities for researchers, policymakers, and civil society organizations attempting to understand and respond to rapidly evolving social dynamics [2].

In our work as researchers, we have observed the significant barriers that exist between the complexity of demographic datasets and their practical application by professionals in the field.

© The Author(s), under exclusive license to Springer Nature Switzerland AG 2026
B. M. Suárez et al. (Eds.): R Day 2025, CCIS 2824, pp. 139–152, 2026.
https://doi.org/10.1007/978-3-032-18455-9_8

The Comprehensive Household Survey (GEIH) of Colombia, administered by the National Administrative Department of Statistics (DANE), represents the country's primary source for labor market statistics and demographic information [5]. Despite its comprehensive nature and national representativeness, the complexity of its modular structure and considerable scale present a significant obstacle to its utilization by researchers.

To overcome this challenge that limits the usability of such a valuable resource, we are motivated to develop this Shiny application. This initiative is grounded in direct experience from research on Venezuelan migration patterns and labor market integration. Time and again, we found colleagues, students, and policymakers who possessed valuable domain expertise but lacked the technical resources to conduct independent analysis of GEIH data. This gap between data availability and usability inspired us to create a tool that would democratize access to one of Colombia's most important demographic datasets.

To address this gap, the central objective of this article is to present an open-source computational tool designed to facilitate reproducible analysis and visualization of the Comprehensive Household Survey (GEIH). Although the application architecture is generalizable and enables cross-sectional exploration of all survey modules (health, education, housing), we use the characterization of the Venezuelan migrant population as the primary case study to demonstrate the analytical capabilities of the tool.

This dual approach allows us to validate the technical robustness of the platform against complex and urgent research questions, while providing the academic community with a reusable framework for other domains of study. The application not only offers an intuitive interface for exploring demographic patterns without requiring advanced programming, but also supports reproducible research practices by making calculation methodology transparent and enabling export of processed microdata.

To operationalize this objective, our technical approach leverages the power of R's Shiny framework to create an interactive web application capable of efficiently processing over 200,000 household records annually. The architecture is structured around five main modules (Demographics, Education, Labor Market, Housing, and Health), supporting both national-level analysis and regional disaggregation. This modularity allows users to explore migration patterns through cross-cutting filtering mechanisms without losing the ability to examine other sociodemographic dimensions.

This article makes several contributions to the literature on reproducible research tools and demographic analysis applications. We present a comprehensive methodology for transforming complex survey data into accessible interactive visualizations, demonstrate performance optimization techniques for handling large-scale household survey data in web applications, and provide a replicable framework for creating policy-relevant research tools. Additionally, we contribute to the growing body of work on Shiny applications in academic research, documenting both technical innovations and lessons learned from our deployment experience.

2   Related Work

2.1   Shiny Applications in Academic Research

The use of Shiny for academic research has grown substantially since its introduction in 2012, with applications spanning disciplines from biomedical research to social sciences [1]. Recent systematic reviews have documented over 445 unique Shiny applications across 229 peer-reviewed journals, demonstrating the framework's versatility for creating research tools [11]. These applications range from simple visualization tools to complex analytical workflows that guide users through multi-step analysis processes.

Our work builds on established principles of reproducible research in R, particularly the integration of interactive applications with traditional research workflows [10]. The literature emphasizes the importance of maintaining transparency in analytical processes while making tools accessible to non-technical users [13]. This balance between accessibility and methodological rigor has been central to our design philosophy.

2.2   Visualization of Demographic Data

Interactive demographic visualization has become increasingly important as migration patterns grow more complex globally. Previous work on migration data dashboards has focused primarily on aggregated flows and basic demographic breakdowns [9]. However, few tools have specifically addressed the intersection of migrant status with detailed socioeconomic indicators, particularly in the context of large-scale regional migration crises.

The Comprehensive Household Survey (GEIH) has been the subject of extensive academic analysis by researchers who have documented its fundamental value for understanding Colombian labor market dynamics, the country's demographic transitions, and the sociodemographic characteristics of the population [7]. This statistical operation, which covers the largest national survey with approximately 315,000 households annually, has established itself as the primary source of information for characterizing labor market behavior and is an essential input for evidence-based public policy formulation.

However, despite the informational richness of GEIH microdata and its public availability through the Open Data portal, most existing analyses require substantial technical expertise in specialized statistical software for data analysis, which significantly limits broader access to the insights contained in this data.

Although there are important institutional precedents, such as the DANE's Geovisor and official labor market visualization dashboards, these tools are typically closed-source and offer pre-aggregated visualizations that limit researchers' ability to define custom subdomains. Unlike these solutions, our application distinguishes itself by offering an open-source R/Shiny architecture that enables, in an unprecedented way, dynamic manipulation and real-time filtering of GEIH microdata, overcoming the limitations of traditional static approaches. This characteristic is fundamental, as it enables not only the visualization of standard indicators but also the ad-hoc disaggregation of specific populations (such as recent migrants by educational level) and transparent auditing of calculation algorithms, thereby filling a critical gap in the current ecosystem of digital demography tools in Colombia.

2.3 Reproducible Research Platforms

The broader movement toward reproducible research has generated numerous tools and platforms designed to bridge the gap between complex analysis and broader user communities [8]. Shiny applications represent one approach to this challenge, offering interactive access to underlying analytical capabilities while maintaining transparency about methodological decisions [4].

Our application contributes to this ecosystem by providing a domain-specific tool that addresses particular needs in migration and labor market research. The emphasis on data export capabilities and methodological transparency aligns with broader principles of open science and reproducible research practices.

3 Data and Methodology

3.1 Structure and Complexity of the GEIH Survey

The Comprehensive Household Survey (GEIH) represents Colombia's most integral source for demographic information and labor market data. Conducted monthly by DANE, the survey employs a probabilistic, multi-stage sampling design that ensures national representativeness across 24 capital cities and their metropolitan areas [6]. The survey universe includes the civilian population not residing in institutions within Colombia's national borders, with coverage extending to both urban and rural areas.

The multi-module structure of GEIH presents both opportunities and challenges for researchers. Each monthly wave of the survey consists of several interconnected modules: a basic demographic module capturing household composition and individual characteristics, a detailed labor market module following International Labor Organization guidelines, an income module tracking multiple income sources, and specialized modules addressing migrant status, education, health access, and housing conditions. This comprehensive structure enables rich analysis but creates significant complexity for data users.

Beginning in 2021, DANE implemented methodological updates to GEIH that incorporated enhanced migration measurement capabilities [3]. These changes, designed to improve statistical visibility of vulnerable populations including Venezuelan migrants, introduced new variables for migrant status, country of origin, and migration trajectory. However, the increased complexity has made the data even more challenging for non-specialist users to navigate.

We process GEIH data at the individual level, maintaining the survey's complex weighting structure while creating analytical variables suitable for interactive visualization. Our data processing workflow begins with raw monthly files provided by DANE, which we systematically merge and clean to create consolidated datasets that preserve survey representativeness while optimizing for web application performance.

3.2 Data Processing Architecture

Our data processing pipeline addresses multiple technical challenges inherent in handling large-scale household surveys. GEIH's hybrid architecture, combining longitudinal panel components and monthly cross-sections, requires specialized temporal harmonization techniques and robust algorithms to integrate multi-stage data while preserving national representativeness. The incorporation of the migration module since 2019 presents additional variable standardization challenges, requiring specific mapping protocols, cross-validation, and automated quality control checks that detect inconsistencies in unique identifiers and merge issues across different sources to ensure coherence with existing historical series.

We developed two primary processing scripts that use the data.table package to handle survey scale efficiently. The national-level script processes approximately 232,000 household records annually, generating summary statistics and analytical variables optimized for web application visualization. The regional script provides similar functionality while maintaining disaggregation across Colombia's 32 departments plus Bogotá (see Fig. 1).

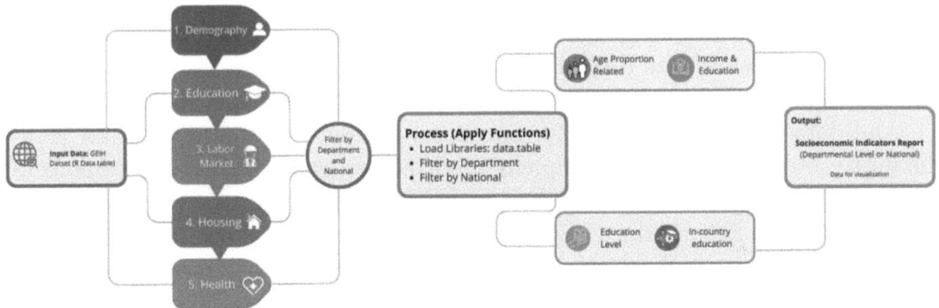

Fig. 1. Pipeline Flow

Our processing approach prioritizes performance while maintaining analytical flexibility. We pre-calculate frequently requested summary statistics and store them in optimized formats that minimize real-time computation requirements. This approach enables responsive user interfaces while preserving the ability to conduct detailed subgroup analysis.

The process of identifying migrant status deserves particular attention given its importance to our target user community. We implement DANE's official migration classification criteria, identifying Venezuelan migrants through multiple variables including country of birth and previous residence. This multivariable approach provides robust identification while acknowledging the complexity of migration experiences and documentation status.

3.3 Technical Architecture and Implementation

Our Shiny application employs a modular architecture that balances functionality with maintainability. The main application file (app.R) coordinates five specialized modules, each addressing distinct analytical domains (Demographics, Education, Labor Market, Housing, and Health) while maintaining consistent user interface conventions and data handling approaches. To facilitate community validation and public access to these analyses, we have deployed a functional version of the tool at https://jsidte-daniel-molina.shinyapps.io/shiny-app/, enabling real-time interaction with processed microdata.

The user interface employs `shinydashboard` to create an intuitive navigation structure that guides users through analytical options without overwhelming them with technical complexity. We implemented conditional panels that dynamically adjust available options based on user selections, reducing cognitive load while maintaining analytical depth.

Our server-side implementation leverages reactive programming principles to ensure efficient computation and responsive user interactions. We employ reactive values to coordinate across modules while minimizing unnecessary recalculations. This approach proves particularly important when handling large datasets and complex visualizations.

Visualization components integrate `plotly` to generate static and interactive graphics as needed for analytical purposes. Population pyramids, for example, leverage `plotly`'s interactive capabilities to facilitate detailed exploration of age and gender distributions, while summary graphs are designed to communicate key indicators clearly and directly. This visualization architecture efficiently balances computational performance with an intuitive user experience.

The data export functionality represents a crucial component of our reproducibility strategy. We have implemented export capabilities using `openxlsx` and base R functions to provide data in CSV and Excel formats. These exports include both the filtered data requested by users and metadata documenting the filtering criteria applied, supporting reproducible research practices.

3.4 Statistical Specification of Labor Indicators

To ensure consistency with DANE's methodological standards and International Labor Organization (ILO) guidelines, the application implements indicator calculation through design-based ratio estimators accounting for GEIH's complex sampling structure.

Let U be the survey universe and s the observed sample. For each individual $i \in s$, we define w_i as the expansion factor (variable `fex_c18` in the microdata), which represents the number of people in the population that individual i represents. The application allows users to define a subdomain of interest $d \subset U$ through dynamic filters (e.g., Venezuelan migrants in Bogotá).

Key labor market indicators are calculated dynamically over this subdomain d as follows:

1. **Working-Age Population (WAP):** Defined as the population aged 12 or older in urban areas and 10 or older in rural areas.

$$\text{WAP}_d = \sum_{i \in s \cap d} w_i \cdot I(i \in \text{WAP}) \tag{1}$$

Where $I(\cdot)$ is an indicator function.

2. **Global Participation Rate (GPR):** Measures the pressure of labor supply on the working-age population.

$$\text{GPR}_d = \frac{\sum_{i \in s \cap d} w_i \cdot I(i \in \text{EAP})}{\text{WAP}_d} \times 100 \tag{2}$$

Where EAP corresponds to the Economically Active Population (employed + unemployed).

3. **Employment Rate (ER):**

$$\text{ER}_d = \frac{\sum_{i \in s \cap d} w_i \cdot I(i \in \text{Employed})}{\text{WAP}_d} \times 100 \tag{3}$$

4. **Unemployment Rate (UR):** Represents the proportion of the labor force actively seeking work but unable to find it.

$$\text{UR}_d = \frac{\sum_{i \in s \cap d} w_i \cdot I(i \in \text{Unemployed})}{\sum_{i \in s \cap d} w_i \cdot I(i \in \text{EAP})} \times 100 \tag{4}$$

The computational implementation of these formulas uses vectorized operations in the `data.table` package to perform weighted summation of expansion factors w_i in real-time, recalculating population denominators each time the user modifies subdomain d filtering parameters.

4 Application Features and User Experience

4.1 Modular Design and Navigation

Our application organizes analytical capabilities into five main modules: Demographics, Education, Labor Market, Housing, and Health. Each module provides specialized visualizations and analytical tools while maintaining consistent navigation patterns and user interface conventions. This modular approach allows users to focus on specific domains while preserving the ability to conduct integrated analysis across multiple dimensions.

The Demographics module serves as the application's foundation, providing population pyramids, gender distributions, and marital status breakdowns. These visualizations support both national and regional analysis, with dynamic filtering by Venezuelan migrant status. Population pyramids, in particular, reveal marked differences between the general population and Venezuelan migrants,

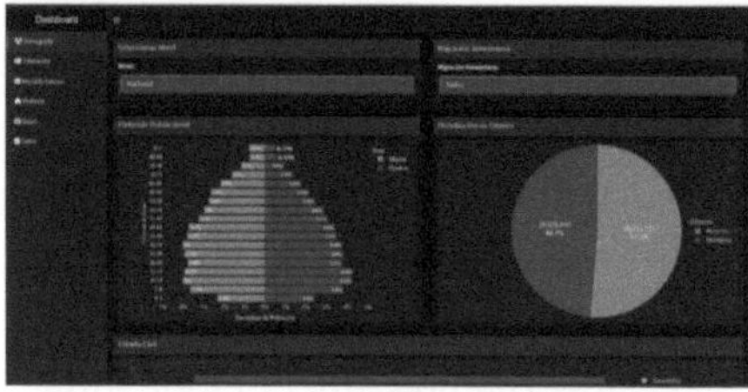

Fig. 2. Demographics Module.

Fig. 3. Education Module.

highlighting the characteristically younger age profile of recent migration flows (see Fig. 2).

The Education module addresses critical questions about human capital and economic integration. Our visualizations reveal patterns of educational achievement and income differentials by educational level, providing insights into labor market segmentation and integration challenges. Interactive graphics allow users to explore how educational credentials translate to economic outcomes differently for Venezuelan migrants compared to Colombian nationals (see Fig. 3).

The Labor Market module constitutes a significant contribution to public policy formulation, as it enables deep and continuous analysis of participation, employment and unemployment rates, as well as detailed study of occupational patterns and structural employment characteristics in Colombia, providing reliable indicators and updated series fundamental to decision-making and the effective design of government interventions. (see Fig. 4)

Fig. 4. Labor Market Module.

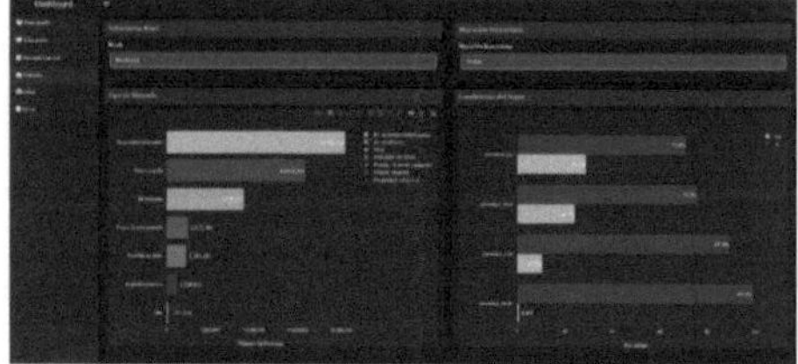

Fig. 5. Housing Module.

The Housing and Health modules complete our comprehensive approach to social integration analysis. Housing condition indicators provide insights into living standards and integration experiences, while health access measures document barriers to essential services. Together, these modules enable holistic assessment of migration impacts and integration outcomes (see Fig. 5).

4.2   Interactive Visualization, Export, and Reproducibility

Our visualization approach prioritizes interpretability without sacrificing analytical rigor. To ensure a consistent and professional user experience, we have implemented custom CSS styles and accessible color palettes (`paletteer` and `viridis`) that ensure readability for users with diverse visual abilities. Interactivity is managed through `plotly`, enabling detailed exploration through hover and drill-down functionalities, supported by pre-computation strategies that maintain fast response times even with complex visualizations of large data volumes.

A distinctive component of the application is its dynamic filtering capability, which enables smooth transitions between national and regional perspectives. In particular, the Venezuelan migration filter facilitates comparative analysis across population groups, revealing integration patterns and socioeconomic gaps. Complementing this visual exploration, the Data module offers direct access to underlying microdata through interactive tables (`DT`), allowing users to examine, search, and filter specific observations for granular investigations.

Recognizing the importance of open science, the tool integrates robust export capabilities that support reproducible research practices. Users can download filtered datasets in both CSV format (for statistical analysis) and Excel, preserving the survey's weighting structure. Each export automatically includes metadata documenting the filtering criteria and analytical decisions applied during the session, ensuring that visual findings can be validated and replicated in external software environments (see Fig. 6).

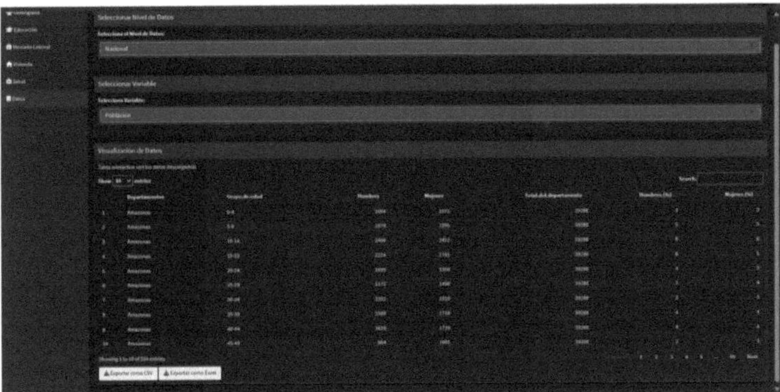

Fig. 6. Data Module

5   Technical Challenges and Implementation Solutions

5.1   Processing Optimization and Usability

Managing GEIH's scale, with over 200,000 annual records, presents significant performance challenges. To address these, our primary approach involves pre-computing summary statistics and intensive use of the `data.table` package.

This strategy, which reduces computation times by orders of magnitude compared to base R operations, allows us to trade storage space for response speed, guaranteeing smooth interactivity even under the typical memory constraints of free hosting.

Complementing this backend optimization, interface design addresses the challenge of visual complexity through progressive disclosure. We implement conditional panels that dynamically adapt visible options based on user context (for example, hiding regional filters in the national view), thereby reducing cognitive load. This approach is reinforced by responsive design that ensures functionality on both desktop and mobile devices, enabling the analytical depth of microdata to be accessible without overwhelming the end user.

5.2 Deployment and Hosting Considerations

Public deployment of the application on free platforms such as ShinyApps.io imposes strict memory and concurrency restrictions. To mitigate these limitations, our architecture minimizes resource consumption through the precomputation strategies already described, enabling stable performance without infrastructure costs. To facilitate reproducibility of the computing environment, we provide open access to source code, allowing other researchers to audit the libraries used and replicate deployment on their own systems. Although we recognize the potential of Docker containerization to scale the tool on institutional servers, we currently prioritize free hosting to maximize immediate accessibility for our target user community.

6 Results and Impact

6.1 User Participation and Adoption

Since deployment, our application has demonstrated significant reach across our target user communities. The combination of academic institutions and government agencies accessing the platform reflects successful achievement of our goal to democratize access to GEIH data analysis capabilities.

User behavior analysis reveals engagement patterns that validate our design decisions. The modular structure enables focused analysis, with users typically engaging deeply with one or two modules per session rather than attempting comprehensive analysis across all domains. This pattern supports our decision to organize functionality around distinct analytical domains.

The implementation of export functionality reinforces the project's commitment to reproducible research. By enabling direct download of processed microdata, the application transcends its purely visual function to consolidate as a data preparation platform, offering researchers an efficient mechanism to obtain valuable inputs for econometric models or external statistical analysis.

The geographic distribution of users reflects both Colombian domestic use and participation from the international research community. The application has supported research projects across multiple universities and provided data access for policy analysis by government agencies at national and regional levels.

6.2   Research Applications and Results

Our application has supported diverse research demonstrating its potential to deepen understanding of migration dynamics and labor market behavior. The incorporated interactive tools have facilitated dynamic exploration of patterns and relationships between variables, enabling identification of emerging trends and contrasting scenarios intuitively and efficiently. Likewise, the export functionality has enhanced detailed statistical analysis, enabling users to conduct advanced modeling and validation in specialized environments without sacrificing data traceability or consistency.

Comparative analysis capabilities, particularly the ability to examine Venezuelan migrants separately from the general population, have provided insights into integration patterns and policy needs. Researchers have documented differences in educational achievement, labor market participation, and housing conditions that inform both academic understanding and policy development.

The application's accessibility has enabled participation of researchers and professionals who previously lacked access to GEIH data analysis capabilities. This democratization has broadened the research community engaging with migration and demographic topics, contributing to more diverse perspectives and approaches in the literature.

6.3   Lessons Learned and User Feedback

User feedback has highlighted several successful design elements while identifying areas for potential improvement. The intuitive navigation structure receives consistent praise, validating our emphasis on user experience design. The quality of visualization and interactivity have enabled users to generate insights that would be difficult to achieve through traditional statistical software alone.

Performance considerations have proven crucial to user adoption. Precomputation strategies successfully maintain responsive interfaces, although users working with multiple simultaneous filters occasionally experience slower response times. This feedback has informed our ongoing optimization efforts and future development priorities.

The modular architecture has proven highly successful, enabling users to focus on specific analytical domains without confronting unnecessary complexity. This approach has been particularly valuable for policymakers and NGO staff who often have specific questions rather than broad exploratory research objectives.

7   Future Directions and Improvements

7.1   Planned Technical Enhancements

Our development plan contemplates improvements oriented toward expanding analytical capabilities without compromising usability. We plan to incorporate new indicators, especially analysis of income disaggregated by economic sector and detailed health condition metrics, along with integration of information covering a greater number of years.

Multilingual support represents a priority development area, with English translation planned to expand international accessibility. This enhancement would enable broader use by the international research community studying migration patterns and labor market integration in Latin America.

Enhanced export capabilities, including PDF report generation with embedded visualizations, would support policy communication and academic dissemination needs. We envision automatic report generation that combines user-selected analyses with contextual information about data sources and methodological approaches.

Performance optimization remains an ongoing priority, focusing especially on reducing response times for complex filtering operations. We are evaluating additional strategies such as pre-computation and implementation of a robust database backend that supports larger datasets and more sophisticated queries, while enabling efficient generation of econometric and statistical reports.

7.2 Methodological Extensions

Future versions will incorporate temporal analysis capabilities, allowing users to examine trends over time rather than focusing on individual survey waves. This enhancement would support longitudinal research applications and policy evaluation while maintaining the accessibility that characterizes our current approach.

Advanced statistical capabilities, including basic regression analysis functionality and statistical tests, could expand the application's utility for users seeking to conduct more sophisticated analysis without leaving the web interface. These enhancements would maintain our commitment to accessibility while supporting more complex research applications.

Integration with other Colombian survey datasets, particularly the Quality of Life Survey and specialized migration surveys, could provide comprehensive analytical capabilities covering broader aspects of social and economic integration. Such integration would position our application as a comprehensive platform for demographic and social research in Colombia.

7.3 Community Engagement and Sustainability

We envision expanding community participation through user workshops and training sessions that build capacity for effective use of demographic data in research and policy applications. These activities would complement the technical platform by building user skills and fostering collaborative research networks.

Sustainability considerations include exploring funding mechanisms that could support ongoing development and hosting while maintaining free access for our target user communities. Partnership with government agencies, international organizations, and academic institutions could provide resources for development and continuous improvement.

7.4   Limitations and Constraints

Despite its advantages, the tool presents inherent limitations. First, server RAM dependence limits the number of concurrent users in the free hosting version. Second, although the application calculates point estimates, the current version does not include automated calculation of confidence intervals through bootstrap or Taylor linearization due to the computational load this would impose in real-time. Finally, the tool depends on DANE's flat file structure, so changes in official collection methodology will require manual code base updates.

8   Conclusion

Our interactive Shiny application for GEIH data analysis represents a successful effort to democratize access to Colombia's most comprehensive demographic and labor market dataset. Through careful attention to user experience design, technical optimization, and reproducible research practices, we have created a tool that serves diverse user communities while maintaining analytical rigor and methodological transparency.

The application's modular architecture, interactive visualization capabilities, and data export functionality address critical needs in migration and labor market research. By making demographic analysis accessible to users without programming experience, we have expanded participation in evidence-based research and policy development. The emphasis on Venezuelan migration patterns addresses timely policy questions while demonstrating broader principles for demographic data visualization.

The technical innovations in data processing, performance optimization, and interface design provide lessons for other researchers seeking to create accessible research tools. Our experience with deployment challenges and solutions offers practical guidance for similar projects, while our emphasis on reproducible research practices contributes to broader efforts to improve scientific transparency and replicability.

Positive user engagement and research applications demonstrate successful achievement of our goals to democratize data access and support evidence-based analysis. The application has enabled research projects, supported policy development, and facilitated broader community participation with demographic data analysis. These results validate the investment in developing accessible research tools while highlighting opportunities for continued improvement and expansion.

Looking forward, we see this work contributing to a broader ecosystem of accessible research tools that bridge the gap between complex datasets and the diverse communities that need to use them. The success of our approach suggests potential for replication in other contexts, supporting the movement toward open science. In consonance with this commitment to transparency, the microdata used in this study are publicly accessible through DANE and the complete application source code is available in the GitHub repository: https:// github.com/dmgsjj/first_app. We distribute this code under the MIT license to encourage its reuse and adaptation, thereby ensuring that our methodology can serve as a foundation for future research.

Our experience developing and deploying this application reinforces the value of collaborative approaches that bring together technical expertise, domain knowledge, and insights from the user community. The intersection of these perspectives has been essential to creating a tool that successfully balances analytical capability with accessibility, serving as a model for future research tool development initiatives.

As migration patterns continue to evolve and demographic data become increasingly important for policy development, tools such as ours will play crucial roles in ensuring that evidence-based analysis informs decision-making across all levels of government and civil society. We are committed to continuing this work and supporting the broader community of researchers and professionals working to understand and address the challenges and opportunities presented by migration and demographic change in Latin America.

The authors declare that they have no competing financial interests nor known personal relationships that could have influenced the work presented in this article.

References

1. Beeley, C.: Web application development with R using shiny. Packt Publishing, Birmingham (2016)
2. Banco de la república: migration waves from venezuela in colombia: phenomenon characterization and macroeconomic effects. Estudios Económicos Regionales 97, Banco de la República, Bogotá (2020)
3. de la República, B.: Methodological updates and primary changes to the comprehensive household survey (GEIH). Monetary Policy Report, Banco de la República, Bogotá (2022)
4. Chang, W., et al.: Shiny: web application framework for R. R package version 1.7.1 (2021)
5. DANE: gran encuesta integrada de hogares - geih. national administrative department of statistics, bogotá (2022)
6. DANE: great integrated household survey (GEIH) methodology. technical documentation, national administrative department of statistics, bogotá (2022)
7. Delgado-Prieto, L.: dynamics of local wages and employment: evidence from the venezuelan immigration in colombia. IZA Conference Paper AMM_2021 (2021)
8. Gandrud, C.: Reproducible research with R and R studio. Chapman and Hall/CRC, Boca Raton (2020)
9. International Organization for Migration: World Migration Report 2020. IOM Publications, Geneva (2020)
10. Peng, R.D.: Reproducible research in computational science. Science **334**(6060), 1226–1227 (2011)
11. Perkins, M.A., Carrier, J.W.: Protocol for programming a degree-to-careers dashboard in R using Posit and Shiny. STAR Protocols **5**(1), 102956 (2024)
12. UNHCR: global trends: forced displacement in 2020. UN High Commissioner for Refugees, Geneva (2020)
13. Wickham, H.: Mastering shiny: build interactive apps, reports, and dashboards. O'Reilly Media, Sebastopol (2023)

Applications of R in Multiple Disciplines

Potential Contributions of R Packages to Statistical Quality Control in Industry 4.0

Laura V. Pena-Sanchez[✉] [iD] and Carmen E. Patiño-Rodríguez [iD]

Universidad de Antioquia, Medellín, Colombia
`valentina.pena1@udea.edu.co`

Abstract. The growing adoption of Industry 4.0 technologies has increased the need for flexible and data-driven tools to enhance Statistical Quality Control (SQC). In this context, this study aims to analyze the potential contributions of four R packages—`qcc`, `qcr`, `qicharts2`, and `SixSigma`—to modern quality management systems. The purpose is to describe how these existing tools, when used together, can support real-time monitoring, process characterization, and capability analysis within digitalized industrial environments. The research follows an analytical and descriptive approach, comparing the functionalities, outputs, and complementarities of each package. `qcc` is identified as the foundational tool for classical control chart construction; `qcr` extends its scope through enhanced graphical flexibility and integrated capability analysis; qicharts2 introduces automated detection of instability signals based on Shewhart rules; and `SixSigma` focuses on structured process characterization and capability evaluation under the DMAIC framework. Results show that combining these packages provides a comprehensive environment for process monitoring and decision support, enabling automation, interoperability, and reproducibility. The study concludes that these R-based tools collectively bridge traditional SQC techniques with Industry 4.0 principles, reinforcing connectivity, transparency, and continuous improvement, while facilitating the transition from reactive quality control toward intelligent and adaptive process management.

Keywords: Statistical Quality Control · R Packages · Industry 4.0

1 Introduction

Statistical Quality Control (SQC) is a valuable methodology that employs statistical techniques to evaluate variability within manufacturing processes. Control charts and related analytical tools are used to assess the process state and to implement preventive actions through corrective measures before the process becomes out of control [1].

With the advent of digitalization, SQC has become integrated into Industry 4.0, commonly referred to as the Fourth Industrial Revolution. This new paradigm is grounded in principles such as interoperability, virtualization, decentralization, real-time capability, service orientation, and modularity [2]. These principles enable production systems to

B. M. Suárez et al. (Eds.): R Day 2025, CCIS 2824, pp. 155–190, 2026.
https://doi.org/10.1007/978-3-032-18455-9_9

evolve into flexible and autonomous environments capable of rapidly adapting to market demands. Moreover, advanced technologies such as robotics, 3D printing, and augmented reality enhance customization and worker support, while cybersecurity remains essential to ensure process reliability. Collectively, these elements not only modernize production systems but also position Big Data as a strategic driver for process optimization, fault prediction, and real-time decision-making based on massive information flows [3].

Big Data is characterized by its volume and velocity, referring to the magnitude of data generated and the speed at which it must be captured, processed, and analyzed. In addition to these, five further dimensions broaden its analytical scope: variety, encompassing both structured and unstructured data; veracity, ensuring data reliability; value, distinguishing useful from irrelevant information; viability, reflecting the organizational capacity to exploit data; and visualization, facilitating the clear representation of patterns and inefficiencies [4]. These dimensions highlight both the potential and the challenges of Big Data, as information arises from multiple heterogeneous sources such as social networks, industrial sensors, transactional records, biometric data, and digital documents [5].

In this context, R statistical packages provide a robust foundation for implementing SQC in digital environments. One of the most widely used is the qcc package, developed by Luca Scrucca (2004), which includes a broad range of control charts—such as Shewhart, CUSUM, and EWMA—for continuous, attribute, and count variables, along with process capability analysis and cause–effect diagrams, making it one of the most comprehensive tools available [6]. Similarly, other useful though less disseminated packages exist, such as SixSigma, developed by Cano, Moguerza, and Redchuk (2012), which provides a methodological framework aligned with the DMAIC cycle: Define (formulate a question), Measure (conduct preliminary research), Analyze (develop a hypothesis), Improve (test the hypothesis through experimentation), and Control (analyze data and draw conclusions). This package also integrates complementary tools such as repeatability and reproducibility (R&R) studies, capability indices, Pareto charts, process maps, loss function analysis, and variable control charts [7].

The qicharts2 package, developed by Jacob Anhøj (2018), focuses on the early detection of process variation using Shewhart statistical rules for common and special causes, providing high flexibility for quality control applications [8]. Additionally, the qcr package, introduced by Cano, Moguerza, and Redchuk (2017), extends the approach to multivariate control charts such as Hotelling's T2, MCUSUM, and MEWMA, and incorporates automatic Shewhart rules, process capability analysis, and an educational orientation through compatibility with R Markdown for reproducible reporting [9]. More recently, Violante et al. (2024) proposed the B-ATTRIVAR SS S2 chart, developed in R using the Shiny package by Joe Cheng (2012), which enables interactive interfaces to facilitate process visualization and monitoring. This tool combines attribute and variable inspections, offering operational simplicity alongside a more comprehensive analysis of variability—an example of how applied statistics can integrate with digitalization and the principles of Industry 4.0 [10].

Nevertheless, these packages face significant challenges in the Big Data era, where the analysis of large-scale and real-time information requires not only greater computational power but also new methodological approaches. The diversity of data sources exposes limitations in memory, processing speed, and scalability within traditional tools. As noted by Aykroyd, Leiva, and Ruggeri (2019), the classical assumptions of normality and independence become insufficient under these conditions, prompting the development of more robust methods such as asymmetric distributions, Bayesian models, machine learning techniques, and advanced transformations like wavelets. These developments indicate that SQC must not only adapt to digitalized industrial environments but also integrate its tools within the dynamics of Big Data and Industry 4.0 [11].

This article focuses particularly on Shewhart control charts and the various tools used for statistical quality control implemented in different R packages, with the aim of analyzing their functionalities, advantages, and limitations within the framework of current Big Data challenges. The selection of these tools is based on the fact that, despite their relevance in process quality control practice, relatively few packages are available and the existing documentation remains limited, creating a gap in their use and understanding within advanced and modern production environments. This study seeks to systematize the existing information, provide practical application examples, and offer a critical perspective on the role of these tools in the Big Data era. In doing so, it aims to contribute to both the academic and professional fields, highlighting the importance of adapting control charts particularly Shewhart charts and other related tools to new scenarios characterized by massive and heterogeneous data.

2 Contextual Framework

The application of statistical quality control within Industry 4.0 necessitates advanced analytical tools for real-time process monitoring and optimization [12]. This paradigm shift demands robust and adaptable software solutions capable of handling extensive data streams generated by interconnected systems and devices [13]. Control charts, a fundamental tool in statistical process control, are critical for detecting deviations and maintaining process stability in these complex environments [14]. Many R packages have been developed to address these needs, offering various functionalities for both Phase I and Phase II control chart analyses, encompassing both univariate and multivariate data [15]. The diverse applications of control charts extend beyond traditional manufacturing to include areas such as healthcare monitoring, where cost-effectiveness of patient observation is paramount [16].

Optimization packages often involve selecting appropriate chart types, sampling intervals, and control limits to minimize costs associated with false alarms and undetected out-of-control conditions, a concept explored in various statistical process control models. This section present main theorical frameworks for three R packages—`qcc`, `SixSigma, qicharts2, and qcr`—for their applicability in statistical quality control within the framework of Industry 4.0.

The `qcc package` is frequently cited and compared alongside other significant R packages in the SQC landscape, such as spc, edcc, qicharts, and ggQC, underscoring its prominence and contribution to the statistical computing ecosystem for quality management (Dobi & Zempléni, 2021). Its design emphasizes ease of use, with functions that allow for straightforward data input, chart generation, and interpretation, making it a valuable resource for statistical quality control applications.

2.1 qcc R Package

The `qcc package` is a widely used R tool for statistical quality control, providing a structured framework for process monitoring and improvement [6, 17]. As outlined in its typical documentation, such as vignettes and tutorials, the package is designed to be accessible for teaching, practitioners and researchers alike, enabling the application of various statistical methods for quality assurance.

The `qcc package`'s central functionality is its comprehensive support for a diverse range of control charts. These charts detect process deviations and maintaining stability in traditional and can adapt to advanced manufacturing environments, such as Industry 4.0 [9, 18]. The package facilitates the creation of numerous standard univariate control charts, including but not limited to `X-bar`, `S`, `R`, `p`, `np`, `c`, `u`, `EWMA`, and `CUSUM` charts [19]. These charts are crucial for analyzing individual measurements and subgroup data.

Beyond univariate analysis, `qcc` also offers capabilities for multivariate statistical process control, which is increasingly relevant with the complex data streams generated by interconnected systems in Industry 4.0 [16, 20] (see Table 1).

Table 1. Main Functions of the qcc R Package

Category	Function	Purpose
Control Chart Construction	`qcc()`	Chart control for variables or attributes X-bar, R, S, p, np, c, u charts
	`ewma()`	Generates EWMA average charts
	`cusum()`	Constructs CUSUM
	`mqcc()`	Constructs multivariate control charts such as Hotelling's T^2
Data Preparation and Grouping	`qcc.groups`	Groups data into subgroups for control charting
Capability Analysis	`process.capability()`	Calculates and visualizes process capability indices
	`process.capability.normal()`	Performs capability analysis assuming normal distribution

(continued)

Table 1. (*continued*)

Category	Function	Purpose
	`process.capability.nonnormal()`	Handles capability analysis for non-normal data
Descriptive Analysis	`pareto.chart()`	Displays the most frequent sources of variation
	`cause.and.effect()`	Generates cause-and-effect diagrams
Visualization and output	`summary.qcc()`	Summarizes key control chart statistics
	`plot.qcc()`	Visualizes charts with customizable options

2.2 `Sixsigma` R Package

The `SixSigma package` in R offers a comprehensive suite of tools specifically tailored for implementing the Lean Six Sigma methodology, providing functionalities for process improvement, statistical analysis, and quality management. This includes functions for various control charts, process capability analysis, and design of experiments, essential components of the DMAIC cycle [21]. It further extends its utility by providing functions for defining and solving optimization problems, allowing users to fine-tune process parameters for enhanced efficiency and reduced variability.

The `Sixsigma package` in R is a comprehensive toolkit designed for the implementation of the Lean Six Sigma methodology, providing functionalities essential for process improvement, statistical analysis, and quality management [21] (see Table 2).

Table 2. Main Functions of the `qcc` R Package

Category	Function	Purpose
Control Chart Construction	`ss.cc, ss.cc.constants, ss.cc.getc4, ss.cc.getd2, ss.cc.getd3`	Chart control for variables, limits and constants

(continued)

Table 2. (*continued*)

Category	Function	Purpose
Databases	`ss.data.batteries, ss.data.bills, ss.data.bolts, ss.data.ca, ss.data.density, ss.data.doe1, ss.data.doe2, ss.data.pastries, ss.data.pb1, ss.data.pb2, ss.data.pb3, ss.data.pb4, ss.data.pc, ss.data.pc.big, ss.data.pc.r, ss.data.rr, ss.data.strings, ss.data.thickness, ss.data.thickness2, ss.data.wbx, ss.data.wby`	Provide benchmark datasets for testing and validating control chart procedures
Capability Analysis	`ss.ca.yield()ss.ca.z(),ss.study.ca()`	Calculates and visualizes process capability indices
Descriptive Analysis	`pareto.chart()`	Displays the most frequent sources of variation
	`ss.ceDiag()`	Generates cause-and-effect diagrams
	`ss.pMap()`	Generates Process Map
	`ss.lf(), ss.lfa()`	Evaluates the Loss Function for a process
	`ss.summary()`	Computes basic descriptive statistics
Visualization and output	`summary.qcc()`	Summarizes key control chart statistics
	`plot.qcc()`	Visualizes charts with customizable options
Diagnostic and Auxiliary Tools	`climProfiles(),outProfiles()`	Highlights data points outside control limits
	`ss.ci ()`	Computes a confidence interval
	`ss.rr()`	Evaluate Gage R&R studies

2.3 `qichart` and `qichart2` R Package

The `qicharts package`, later updated as `qicharts2` [8], provides a structured framework for generating quality control charts in R. The package supports the creation and interpretation of multiple control charts, essential for process analysis and deviation detection in quality improvement contexts [22].

While `qicharts` [23] relies on base R graphics and manual parameter settings, `qicharts2` [8] extends the functionality through the integration of ggplot2 and the tidyverse ecosystem, offering improved automation, reproducibility, and visual clarity. This evolution enhances the package's usability in data-driven environments, allowing users to manage grouped data, apply standard SPC rules automatically, and produce publication-quality charts (see Table 3).

Table 3. Main Functions of the `qichart` and `qichart2` R Packages

Category	Function		Purpose
	`qichart`	`qichart2`	
Control Chart Construction	`qic.chart()`	`qic()`	Chart control for variables or attributes X-bar, R, S, p, np, c, u charts
		`bchart`	Bernoulli CUSUM chart for binary data
Data Preparation and Grouping	`qic.data()`	`aggregate() +` `qic()`	Prepares input data in the correct format for control chart generation
Diagnostic and Auxiliary Tools		`limits()`	Allows extraction and customization of calculated control limits and center lines
	`qic.add()`	`highlight()`	Highlights data points outside control limits
Visualization and Output	`plot.qic`	`Integration with ggplot2 dplyr, tidyr`	Visualizes charts with customizable options
		`summary.qic`	

The primary distinction between `qicharts2` and its predecessor, `qicharts`, lies in `qicharts2`'s enhanced functionality, broader chart type support, and improved graphical capabilities [8] qicharts is a base-R implementation focused on classical Shewhart charts, suited for legacy or lightweight analysis. In contrast, `qicharts2` is a modern reimplementation using `ggplot2`, offering more flexibility, automation, and visual quality, making it ideal for healthcare quality monitoring and continuous improvement initiatives. In addition, `qicharts2` is applied in healthcare, where control charts are used to monitor patient care and treatment outcomes [16, 24]. The package enables clear visualization of process data over time, facilitating the distinction between common and special cause variation, a key requirement for sustained quality improvement [22, 24].

2.4 `qcr` R Package

The `qcr package` implements a broad set of univariate and multivariate statistical quality control methods and nonparametric alternatives designed for practical process monitoring and capability assessment. It complements existing R tools by adding depth in nonparametric multivariate control charts, functional-data control charts, and an extended set of capability indices that address both parametric and nonparametric conditions [9].

Functionally, `qcr` provides Shewhart-type charts for individual and subgroup data, attribute charts, and distribution-sensitive control procedures such as EWMA and CUSUM. For multivariate monitoring it supports Hotelling's T2, MEWMA and MCUSUM charts and, critically, nonparametric multivariate charts based on data depth (see Table 4). These nonparametric alternatives are particularly relevant when Gaussian assumptions are violated or when data display complex dependence structures common in digitalized manufacturing environments. The package also includes Phase I and Phase II methods for functional data, enabling charting of profile or curve-type outputs from automated sensors or process scanners.

Table 4. Main Functions of the `qcr` R Package

Category	Function	Purpose
Control Chart Construction	`qcs.xbar, qcs.R, qcs.S, qcs.p, qcs.np, qcs.c, qcs.u, qcs.one`	Create classical Shewhart-type control charts for variables and attributes in Phase I and II analysis
	`qcs.cusum, qcs.ewma, qcs.pcr`	Generate cumulative sum and exponentially weighted moving average charts for detecting small or gradual process shifts
	`mqcs.t2, mqcs.mewma, mqcs.mcusum, npqcs.Q, npqcs.r, npqcs.S`	Construct nonparametric multivariate control charts based on depth and rank statistics for non-normal data
	`fdqcs.depth, fdqcs.rank`	Develop functional data control charts using depth or rank functions for continuous profile monitoring
Data Preparation and Grouping	`qcd, mqcd, npqcd, fdqcd`	Define and preprocess quality control data objects for univariate, multivariate, nonparametric, or functional analysis

(continued)

Table 4. (*continued*)

Category	Function	Purpose
	`state.control, mstate.control, npstate.control`	Specify process state parameters and grouping structures for simulation or state-transition analysis
	`qcs.add, mqcs.add, npqcs.add`	Add new subgroups or observations to existing control chart objects for Phase II monitoring
Databases	`orangejuice, oxidation, plates, pistonrings, pcmanufact, presion`	Provide benchmark datasets for testing and validating control chart procedures
Capability Analysis	`qcs.ca, qcs.cp, qcs.cpn, qcs.hat.cpm`	Calculates and visualizes process capability indices
Descriptive Analysis	`summary.qcs, summary.mqcs, summary.npqcs`	Displays the most frequent sources of variation
Visualization and output	`plot.qcs, plot.mqcs, plot.npqcs`	Produce visual control charts for univariate, multivariate, and functional data analysis
Diagnostic and Auxiliary Tools	`summary.qcs, summary.mqcs, summary.npqcs`	Summarize results with control limits and violations

On process capability, `qcr` implements first- to fourth-generation capability indices and their nonparametric counterparts, together with graphical capability outputs. This comprehensive capability suite allows evaluation of process conformity under non-normality, skewness, and multimodality—conditions increasingly encountered in high-frequency, sensor-derived datasets.

From a theoretical standpoint, `qcr` integrates classical SQC principles (variance decomposition, control limits, and signal detection rules) with modern robust and depth-based statistics to maintain interpretability while increasing applicability to nonstandard data. This makes it suitable for Industry 4.0 contexts where streaming, correlated, or functional data are normative.

Limitations include relatively limited support for streaming implementations out-of-the-box and potential computational cost for depth-based multivariate methods on very large datasets; both issues can be mitigated by preprocessing, subsampling, or coupling `qcr` with scalable data pipelines. Overall, `qcr` extends the SQC toolbox by providing robust options for nonparametric and functional monitoring, bridging classical methods and modern data types.

3 Methodology

This study employed a comparative analysis of statistical packages designed for quality control applications in industrial environments. The methodology began with the Data and Package Selection phase, which involved the careful selection of representative datasets and the identification of R packages specifically focused on statistical quality control. Following this, the Definition of Evaluation Dimensions established the core criteria for assessment, encompassing data input requirements, basic quality control tools, attribute charts, variable charts, and process capability analysis. The Testing Procedure then involved the systematic execution of relevant functions for each defined dimension, coupled with a thorough analysis of computational accuracy and visualization quality. Finally, an Integration & Comparative Analysis was conducted to summarize the technical performance of each package and assess its industrial applicability and usability. This rigorous evaluation framework allowed for an unbiased comparison of functionalities, highlighting the strengths and weaknesses of each package for real-world manufacturing scenarios.

For each package, five core aspects were evaluated: (i) data input requirements, (ii) basic quality control tools for process characterization and visualization, (iii) attribute charts, (iv) variable charts, and (v) process capability analysis. We executed a dataset, belonging to injection process, to assess the performance of any package that included functions relevant to these aspects. The evaluation examined each package's ability to handle different data structures, the accuracy of its statistical computations, and the clarity and quality of its graphical outputs (see Fig. 1).

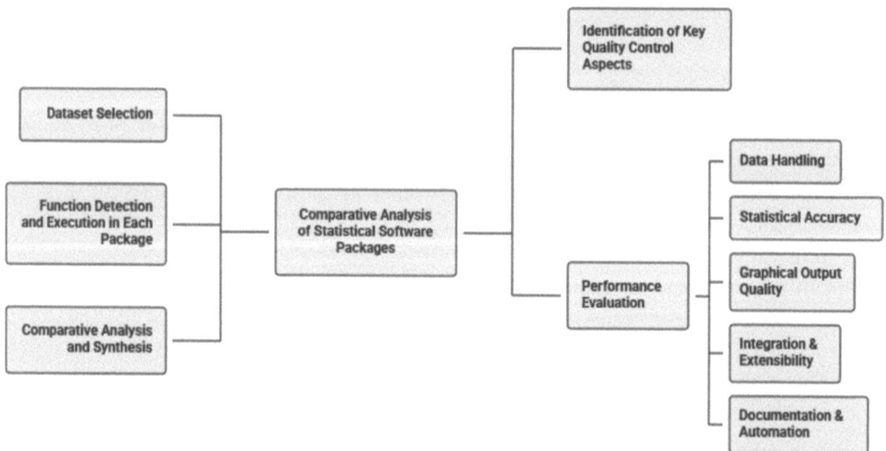

Fig. 1. Methodology framework

Furthermore, practical aspects critical for industrial deployment were examined, including documentation quality, community support, and capabilities for automation and report generation [21]. This systematic approach established an objective framework for comparing the technical performance and practical applicability of each package within the context of statistical process control and Industry 4.0 quality management.

4 Results and Technical Analysis of Quality Control R Packages

This section presents a comprehensive evaluation of the main R libraries applied to Statistical Process Control (SPC), focusing on their performance, analytical precision, and usability when implemented on real manufacturing data. The dataset corresponds to a plastic injection molding process aimed at analyzing the variability in the weight of Dispenser Pump-Type Caps throughout the production stages. Two types of variables were considered: a continuous variable representing the injection weight (g) and an attribute variable referring to the number of defects per lot, both derived from the same production samples. The analysis was conducted using data from 109 subgroups of 24 observations each, ensuring statistical robustness (See Table 5).

Table 5. Description of variables

Variables	Definition	Subgroups	n
Continuous	Injection weight (g)	109	24
Attribute	Defects per lot	109	24

The weight measurements were obtained using a SARTORIUS analytical balance with 0.0001 g resolution, while defect was classified according to specific nonconformities such as deformations, incomplete cavities, absence of cavity numbering, bubbles, material contamination, and burrs. This dual-variable configuration allows the assessment of both process stability and product conformity under controlled production conditions.

In the following subsections, the R packages—qcc, SixSigma, qicharts2, qcr—is analyzed. The comparative study examines data input structures, basic diagnostic tools, attribute and variable control charts and process capability indices, complemented by specific commands and visual outputs generated by each R package. Finally, the discussion integrates aspects of usability, visualization, applicability, and industrial relevance to highlight the strengths, limitations, and distinctive features of each R package in supporting data-driven quality management.

The analysis emphasizes how these R Packages differ in data handling, chart generation, diagnostic accuracy, and interpretability, considering both univariate and attribute-type process variables. Furthermore, the study highlights specific functionalities that facilitate the identification of out-of-control situations, as well as limitations that may affect their adaptability to complex industrial contexts. This section approach ensures a critical and comparative examination of R-based SPC implementations, identifying strengths and methodological limitations of each R package in the context of real-world industrial data, and establishing a technical foundation for future applications in automated and data-driven quality management systems.

4.1 Data Input Requirements

The correct structuring and configuration of datasets represent the first critical step for implementing Statistical Process Control (SPC) tools in RStudio. Each R package—qcc, SixSigma, qicharts2 and qcr—requires specific data formats, parameter definitions, and object structures that directly influence the accuracy of the resulting control charts and diagnostic outputs. Therefore, understanding the syntax and data input conventions of each package is essential to ensure the correct interpretation of control limits, subgroup definitions, and violation detection. In this study, the comparison focuses on how this R packages processes continuous and attribute-type variables derived from the same injection molding process dataset, emphasizing the preparation of grouped data, labeling of subgroups, and specification of statistical parameters such as mean, range, and standard deviation. This analysis establishes a standardized reference for the practical configuration of SPC data in RStudio, enabling reproducible and reliable process evaluations.

The qcc package serves as a fundamental tool for the implementation of Statistical Process Control (SPC) within R, providing a structured and efficient environment for process analysis and monitoring. Its design facilitates the practical application of quality control techniques in both industrial and service contexts, enabling users to detect deviations, assess process stability, and ensure product conformity [25].

When using this package correctly, the data corresponding to the continuous variable must be arranged in a matrix form (m × n), where each row represents a subgroup and each column corresponds to an observation (x) within the subgroup.

In this case, the plastic injection molding process dataset is structured with m = 109 and n = 24. Equation 1 shows the structure of the matrix.

$$\begin{matrix} x_{11} & x_{12} & x_{13} & \dots & x_{124} \\ x_{21} & x_{22} & x_{23} & \dots & x_{224} \\ x_{31} & x_{32} & x_{33} & \dots & x_{324} \\ \dots & \dots & \dots & \dots & \dots \\ x_{1091} & x_{1092} & x_{1093} & \dots & x_{10924} \end{matrix} \tag{1}$$

For the attribute variable, the data are structured in a DataFrame with two columns: Injection, representing the subgroup identifier, and Defects, indicating the total number of defects found per lot. Table 6 shows the first five rows of the DataFrame.

Table 6. Structure of the attribute variable dataset

Injection	Defects
1	16
2	22
3	18
4	8
5	14

The `SixSigma package` presents a methodological approach different from the other packages analyze, as its main purpose is the characterization and initial diagnosis within the DMAIC cycle, supporting the application of the Six Sigma methodology.[7].

To correctly use the functions of this package that require the dataset, it is necessary to first calculate the mean of each subgroup (m) and group them in a DataFrame with a column labeled Mean. Table 7 shows the first five rows of the DataFrame.

Table 7. Structure of the DataFrame of subgroup means

Mean
2.663596
2.677387
2.692408
2.693271
2.692488

The `qicharts2 package` is distinguished by its practical approach to the construction and interpretation of control charts based on Shewhart rules. For the continuous variable, this package works with data in a "long" format, where each observation (x) is represented by two columns in a DataFrame: Injection, which identifies the subgroup, and Mass, which corresponds to the measured value. Table 8 shows the first ten rows of this dataframe. The data were transformed into the required long format using the R script provided in Appendix – Supplementary R Scripts.

Table 8. Structure of the `qicharts2` and `qcr` input dataset

Injection	Mass
1	2.7128
1	2.6504
1	2.6858
...	...
2	2.7033
2	2.6474
2	2.6574
...	...
3	2.7200
3	2.6550
3	2.6807
...	...

(continued)

Table 8. (*continued*)

Injection	Mass
...	...
109	2.70050
109	2.73550
109	2.74660

`qcr package` requires the data to be provided as an R data frame, since this format ensures consistency in variable naming, indexing, and data manipulation. The data frame structure allows the package to internally reference variables by name or position (var.index and sample.index), and to store additional attributes such as subgroup sizes, data type, and source name. Each record should represent one observation, and the dataset must include at least two key columns: one containing the measured variable and another identifying the subgroup or sample to which the measurement belongs, (see Table 8).

When properly created, a `qcd` object includes both the raw data and metadata, which specifies the nature of the observations. This standardized object-oriented structure enables seamless integration with downstream functions, ensuring reproducibility, clear data provenance, and compatibility with visualization and reporting functions within the `qcr` workflow. This organization ensures that the software can correctly aggregate observations within each subgroup, enabling the computation of subgroup statistics such as means and ranges. In the case of the `qcr package`, this structure is essential for generating `qcd` objects, which serve as the foundational input for variable control charts.

An important consideration in preparing variable-type data is the distinction between data.frame and tibble objects. In R, tibbles—common when using packages such as dplyr or tidyverse—are stricter than base data frames and do not allow automatic column recycling. The function `qcd()` in the `qcr` package was originally written to operate on base data frames, where assignments like `data$sizes<- sizes` are permitted. However, when the input is a tibble, this assignment generates the recycling error observed: *"Assigned data sizes must be compatible with existing data."*

To prevent this issue, the tibble should first be converted into a base data frame using `Injection<- as.data.frame(Injection)` before calling `qcd()`. This ensures full compatibility with the internal operations of `qcr package` and the successful creation of the `qcd` object.

For the attribute variable, as in the `qcc package`, the data are structured in a DataFrame with two columns: Injection, representing the subgroup identifier, and Defects, indicating the total number of defects found per lot (see Table 6).

For the construction of control charts based on attribute data using `qcr package`,, the dataset must capture both the count of defects and the corresponding sample size inspected. In `qcr package` is necessary that each record represents a single subgroup, typically including the number of defects (x), the sample identifier (sample), and the number of units inspected (sizes). Unlike variable-type datasets, which are organized purely as data frames, attribute-type datasets in the `qcr` package are internally managed as lists that combine numeric vectors with metadata. The distinction between data structures is crucial for accurate chart generation in the `qcr` package. Variable-type data must be provided as a data frame to preserve the one-to-many relationship between subgroups and individual measurements. In contrast, attribute-type data must be structured as a list, ensuring that the vectors representing defect counts and sample sizes are correctly paired and numerically compatible. When a qcd object is created for attribute data, it combines these components along with key attributes, such as `"type.data"=` `"attribute"`. It is important to note that, in last version of the `qcr` package. Once the object is correctly structured, it becomes the standardized input for functions such as `qcs.u()` or `qcs.c()`, which compute control limits and generate the corresponding charts.

4.2 Basic Quality Control Tools for Process Characterization and Visualization

Process characterization is essential to understand each stage of a process, improve efficiency, standardize procedures, and ensure regulatory compliance. By analyzing each stage, organizations can identify weaknesses, find improvement opportunities, and reduce errors. The main goal of process characterization is to identify and correct the causes of quality problems [26].

The R packages analyzed in this study—`qcc`, `SixSigma`, and `qicharts2`—include several tools for process characterization, such as:

Pareto Chart: a histogram that orders process causes or problems from highest to lowest frequency, helping to prioritize improvement actions [27].

Cause-and-Effect Diagram: links effects with their potential causes; useful during brainstorming sessions or quality improvement meetings [27].

Process Map: a visual representation of process stages that provides an overview of workflows without requiring technical details [26].

The `qcc package` stands out in process characterization because it includes two main functions for generating Pareto charts and cause-and-effect diagrams, namely `paretoChart()` and `causeEffectDiagram ()`, respectively. These tools provide a clear and concise overview of the process and help identify the main sources of variation affecting quality. The Pareto chart was generated using the R code provided in Appendix – Supplementary R Scripts.

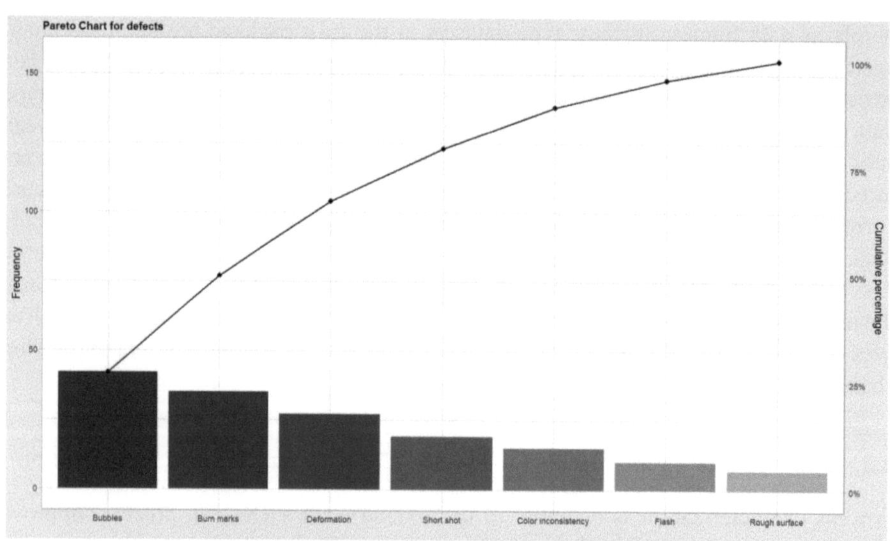

Fig. 2. Pareto Diagram using the `qcc` Package

Figure 2 shows the Pareto chart produced using the `paretoChart()` function from the `qcc package`. The resulting graph is visually clear and easy to interpret. Its execution requires creating two vectors in advance: one for the types of defects (x-axis) and another for their frequency (y-axis). Although the procedure is straightforward, it can become somewhat tedious when analyzing multiple defect categories, since these vectors must be defined manually before running the code. Additionally, it was observed that the ylab parameter—intended to label the vertical axis—was not applied correctly, which can limit the customization of the chart.

Subsequently, the cause-and-effect diagram was generated using the R code provided in Appendix – Supplementary R Scripts.

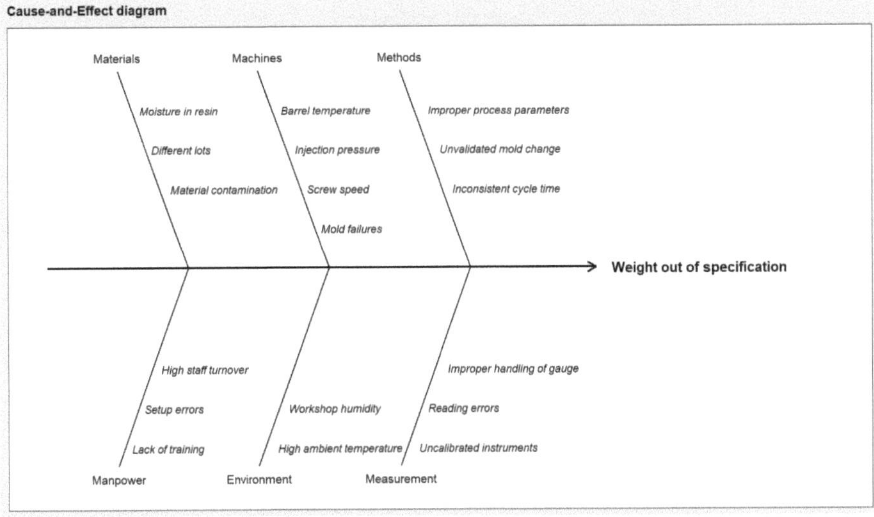

Fig. 3. Cause and Effect diagram qcc package

Figure 3 presents the cause-and-effect diagram produced with the causeEffect-Diagram () function. Implementing this diagram is relatively straightforward, as the vectors are defined directly within the function rather than separately. However, when dealing with large datasets, entering all the information manually can become time-consuming.

The function may also present visualization issues when the cause descriptions are too long, causing text overlap and reducing readability. To mitigate this, the text size can be adjusted using the cex= 0.8 parameter, as shown in this case. However, the value of this parameter must be determined empirically, since the adjustment process is not automated.

On the other hand, the SixSigma package also includes tools for process characterization. To create a cause-and-effect diagram, the ss.ceDiag() function was used, as detailed in Appendix – Supplementary R Scripts.

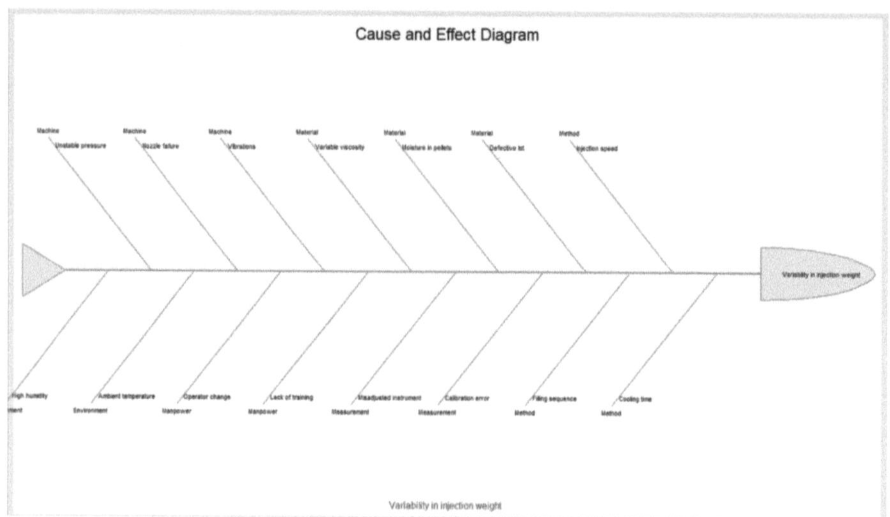

Fig. 4. Cause and Effect Diagram `SixSigma package`

Figure 4 shows the diagram generated with the `ss.ceDiag()` function from the `SixSigma package`. Unlike `qcc`, this function requires the prior creation of three vectors: one for the effect, another for the causes, and a third for the groups in which these causes are classified. This requirement reduces the level of automation. In addition, the chart presents the same limitation observed in qcc: when the text is too long, it tends to overlap, affecting readability.

In this case, the function does not accept the `cex` argument, so the `windows(width= 10, height= 8)` command was used to enlarge the display area and improve visual clarity. It is worth noting that, while the `SixSigma package` offers a more comprehensive approach to quality analysis—covering capability studies, R&R analysis, and DMAIC-based evaluations—its graphical functions tend to be less flexible compared to libraries such as `qcc` or `qicharts2`.

Additionally, the `SixSigma` package includes a distinctive function for creating process maps: `ss.pMap()`, as detailed in the Appendix – Supplementary R Scripts.

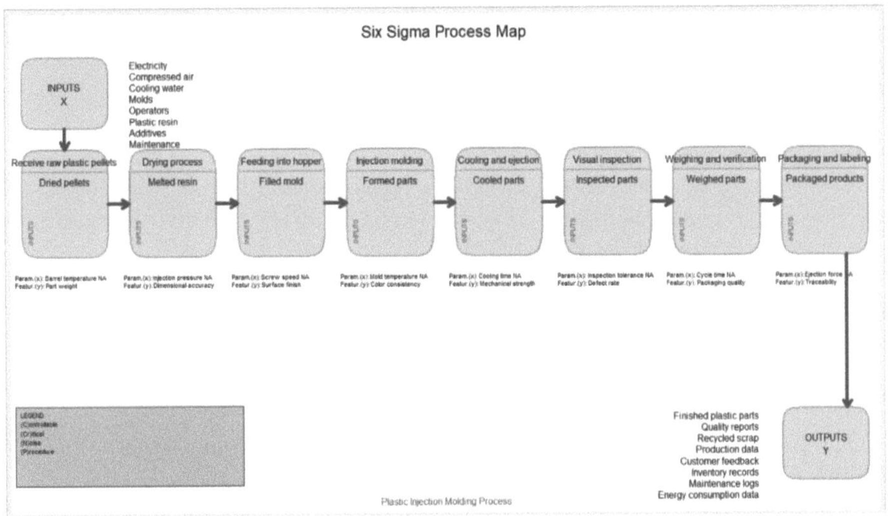

Fig. 5. Process Map `Six Sigma` package

Figure 5 shows the diagram generated by this function. Although `ss.pMap()` enables the graphical representation of process stages, its implementation is rather cumbersome. The function requires defining multiple parameters—such as inputs.overall, steps, and input.output—which makes its use tedious and far from automated.

From a visual standpoint, the resulting diagram does not effectively convey the sequence and relationships among process stages, deviating from the conventional format of a process map. Furthermore, the text size within the chart can affect readability, requiring manual adjustment of the display window dimensions to improve visualization.

Overall, while the `ss.pMap()` function represents a valuable attempt to integrate process characterization tools within the `SixSigma package`, its complexity and limited visual clarity reduce its practical applicability in real-world settings.

In contrast to the `qcc` and `SixSigma` packages, `qicharts2 and qcr` packages does not include specific functions for constructing Pareto charts, cause-and-effect diagrams, or process maps. Its primary focus is on statistical analysis and data visualization through control charts and run charts. This reflects its design, which is centered on continuous process performance monitoring rather than structural process characterization.

4.3 Attribute Charts

Attribute control charts constitute a fundamental component of Statistical Process Control (SPC), designed to monitor the quality performance of a process based on discrete or count data. These charts are applied when the quality characteristic of interest can be classified as conforming or nonconforming, or when defects are counted per inspection unit.

They allow practitioners to detect variations in defect rates or counts, facilitating the timely identification of assignable causes and supporting process stabilization. Depending on the nature of the data and the sampling plan, different types of attribute charts— such as p, np, c, and u charts—can be applied to evaluate process behavior and ensure compliance with quality standards.

In this study, attribute control charts are employed to assess the stability of the injection molding process by analyzing the number of defects per lot, where each lot corresponds to a group of 24 injections (n = 24).

Specifically, the c-chart is used to monitor the number of defects per inspection unit when the sample size (n) remains constant. This type of chart assumes that the number of defects follows a Poisson distribution and allows for the evaluation of process stability by comparing observed defect counts against statistically derived control limits. The c-chart thus focuses on discrete defect counts rather than proportions [28].

The qcc package enables the construction and visualization of the c-chart, which is suitable for monitoring defect counts when the sample size (n) remains constant. The dataset used for this analysis corresponds to the attribute data described in the Data Input section and follows the structure presented in Table 6.

The c-chart was generated using the qcc() function, which automatically computes the process center line and control limits assuming a Poisson distribution for defect counts (see Appendix – Supplementary R Scripts).

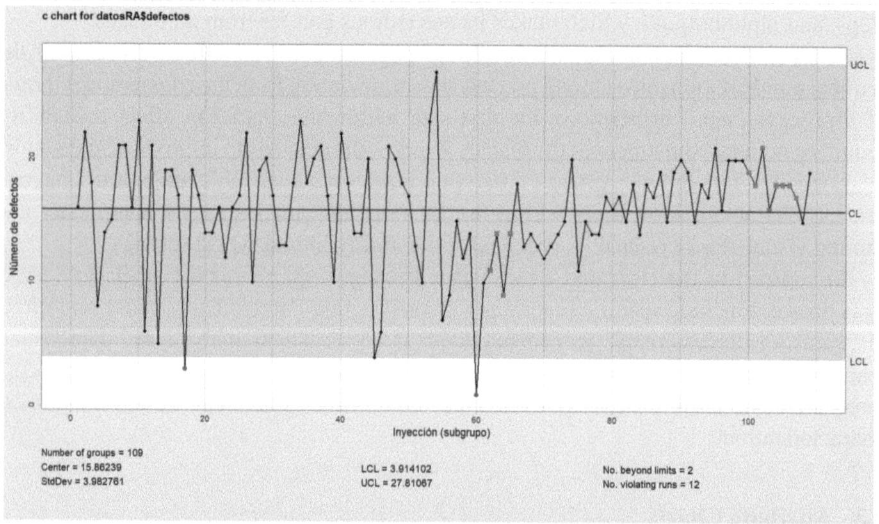

Fig. 6. C-chart for defect counts per subgroup generated with the qcc package

The Fig. 6 shows the c chart generated using the qcc package. The chart displays the number of defects per subgroup (injection) along with control limits calculated under the assumption of a Poisson distribution. The central line represents the process mean, while the upper and lower control limits (UCL and LCL) define the expected range of variation.

From the six classical indicators of an out-of-control process, the `qcc packages` identifies two main behaviors: shifts in the process mean (highlighted in yellow) and out-of-control points (highlighted in red), which indicate potential process instability.

These visual signals provide a clear diagnostic overview, facilitating the detection of special causes and supporting timely corrective actions. This approach offers a simple and efficient way to monitor attribute data and assess process stability in manufacturing environments.

It is also worth noting that while this function produces a clear and visually appealing chart, it presents a minor limitation when assigning a title. The `xlab` and `ylab` arguments correctly label the axes; however, the `main` argument does not display the chart title as expected.

The `qicharts2` package provides a flexible and automated approach to constructing control charts in R, designed to support quality improvement and process monitoring applications. Unlike classical SPC libraries, `qicharts2` is built on top of the `ggplot2` visualization framework, allowing the creation of highly customizable and publication-ready charts.

For attribute-type data, the package automatically selects and configures the appropriate chart type according to the specified parameters in the main function `qic ()`, which serves as the core of the package. This function simplifies chart creation by internally handling the calculation of control limits, central lines, and statistical assumptions based on the selected chart family.

In this study, the c-chart was constructed using the `qic ()` function to monitor the number of defects per inspection lot when the subgroup size remains constant. The data used follows the structure described in Table 6 and the corresponding R implementation is provided in the Appendix – Supplementary R Scripts.

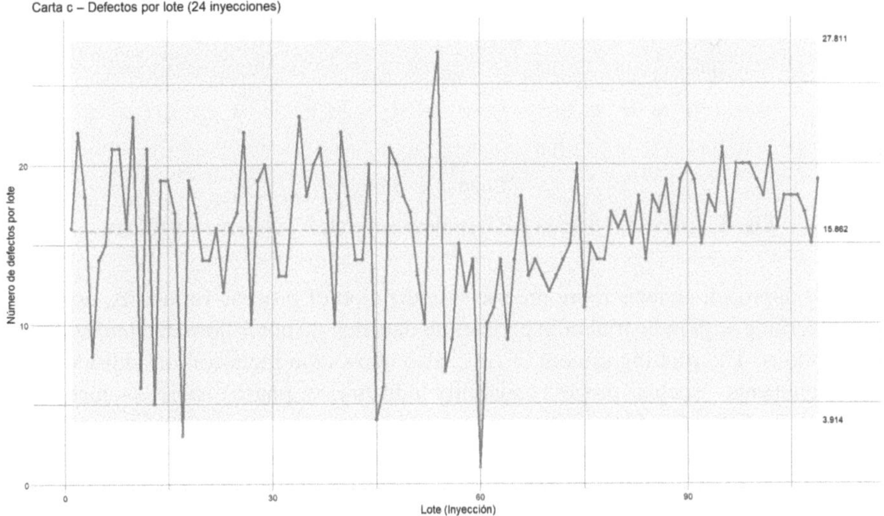

Fig. 7. C-chart for defect counts per subgroup generated with the qicharts2 package.

Figure 7 shows the c-chart generated using the `qicharts2 package`. The `qic ()` function automatically computes the control limits based on a Poisson distribution and produces the chart using `ggplot2` visualization framework. The central line, along with the upper and lower control limits, is shown by default, and the chart dynamically adjusts to the size and type of the data.

Compared to `qcc`, the `qicharts2` package requires fewer arguments to generate the chart, as it does not need additional functions to produce a clear and visually appealing output. Moreover, qicharts2 provides more effective handling of axis and title labeling, since each argument is executed correctly and accurately reflected in the final control chart.

The `SixSigma package` does not include specific functions for constructing attribute control charts (p, np, c, or u charts). In summary, among the evaluated packages, `qcc` and `qicharts2` provide comprehensive tools for monitoring attribute data, while `SixSigma` focuses primarily on continuous variable analysis within the DMAIC framework.

The type C control chart generated with the `qcr R package` is shown in Fig. 8 built on the structure of `qcc R package`. While `qcc R package` offers a solid framework for standard control charts, its visualization options are limited. In contrast, `qcr R package` uses the `plot ()` function, which allows users to adjust chart elements such as axis scales, control limits, annotations, and colors.

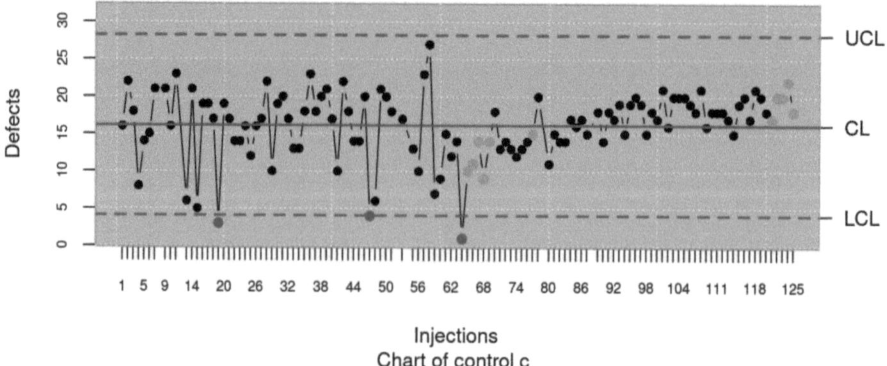

Fig. 8. c chart for defects per injections generated with `qcr R package`

This approach enables more precise visualization of process variability, especially when working with industrial or experimental data that do not follow standard statistical assumptions. The plotting system in `qcr` also allows the inclusion of additional analytical elements—such as process capability indicators or phase divisions—making the chart more informative for quality assessment.

In addition, qcr improves the analytical functions of `qcc` by using S3 methods, which enhance data handling and compatibility with other R packages. This makes it easier to integrate control charts into broader quality control analyses or real-time monitoring processes.

4.4 Variables Charts

Variable control charts are of great importance within Statistical Process Control (SPC) for monitoring process stability and performance when the quality characteristic of interest is measured on a continuous numerical scale. Unlike attribute charts, which handle count or classification data, variable charts allow for the evaluation of the process mean and variability using quantitative measurements such as dimensions, weight, or temperature.

These charts are particularly valuable for identifying both the accuracy and precision of the data, enabling a more accurate assessment of process performance. Depending on the sampling plan and the nature of the data, several types of variable control charts can be employed, including the $\bar{X}$–R chart, the $\bar{X}$–S chart, and the individual (I–MR) chart.

Specifically, the X–R chart (Mean–Range Chart) is one of the most widely used tools for variable data. In this chart, the sample mean is monitored through the X chart, while the process variability is controlled through the R chart, which plots the range (R) of each subgroup. The range represents the difference between the largest and smallest values within a sample and serves as a direct indicator of within-subgroup variation [29].

In this study, variable control charts are applied to the injection molding process to monitor the weight of the injections, which serves as the continuous quality variable. Each subgroup consists of 24 observations, corresponding to the same lot structure used in the attribute analysis. The X–R chart provides a comprehensive view of process performance by allowing the simultaneous evaluation of process mean and variability under the assumption of normality.

The qcc package allows the creation of various types of control charts for continuous variables and includes two specialized charts: CUSUM and EWMA. The CUSUM chart plots the cumulative sums of deviations of sample values from a target value, while the EWMA chart uses an exponentially weighted moving average of process measurements to detect small shifts in the process mean [30].

However, for the analysis of the injection weight process, the qcc package was used to implement the X (mean) and R (range) control charts. The corresponding R script is provided in the Appendix – Supplementary R Scripts.

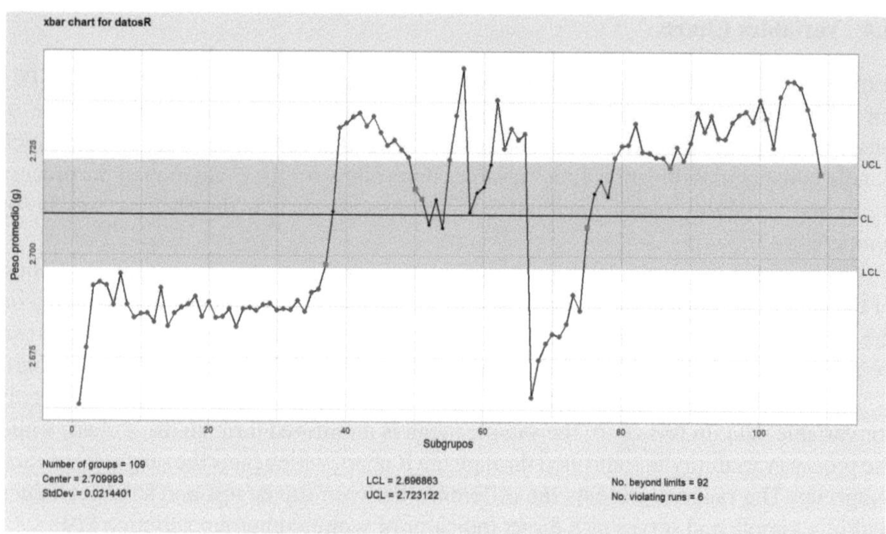

Fig. 9. Xbar chart qcc package

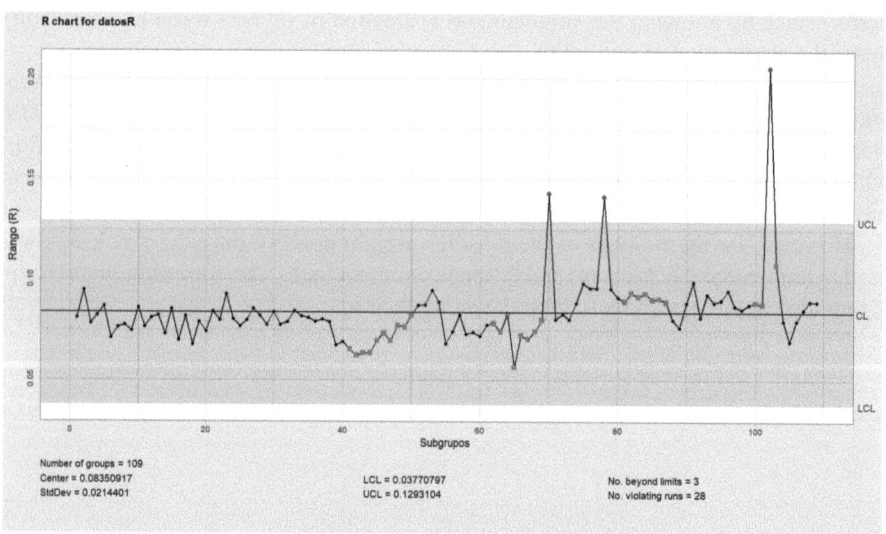

Fig. 10. R chart qcc package

Figure 9 and Fig. 10 show the charts generated using the qcc() function. These charts are visually clear and automatically display the lower, central, and upper control limits. Process shifts are highlighted in yellow, while out-of-control points are shown in red, allowing easy identification of potential process instability. The creation of these charts requires a simple code structure, and the axis labels (xlab and ylab) work correctly.

However, a limitation remains when assigning a title to the chart, as the corresponding argument is not reflected in the final output.

On the other hand, the `qicharts2` package includes the `qic()` function, which allows the generation of various control charts for variables. However, it is important to note that `qicharts2` does not provide a range (R) control chart, unlike `qcc`. Instead, it offers the S chart, which evaluates within-subgroup variability using the standard deviation of the data, providing a more robust alternative for assessing process dispersion. The corresponding R implementation is provided in the Appendix – Supplementary R Scripts.

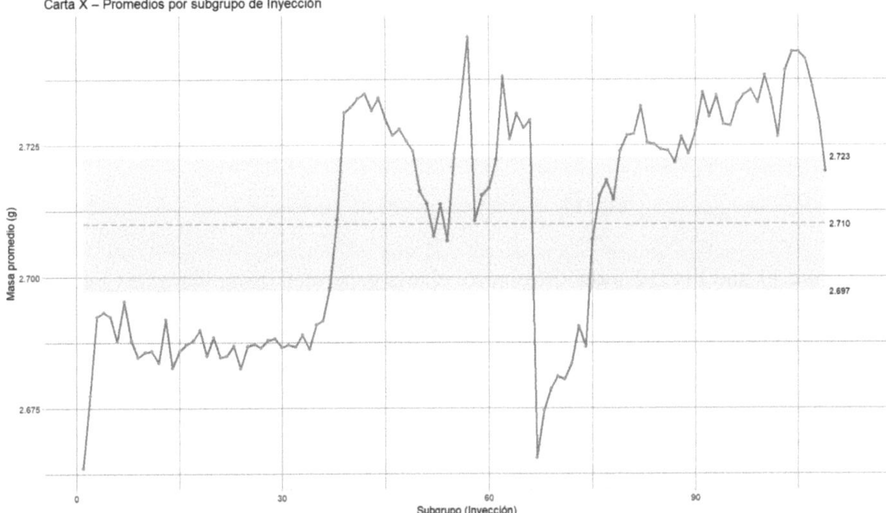

Fig. 11. Xbar chart `qicharts2` package

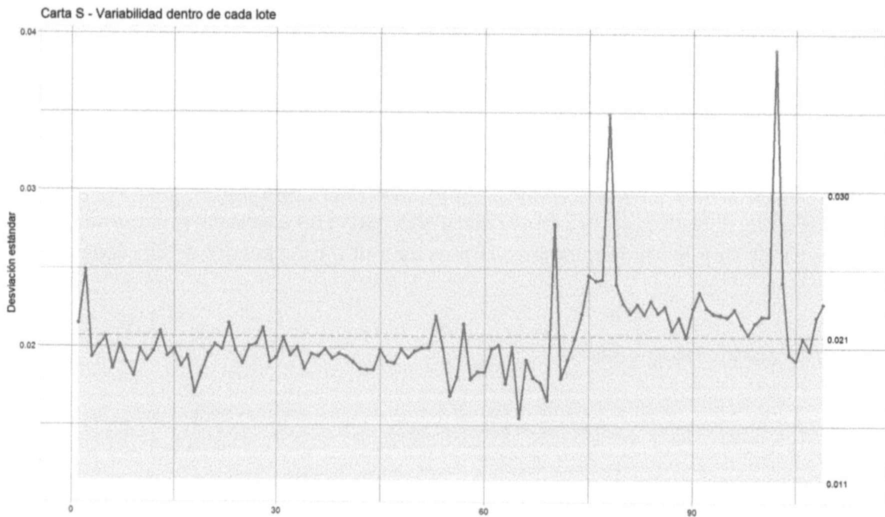

Fig. 12. S chart `qicharts2` package

Figure 11 and Fig. 12 show the control charts generated using the `qic()` function. These charts present a clear and visually appealing display; however, unlike `qcc`, they only highlight the out-of-control points in yellow. This means that users without prior knowledge of statistical process control may find it difficult to identify process changes, whereas `qcc` facilitates this interpretation by visually distinguishing different types of signals through color coding.

The code required to produce these charts is straightforward, as demonstrated earlier, and does not require additional functions to enhance the visual presentation, unlike `qcc`. Moreover, the axes and title can be labeled correctly, allowing for proper customization. However, a notable limitation is that `qicharts2` does not display on the chart the total number of violations or out-of-control points detected.

To conclude these analyses, the `SixSigma package` includes among its tools a function that allows the creation of an individuals and moving range control chart using:

```
ss.cc(
    type = "I-MR",
    data = datos)
```

However, this chart type is not suitable for the data used in this study, as it requires individual observations. Therefore, `SixSigma` proves to be of limited usefulness when implementing control charts for different types of data, particularly when dealing with subgroups or attribute data.

The mean control chart generated with the `qcr` R package provides a clear representation of process stability by monitoring the behavior of subgroup averages over time. Using the `qcr` framework, the chart displays the center line (CL) corresponding to the process mean, along with the upper and lower control limits (UCL and LCL) derived from sample variability. The visualization enables the identification of deviations from statistical control, such as points exceeding the limits or sustained trends suggesting process shifts. Compared to the traditional implementation in qcc, the qcr version allows greater flexibility in chart customization and output interpretation. Through its `summary()` function, users can access key performance information, including the number of points beyond control limits and the presence of violating runs, which support early detection of systematic variation (see Fig. 13).

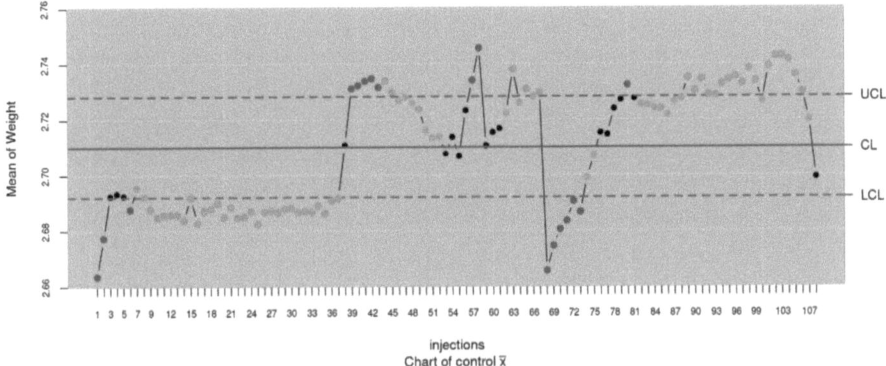

Fig. 13. Means chart qcr Package

Overall, the mean chart in qcr combines statistical rigor with adaptable visualization, offering an efficient tool for routine process monitoring and diagnostic analysis within statistical quality control.

4.5 Process Capability Analysis

The process capability analysis represents a fundamental stage within the framework of Statistical Process Control (SPC), as it determines the actual variability of a manufacturing process in relation to the product design specification values. In this study, the process capability analysis is performed for the continuous variable using the control limits obtained from the measured injection weight data, while considering the specification limits provided by the customers (see Table 9). These limits serve as reference values for calculating the process capability indices, allowing the evaluation of whether the process consistently produces results within acceptable tolerance ranges. Two main indices are considered: the potential process capability index (Cp), which evaluates the potential variability of the process with respect to the specification limits, and the process capability index with respect to process centering (Cpk), which measures the degree of centering of the process within those limits. In this section, only the capability analysis functions available in the qcc, SixSigma, qcr packages are discussed, since the remaining packages do not include specific functions for process capability analysis. Together, these indices provide a quantitative assessment of both the precision and accuracy of the manufacturing process and serve as key indicators for effective statistical quality control and continuous process improvement[31].

Table 9. Specification Limits

LSL	USL
2.65	2.75

For the `qcc package`, the `processCapability()` function is used to perform the process capability analysis considering the established specification limits. This function not only calculates the process potential index (`Cp`) and the process capability index (`Cpk`), but also estimates the proportion of products that fall outside the specification limits—both theoretically (under the assumptions of process normality and stability) and empirically, based on the observed data. In this study, the lower and upper specification limits (LSL and USL) were defined according to the client's requirements, while the control limits were derived from the process data using the mean control chart (`qcc_xbar`). The complete R implementation is provided in the Appendix – Supplementary R Scripts.

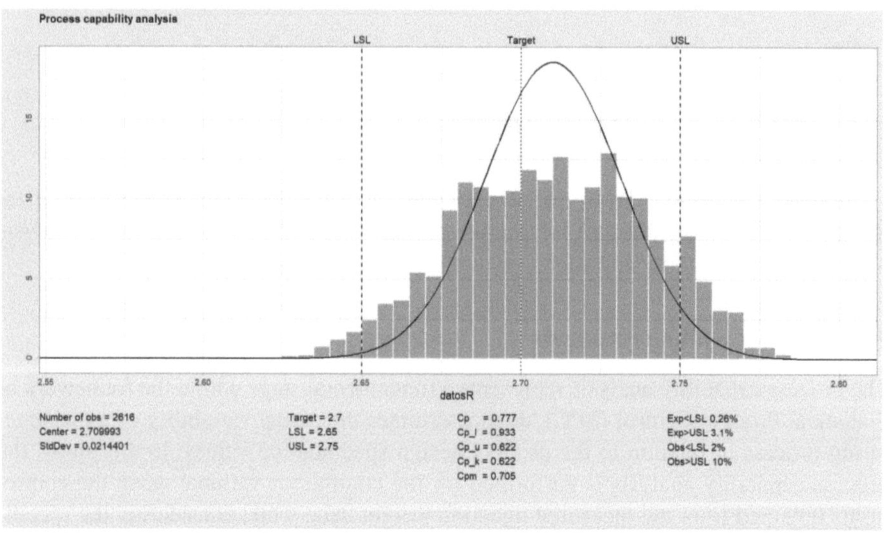

Fig. 14. Process capability analysis `qcc` package.

Figure 14 presents the graphical output generated by the `processCapability()` function from the `qcc` package. The histogram shows the distribution of the observed process data relative to the specified lower and upper specification limits (LSL and USL). The superimposed normal curve represents the theoretical process distribution assumed under conditions of normality. The vertical lines indicate the specification limits and the process mean, allowing a visual assessment of the process centering and dispersion.

The graphical output complements the numerical results provided by the function, which include the process potential index (Cp) and the process capability index (Cpk). These indicators quantitatively describe whether the process variation remains within the specification range and how well the process is centered between the limits.

The `ss.study.ca()` function from the `SixSigma package` performs a complete process capability analysis by comparing the process output with the customer's specification limits.

This function generates a comprehensive report that includes:

- A process histogram with observed and theoretical distributions.
- The specification limits (LSL, USL) and the target value.
- Normality tests (Shapiro–Wilk and Lilliefors K–S).
- Capability indices (Cp, Cpk).
- Defects per million opportunities (DPMO).

This function provides a major advantage to the `SixSigma package`, as it has the ability to integrate these analyses—normality, independence, and process capability—into a single graphical and numerical report, whereas in other packages, such as qcc orqcr, these tasks must be performed separately (see Appendix – Supplementary R Scripts).

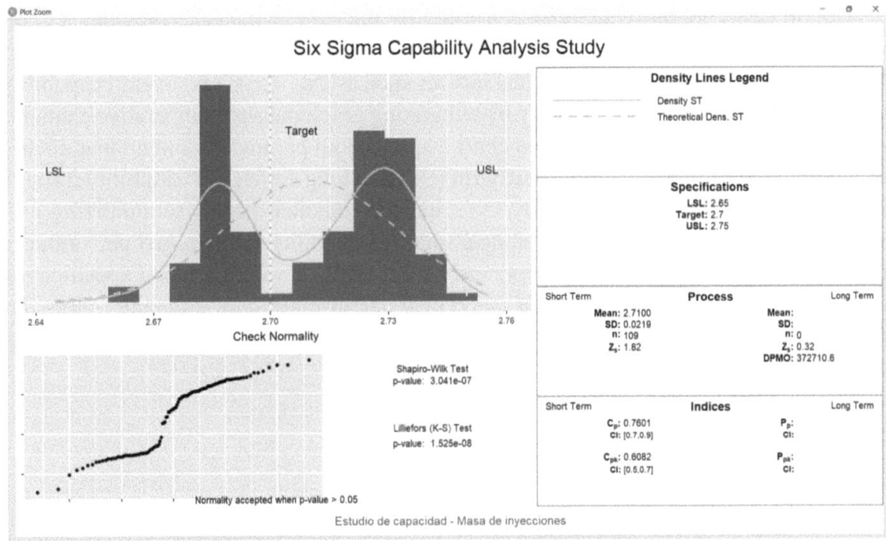

Fig. 15. Process capability analysis `SixSigma` package.

Figure 15 presents the graphical report generated by the `ss.study.ca()` function from the `SixSigma package`. The histogram displays the observed process distribution along with the theoretical normal distribution, allowing a direct comparison of the empirical and expected behaviors. The vertical lines represent the lower specification limit (LSL), the target value, and the upper specification limit (USL), enabling a visual assessment of process centering and dispersion.

The report also includes the results of normality tests Shapiro–Wilk and Lilliefors (K–S) which evaluate whether the process data follow a normal distribution. In the lower section, the calculated capability indices (Cp, Cpk) and defect rate (DPMO) are presented for short-term and long-term perspectives.

This output provides a comprehensive overview of process performance, integrating capability metrics, normality verification, and specification compliance within a single visual summary.

The `SixSigma` package also includes two additional functions for process capability analysis:

```
Cp  <- ss.ca.cp(datosSix$Media, LSL, USL)

Cpk <- ss.ca.cpk(datosSix$Media, LSL, USL)

## Cp = 0.7601443
## Cpk = 0.6082276
```

These individual functions return the values of the Cp and Cpk indices, respectively. They are particularly useful when only the numerical results of these indices are required, without generating the complete graphical report provided by the `ss.study.ca()` function.

The process capability analysis generated by the `qcr` R package provides a comprehensive and visual summary that extends the basic functionality available in `qcc`. While `qcc` primarily reports numerical indices such as Cp, Cpk, and overall capability values, `qcr` enhances interpretability by integrating graphical and comparative elements within a single output. The generated chart displays both parametric and nonparametric capability results, distinguishing short-term (ST) and long-term (LT) variability through overlaid density curves. Additionally, `qcr` includes detailed panels summarizing process data statistics, parametric and nonparametric capability indices, and performance metrics such as the expected and observed percentages of values beyond specification limits (see Fig. 16). The inclusion of theoretical versus empirical probability plots further allows users to visually assess the distributional fit of the process data—an aspect not present in `qcc`.

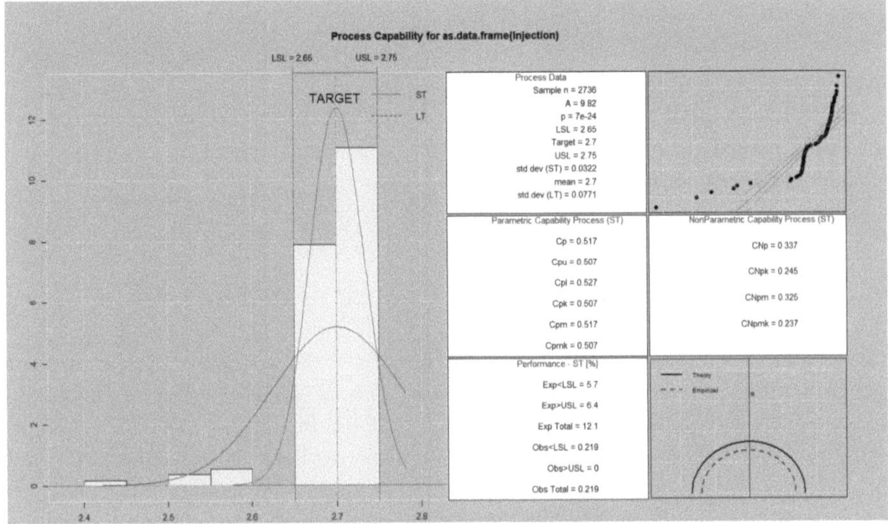

Fig. 16. Capability Process `qcr` package

Unlike the previous packages, qicharts2 does not include specific functions for process capability analysis. Instead, it focuses on the visualization and monitoring of process performance over time.

4.6 Additional Functionalities of Quality Control Libraries

Beyond the generation of control charts, the evaluated packages incorporate complementary functions that enhance statistical interpretation and enable the automatic identification of anomalous patterns in processes.

In the case of `qcc`, the commands `qcc_xbar$violations` and `qcc_r$violations` make it possible to determine in which subgroups of the $\bar{X}$ and R charts violations of Shewhart rules occur, such as points outside the control limits or sustained process shifts. This functionality allows for early detection of instabilities and contributes to more precise quality monitoring.

The `SixSigma` package, through the function `ss.ca.yield(defects= 0, rework= 0, opportunities= 1)`, computes key performance indicators such as Yield, First Time Yield, Rolled Throughput Yield, and Defects per Million Opportunities (DPMO). These metrics provide a quantitative assessment of process capability and efficiency, aligning with the continuous improvement principles of Industry 4.0.

Meanwhile, `qicharts2` offers an automated approach for detecting process instability signals based on Shewhart rules. The function identifies subgroups with anomalous behaviors: runs.signal indicates process shifts (six or more consecutive points above or below the mean), while sigma.signal denotes the presence of special causes (points beyond control limits). These signals can be extracted directly from the results generated by `qic ()`, facilitating integration with automated monitoring systems and real-time analytics.

```
anomalos <- subset(resultado$data, runs.signal == TRUE |
sigma.signal == TRUE)
```

Collectively, these functions serve as a bridge between traditional statistical control and advanced analytics in cyber-physical environments, where automatic anomaly detection and variability traceability are fundamental pillars of intelligent quality management.

5   Comparative Evaluation in the Context of Industry 4.0

The integrated evaluation of the qcc, qicharts2, SixSigma, and qcr packages shows that each provides distinctive capabilities aligned with different operational requirements in digitally industrialized environments. In light of the core principles of Industry 4.0 and the data-driven "V" dimensions—volume, velocity, variety, veracity, and value—these packages exhibit specific strengths that address the challenges of advanced monitoring, automation, and continuous analysis characteristic of modern industrial operations.

The qcc package stands out for its conceptual maturity and adherence to classical statistical process control methodologies. Its robustness and stability make it particularly suitable for contexts where measurement veracity and process consistency are essential. For organizations progressing gradually toward digitalization, qcc maintains technical rigor and reliability, serving as a bridge between traditional SPC practices and the emerging requirements of Industry 4.0.

In contrast, qicharts2 offers greater methodological flexibility, especially in settings involving complex temporal structures and continuous monitoring. Its ability to handle real-time data streams and adapt to shifting patterns aligns directly with the velocity and variety that characterize highly sensored production systems. These features support agile decision-making in environments where rapid analysis and constant updates are operational imperatives.

SixSigma contributes a more holistic perspective by integrating statistical analysis with tools aimed at systemic process understanding, such as process maps, cause-and-effect diagrams, and structured improvement methodologies. This integrative approach enhances value creation by improving traceability, enabling detailed visualization of process flows, and supporting a comprehensive understanding of operational behavior—attributes especially relevant in digital architectures where system transparency is fundamental.

Finally, qcr complements these capabilities by providing a flexible infrastructure aligned with contemporary R-based analytics. Its architecture facilitates the handling of data with greater volume and variety, integrating naturally with distributed environments, smart sensors, and analytical ecosystems typical of Industry 4.0. These features position qcr as a modern alternative for organizations seeking to extend analytical automation and leverage heterogeneous data sources.

Taken together, the comparison indicates that each package addresses different levels of digital maturity and operational needs. While qcc offers stability and methodological rigor, qicharts2 enables adaptability and rapid responsiveness, SixSigma provides an integrative process-oriented perspective, and `qcr` delivers modern flexibility for large-scale and diverse data structures. Rather than competing, these tools form a complementary suite that can be strategically deployed according to an organization's monitoring, analytical, and automation requirements within the industry 4.0 framework.

5.1 Recommended Applications of R Packages within Industry 4.0

To support the practical interpretation of the comparative analysis, Table 10 summarizes the key Industry 4.0 challenges addressed by each package and outlines their recommended industrial applications. This synthesis provides a concise reference for practitioners and researchers seeking to select the most appropriate R tools according to their operational context, digital maturity level, and data-processing requirements in smart manufacturing environments.

Table 10. Recommended Applications of R Packages in Industry 4.0

Package	Industry 4.0 Challenges Addressed	Recommended Industrial Applications
qcc	Data veracity and process stability in environments transitioning toward digitalization	Routine SQC, traditional production lines, stable high-volume manufacturing processes
qicharts2	Velocity and variety in time-dependent data streams	Real-time monitoring, early detection of instability, highly sensored or IoT-enabled lines
SixSigma	Value creation through systemic process understanding and traceability	DMAIC projects, cause-and-effect analysis, capability studies and benchmarking
qcr	Handling large-volume and high-variety data in distributed environments	Automated dashboards, advanced capability analysis, integration with smart-sensor data

6 Conclusions

The comparative analysis of the `qcr`, `qicharts2`, and `SixSigma` packages demonstrates the potential of R-based tools to enhance Statistical Quality Control (SQC) within the industry 4.0 framework. Each package provides specific analytical and visualization functionalities that, when used together, strengthen process monitoring, capability evaluation, and performance analysis. The `qcr` package stands out for its graphical flexibility and its ability to produce detailed visualizations suitable for real-time applications and automated dashboards. `qicharts2` offers tools for detecting process instability based on Shewhart rules, enabling the early identification of deviations. `SixSigma`, in turn, complements these features by supporting standardized capability studies and comparisons aligned with industrial quality standards.

Taken together, these packages illustrate how open-source tools can support the development of connected and data-driven quality systems consistent with the principles of Industry 4.0. Their combined use facilitates dynamic visualization, automated computation, and reproducible analysis, helping organizations transition from traditional control methods toward adaptive and digitally supported quality management. The integration of these tools does not replace classical SQC methodologies; rather, it modernizes their application by linking statistical analysis with digital transformation processes.

The four R packages—qcc, qcr, qicharts2, and SixSigma—offer complementary functions that form a comprehensive toolkit for quality control. SixSigma is particularly valuable for process characterization, providing tools to summarize and visualize production data within the DMAIC framework. qcc remains the core package for constructing control charts for both variables and attributes, delivering reliable tools for routine SQC analysis. The qcr package extends the capabilities of qcc by offering greater graphical flexibility and a more complete capability analysis, combining parametric and nonparametric indices with performance summaries in a single output. qicharts2 provides a practical framework for monitoring process stability over time; although originally developed for healthcare applications, it is highly adaptable to industrial contexts. Its ability to automatically detect violations of Shewhart rules supports the early recognition of process changes. Together, these four packages provide a complementary and interoperable environment for quality control aligned with the objectives of Industry 4.0.

Despite the combined strengths of these packages, the analysis also highlights an important opportunity for the future development of the R ecosystem: the absence of a unified package capable of integrating, within a single environment, the fundamental principles of Industry 4.0, including: real-time data ingestion and processing, management of high-volume and high-variety data, interoperability with cyber-physical systems and smart sensors, advanced visualization and interactive dashboards, incorporation of both classical and adaptive SQC methods, complete data and process traceability.

In summary, an integrated tool designed explicitly for Industry 4.0—capable of combining statistical analysis, real-time monitoring, advanced visualization, and digital connectivity—would significantly contribute to strengthening modern quality systems and would further establish R as a key platform for advanced industrial analytics.

Appendix – Supplementary R Scripts

All R scripts developed and used in this study are compiled in a separate supplementary document titled "Appendices."

This document contains the complete code for data structuring, process characterization, control chart generation, and process capability analysis using the R packages qcc, SixSigma, and qicharts2. Each section within the supplementary file corresponds to the analyses discussed in the main text, organized as follows (Table 11):

Table 11. Appendix

Appendix	Content/Code Description	Related Figure(s)
A	`Data preprocessing and long-format transformation (qicharts2)`	-
B	`Pareto chart (qcc)`	Figure 2
C	`Cause-and-effect diagram (qcc)`	Figure 3
D	`Cause-and-effect diagram (SixSigma)`	Figure 4
E	`Process map (SixSigma)`	Figure 5
F	`c-chart (qcc)`	Figure 6
G	`c-chart (qicharts2)`	Figure 7
H	`X-R control charts (qcc)`	Figure 9 and 10
I	`X-S control charts (qicharts2)`	Figure 11 and 12
J	`Process capability analysis (qcc)`	Figure 14
K	`Process capability study (SixSigma)`	Figure 15

The supplementary file ensures reproducibility of the results and serves as a reference for implementing Statistical Process Control (SPC) methods in RStudio within the framework of Industry 4.0.

References

1. Carro, R., González Gómez, D.A.: Control estadístico de procesos (2012)
2. Marín, J.F., Castillo, S.G.: Competencias STEM de mayor demanda para afrontar los retos de la Industria 4.0. Revisión bibliográfica para América Latina y Costa Rica: STEM skills in greatest demand to face the challenges of Industry 4.0. Bibliographic review for Latin America and Costa Rica. Latam Rev. Latinoam. Cienc. Soc. Humanidades **5**(4), 21 (2024)
3. del Val Román, J.L.: Industria 4.0: la transformación digital de la industria. In: Valencia: Conferencia de Directores y Decanos de Ingeniería Informática, Informes CODDII (2016)
4. Lavalle, A.: Desarrollo de un sistema de Big Data sobre datos abiertos (2018)
5. Marqués, M.P.: Big Data. Téc. Herram. Apl. Alfa Omega Grupo Ed. (2015)
6. Scrucca, L.: qcc: an R package for quality control charting and statistical process control. Dim Pist. **1**(200), 3 (2004)
7. Cano, E.L., Moguerza, J.M., Redchuk, A.: Six Sigma with R: Statistical Engineering for Process Improvement, vol. 36. Springer, Cham (2012)
8. Anhoej, J.: qicharts2: quality improvement charts for R. J. Open Source Softw. **3**(25), 699 (2018)
9. Flores, M., Fernández-Casal, R., Naya, S., Tarrío-Saavedra, J.: Statistical quality control with the qcr package (2021)
10. Violante, J.P.C., Machado, M.A., dos S. Mendes, A., Almeida, T.S.: An interface to monitor process variability using the binomial ATTRIVAR SS control chart. Algorithms **17**(5), 216 (2024)
11. Aykroyd, R.G., Leiva, V., Ruggeri, F.: Recent developments of control charts, identification of big data sources and future trends of current research. Technol. Forecast. Soc. Change **144**, 221–232 (2019)

12. Waqas, M., Xu, S.H., Amin, M.N.U., Masengo, G.: On 100 years of quality control charts. 16 de abril de 2024, In Review. https://doi.org/10.21203/rs.3.rs-4264704/v1
13. Megahed, F.M., Chen, Y.-J., Zwetsloot, I., Knoth, S., Montgomery, D.C., Jones-Farmer, L.A.: AI and the future of work in statistical quality control: Insights from a first attempt to augmenting ChatGPT with an SQC knowledge base (ChatSQC). arXiv Preprint arXiv:230813550, vol. 525 (2023)
14. Naveed, M., Azam, M., Khan, N., Aslam, M., Saleem, M., Saeed, M.: Control charts using half-normal and half-exponential power distributions using repetitive sampling. Sci. Rep. 14(1), 226 (2024). https://doi.org/10.1038/s41598-023-50137-w
15. Atalay, M., Caner Testik, M., Duran, S., Weiß, C.H.: Guidelines for automating Phase I of control charts by considering effects on Phase-II performance of individuals control chart. Qual. Eng. 32(2), 223–243 (2020). https://doi.org/10.1080/08982112.2019.1641208
16. Dobi, B., Zempléni, A.: Markovchart: an R package for cost-optimal patient monitoring and treatment using control charts. Comput. Stat. 37(4), 1653–1693 (2022). https://doi.org/10.1007/s00180-021-01175-3
17. Walter, O.M.F.C., Henning, E., Zvirtes, L.: Controle Estatístico do Processo em R: análise exploratória do pacote qcc
18. Rizkiah, N., Kurniawati, Y.: An application X-bar chart and statistical process control with R package. UNP J. Stat. Data Sci. 3(2), 197–204 (2025). https://doi.org/10.24036/ujsds/vol3-iss2/363
19. Scrucca, L.: qcc: an R package for quality control charting and statistical process control. R News 4(1), 11–17 (2004)
20. Gomon, D., Fiocco, M., Putter, H., Signorelli, M.: SUrvival Control Chart EStimation Software in R: the success package (2023). https://doi.org/10.48550/ARXIV.2302.07658
21. Cano, E.L., Moguerza, J.M., Redchuk, A.: Six Sigma with R: Statistical Engineering for Process Improvement. Springer, New York (2012). https://doi.org/10.1007/978-1-4614-3652-2
22. Winckler, B., McKenzie, S., Lo, H.: A practical guide to QI data analysis: run and statistical process control charts. Hosp. Pediatr. 14(1), e83–e89 (2024). https://doi.org/10.1542/hpeds.2023-007296
23. Mohammed, M.A., Worthington, P., Woodall, W.H.: Plotting basic control charts: tutorial notes for healthcare practitioners. Qual. Saf. Health Care 17(2), 137–145 (2008). https://doi.org/10.1136/qshc.2004.012047
24. Wolfe, H.A., Taylor, A., Subramanyam, R.: Statistics in quality improvement: measurement and statistical process control. Pediatr. Anesth. 31(5), 539–547 (2021). https://doi.org/10.1111/pan.14163
25. Tamayo, L.M.R.: Cartas de control para optimizar el proceso de pintura de láminas de aluminio. Espacios 39(22), 34–40 (2018)
26. Baro, M., Piña, M.R., Valdiviezo, C.J., Amaya, R.M.: El Proceso DMAIC: Herramientas de Calidad en el Desarrollo de Proyectos de Mejora de la Calidad. Ing. E Innov. 12(1) (2024)
27. Rojas, A.R.-F.: Herramientas de calidad. Univ. Pontif. Comillas Madr. (2009)
28. Gutiérrez, H.: Cartas de control Bayesianas para atributos y el tamano de subgrupo grande en la carta p. Rev. Colomb. Estad. 29(2), 163–180 (2006)
29. Mustafa, A.M., Rodríguez, N.L., Chauvet, S.: Control de calidad: Cartas de control por variables. en Congreso Regional de Ciencia y Tecnología NOA (2002)
30. Rodríguez, L.I., Rodríguez, M.I., Rodríguez, M.A.: Análisis Comparativo de los Gráficos de Control Shewhart, CUSUM y EWMA, para Procesos en Condiciones de Control y Fuera de Control (2012)
31. Carrillo Mejía, F.E.: Aumento de la productividad en la línea de producción de baldosa de porcelanato mediante el análisis de Capacidad de Procesos (2023)

From Free Text to Safety Signals: Integrating Structural Topic Model and Embeddings for Drug–Event in Spanish Pharmacovigilance (Colombia).

Gustavo Bruges Morales[1]([✉]) [iD], Juan Manuel Luna[2] [iD],
Francisco Palencia-Sánchez[1,2] [iD], Paula Alejandra Vargas[1] [iD], Shirly Vanesa Melo[1] [iD],
María Laura Morales[1] [iD], and Mariangeles Meza[1] [iD]

[1] Universidad Autónoma de Bucaramanga, Bucaramanga, Santander, Colombia
gbruges@unab.edu.co
[2] Pontificia Universidad Javeriana, Bogotá, Colombia

Abstract. Pharmacovigilance in Colombia faces challenges in analyzing spontaneous adverse drug reaction (ADR) reports written in free text, where clinical terms mix with colloquialisms, synonyms, abbreviations, and spelling errors. This heterogeneity complicates mapping to Medical Dictionary for Regulatory Activities (MedDRA) and delays signal detection. This study applies a semantic text-mining framework to the Colombian National Pharmacovigilance Database, focusing on 1,199 drug-level ADR narratives through multilingual embeddings, dimensionality reduction, and Structural Topic Model (STM). Drug narratives underwent preprocessing for normalization and entity harmonization via Anatomical Therapeutic Chemical (ATC) classifications, leading to a 25-topic optimal solution. The themes identified include systemic toxicities and administration-related errors, demonstrating that Spanish narratives reveal significant clinical distinctions beyond standardized terms. This novel approach facilitates scalable analysis and enhances regulatory workflows, although challenges such as underreporting and the need for validation remain. Future work should involve temporal modeling and expert feedback to improve methodologies in Spanish-speaking regulatory contexts.

Keywords: Adverse Drug Reaction · Text Mining · Structural Topic Modeling

1 Introduction

Pharmacovigilance is a discipline within pharmaceutical sciences aimed at assessing patient safety and enhancing public health systems through the identification, evaluation, and comprehension of adverse drug reactions (ADRs) and associated issues. Most of the research on adverse events originates from hospital environments, where adverse drug reactions (ADRs) are prevalent [1–5]. Nonetheless, there are other accounts from the communal level [1]. Research, like the ENEAS study from Spain and later investigations

© The Author(s), under exclusive license to Springer Nature Switzerland AG 2026
B. M. Suárez et al. (Eds.): R Day 2025, CCIS 2824, pp. 191–205, 2026.
https://doi.org/10.1007/978-3-032-18455-9_10

in Latin America, indicates that adverse drug reactions (ADRs) incur costs, resulting in disabilities, prolonged hospitalizations, and elevated mortality rates, particularly among the elderly [3].

Structural deficiencies, including the absence of reporting systems, uneven hospital policies, and inadequate analytical capacity, have resulted in the suboptimal functioning of pharmacovigilance systems in Latin America [2]. These obstacles impede the swift recognition of signals and the continual assimilation of adverse event data. As treatments grow increasingly intricate, the necessity for sophisticated analytical frameworks capable of effectively managing diverse data sources, particularly unstructured data, becomes increasingly imperative.

In the last two decades, advancements in natural language processing (NLP) and machine learning have significantly enhanced the transformation of free-text reports into relevant pharmacovigilance insights [5–10]. Data have been sourced from various origins, including spontaneous reporting systems, vaccination safety reports, narrative case reports, electronic health records, and online health debates [11–16]. These methods have shown effective for the automatic identification of adverse drug reactions (ADRs), facilitating signal detection, case prioritization, and comprehensive safety monitoring. Systematic evaluations demonstrate that NLP techniques can proficiently detect ADRs in user-generated content, hence improving conventional disproportionality analyses [9].

The majority of NLP research has been performed in English, resulting in clinical NLP for non-English-speaking environments, particularly in Spanish, being in its nascent phase and deficient in several resources. Recent advancements in Spanish NLP include annotated corpora and standards for ADR identification, along with specialized procedures for clinical documentation in Spanish. Although named entity recognition and adverse event extraction have advanced, national pharmacovigilance databases in Spanish remain underutilized for text mining. Innovative tools like MedLexSp and biomedical embeddings in Spanish facilitate the comprehension of medical terminology in the Spanish language [17].

Topic models are a suite of unsupervised machine learning techniques designed to discover latent thematic structures within large collections of unstructured text. By statistically analyzing patterns of word co-occurrence, these models infer a set of "topics" where each topic is represented as a distribution of related words, and each document is represented as a mixture of these topics. This allows for the automatic summarization, organization, and thematic exploration of text data at scale, transforming raw text into a structured semantic map. Latent Dirichlet Allocation (LDA) and the more advanced Structural Topic Model (STM) are two algorithms that help people make sense of a lot of writing by finding the main themes that run through a lot of documents. By statistically modeling the latent thematic structure, these models reveal patterns and conceptual relationships that are not immediately apparent. For instance, the application of STM to drug-related social media and forum content has been particularly fruitful, successfully uncovering significant themes regarding consumer apprehensions, usage patterns, and temporal shifts in public discourse [16, 18, 19].

A substantial need persists in studies concerning the confluence of Latin American national pharmacovigilance systems, thorough ADR reports in Spanish, and innovative methodologies for semantic representations and topic modeling. Despite research indicating the feasibility of analyzing clinical texts in Spanish to identify adverse drug reactions (ADRs) and the efficacy of semantic text mining (STM) for safety data, insufficient evidence exists to illustrate the combined application of embeddings and STM in examining the semantic structure of ADR narratives on a national scale. Moreover, publicly accessible pharmacovigilance databases, essential for transparency and comparative regional studies, are hardly utilized.

We propose a semantic text-mining framework for Colombian pharmacovigilance data that integrates (1) Multilingual embeddings [20], (2) dimensionality reduction and hierarchical clustering, and (3) STM with covariates derived from the embedding space. Using open public data from the Colombian National Pharmacovigilance Database, we aggregate free-text ADR narratives at the drug level, construct a continuous semantic representation of pharmacological experience in Spanish, and estimate topic structures linked to pharmacological domains and adverse event patterns.

2 Structural Topic Modeling

2.1 Ethical Considerations and Data Governance

The data analyzed in this study were obtained from the open government data portal of the Colombian State (Datos Abiertos Colombia). Specifically, we used the publicly available dataset accessible at:

https://www.datos.gov.co/Salud-y-Protecci-n-Social/EVENTOS-ADVERSOS-ASOCIADOS-A-MEDICAMENTOS/my7f-awtq/about_data.

In Colombia, open data are defined as public information made available in formats that allow their use and reuse under an open license and without legal restrictions, in accordance with the Ley 1712 de 2014 on Transparency and the Right of Access to Public Information. Under this framework, *Datos Abiertos Colombia* corresponds to primary or unprocessed data published in standard, interoperable formats that facilitate access and reuse by any citizen [21].

The data set used here is provided in de-identified form and does not include direct personal identifiers of patients or reporters. Our analysis was conducted exclusively on these anonymized records, and all results are reported at aggregated levels (e.g., drug-level semantic profiles, topic distributions, and clusters) that preclude individual re-identification. For the purposes of this study, we further aggregated spontaneous reports by generic drug name before modeling, which increases anonymity and reduces residual disclosure risk.

2.2 Data Acquisition and Preprocessing

Each record contains a free-text narrative describing the suspected drug, the observed reaction, dosage information, and accompanying clinical details. To ensure analytical consistency, all reports were filtered to retain only those with clearly identifiable drug names and reaction descriptions in Spanish. Individual ADR narratives were then aggregated by *generic drug name*, such that all reports referring to the same medication were concatenated into a unified textual representation. After this consolidation process, a total of *1,199 drug-level descriptions* were retained for analysis.

Textual cleaning and normalization followed a multi-stage natural language processing (NLP) workflow (Fig. 1). The pipeline encompassed:

1. Normalization, including lowercasing, diacritic and punctuation removal, and uniform spacing.
2. Tokenization and lemmatization to ensure standardized lexical representation.
3. Stopword filtering, excluding non-informative functional words
4. Entity harmonization, standardizing the nomenclature of drugs, administration routes, and reactions through the Anatomical Therapeutic Chemical (ATC) classification and cross-referencing with DrugBank identifiers.

This systematic cleaning procedure yielded a semantically coherent corpus of pharmacovigilance narratives, ready for subsequent embedding generation and topic modeling. As illustrated in Fig. 1, the overall workflow comprises four principal modules: (i) data acquisition and preprocessing, (ii) embedding generation, (iii) dimensionality reduction and clustering, and (iv) structural topic modeling and network construction. The sequential design ensures both linguistic standardization and methodological transparency across the computational pipeline.

2.3 Conceptual Overview

Structural Topic Modeling (STM) provides a probabilistic framework for discovering latent thematic structures within large collections of text while simultaneously modeling how document-level metadata influence topic prevalence and content [5].

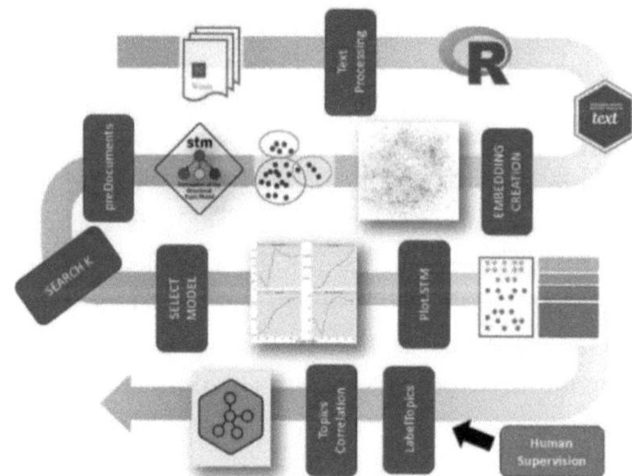

Fig. 1. Workflow for Semantic Pharmacovigilance Using Embeddings and Structural Topic Modeling

Unlike classical topic models such as Latent Dirichlet Allocation (LDA), STM incorporates covariates—quantitative or categorical variables that describe each document—allowing the estimation of both what topics are present and how their frequencies vary across external factors.

In the context of pharmacovigilance, this flexibility makes STM especially valuable for linking linguistic patterns in adverse drug reaction (ADR) narratives to pharmacological properties, drug classes, or embedding-derived semantic structures.

2.4 Model Specification

Formally, for each document, consider a collection of documents $d \in \{1,....,D\}$, where each document corresponds to an aggregated drug-level narrative composed of words $w_{d,n}$ drawn from a vocabulary V, In the Structured Topic Model, the topic proportion vector θ_d for document d follows a logistic-normal distribution.

$$\theta_d \sim \text{LogisticNormal}(\mu = X_d\gamma, \Sigma) \tag{1}$$

Where $\mu_d = Xd\gamma$ is a linear function of document-level covariates Xd with coefficients γ, and Σ is a covariance matrix controlling correlations among topics. For each position n in document d, a latent topic assignment $z_{d,n}$ is drawn from a Multinomial distribution with parameter θ_d

$$w_{d,n} \sim Multinomial(\beta_{zd,n}) \tag{2}$$

Each word $w_{d,n}$ is then generated by first sampling a topic assignment wd,n ~ Multinomial(θ_d) and subsequently sampling a term from the topic-specific word distribution $\beta_{zd,n}$.

STM acts as a bridge between quantitative representations and qualitative interpretation. We incorporated document-level covariates derived from the embedding space and clustering structure so that topic prevalence is partially explained by the position of each drug narrative in the semantic topology.

Specifically, the prevalence component of the model includes (i) the first three principal components (PC1–PC3) of the embedding matrix as continuous covariates and (ii) the HCPC-derived cluster membership as a categorical factor. The prevalence equation is therefore expressed as

$$\text{prevalence} \sim s(\text{PC1}) + s(\text{PC2}) + s(\text{PC3}) + \text{cluster} \tag{3}$$

where s denotes smooth penalized spline functions that capture potential nonlinear associations between semantic coordinates and topic weights. PC1–PC3 summarize major axes of semantic variation, and the cluster variable indexes discrete pharmacological groupings.

2.5 Embedding Generation

Each drug-level narrative was encoded as a dense vector using the multilingual Sentence Transformer model *"paraphrase-multilingual-MiniLM-L12-v2"* implemented via the text package in R [20, 21]. This model captures contextual dependencies between words, producing embeddings that preserve semantic similarity beyond surface co-occurrence. To avoid overrepresenting very common medications, document weights were scaled inversely to the number of reports per drug so that highly reported drugs did not dominate the learned structure.

The resulting matrix of numeric embeddings (one vector per drug-level narrative) served as input for dimensionality reduction procedures and as covariates for the STM. Principal Component Analysis was then applied to reduce dimensionality and extract the main semantic axes explaining most corpus variance, using the *FactoMineR* R package [22]. These components were then used for visual exploration and for subsequent hierarchical clustering analyses.

2.6 Dimensionality Reduction and Clustering

The first three components accounted jointly for 44.6 % of total variance (26.9 %, 12%, and 5.6 % individually), while subsequent ones each contributed less than

5%. The associated eigenvalues—103.46, 46.17, and 21.63—confirmed that most semantic information concentrated along these initial dimensions.

Interpreting the loadings revealed intelligible pharmacological gradients:

- Dim. 1 traced a continuum of *therapeutic complexity*, from symptomatic drugs to specialized biologics and antineoplastics.
- Dim. 2 delineated *systemic versus local* pharmacological action.
- Dim. 3 distinguished *peripherals from central* targets, especially within nervous-system agents.

These three axes together defined a semantic topology in which similar medications occupied contiguous regions. Hierarchical Clustering on Principal Components (HCPC) [22] identified three large clusters corresponding broadly to therapeutic domains. A 2-D projection of these embeddings is shown conceptually in Fig. 2, where clusters appear as distinct clouds colored by cluster group.

2.7 Structural Topic Modeling and Model Selection

The STM estimation was conducted in R using the *stm* package [16]. A sequence of models with:

$$K = \{5,\ 10, 25,\ 20,\ 25,\ 30,\ 40,\ 50,\ 70\}$$

was fitted to identify the most interpretable number of topics.

Model performance was judged by four diagnostics: held-out likelihood (predictive accuracy), semantic coherence (intra-topic consistency), exclusivity (vocabulary uniqueness), and the evidence lower bound (ELBO) indicating convergence stability [16].

Balancing coherence and exclusivity indicated **K = 25** as the optimal trade-off, yielding topics that were both interpretable and pharmacologically meaningful without overfitting the corpus.

Code Availability All scripts used for data preprocessing, model specification, structural topic modeling, embedding generation, dimensionality reduction, and clustering have been deposited in a public repository to facilitate reproducibility and extension of our work (https://github.com/Gustavo-Bruges/rday-2025).

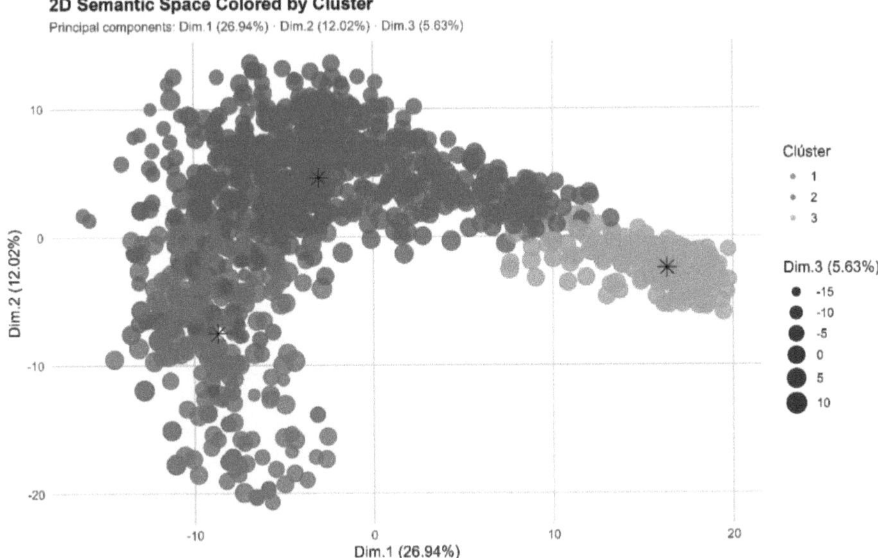

Fig. 2. Two-Dimensional Semantic Projection of Drug Embeddings. A 2D scatterplot of 1,199 drug embeddings projected onto the first two principal components. Colors represent the three HCPC clusters, showing partial overlap but preserving the main semantic organization observed in the 3D space. Point size reflects the third principal component (Dim. 3, 5.63% of variance), highlighting vertical variance gradients related to pharmacological diversity, and therapeutic scope.

3   Results and Discussion

The Structural Topic Model fitted with $K = 25$ yielded a well-differentiated thematic structure, encompassing physiological systems, pharmacological mechanisms, and administration-related phenomena. Rather than being treated as isolated findings, these topics can be understood as recurrent linguistic patterns that reflect pharmacological experience within Spanish-language narratives, consistent with recent applications of topic modeling to health-related corpora [18, 23, 24]. Each topic was reviewed qualitatively through its FREX (Frequency and Exclusivity) [16, 25] terms and representative documents to verify interpretability and alignment with established pharmacology. Across the 1,199 aggregated drug narratives, STM recovered 25 coherent themes (Fig. 3).

Panels show exclusivity, held-out likelihood, residuals, and semantic coherence across K. The intersection where coherence and exclusivity balance corresponds to approximately $K = 25$, considered the best trade-off between interpretability and statistical fit.

The Structural Topic Model uncovered a structural landscape of adverse event narratives ranging from systemic organ-level toxicities to localized and route-specific complaints (Table 1).

Topics such as *Acute Myocardial Ischemia* and *Cardiac Failure* and *Hepatocellular Injury* and *Elevated Liver Enzymes* reflected canonical cardiotoxic and hepatotoxic

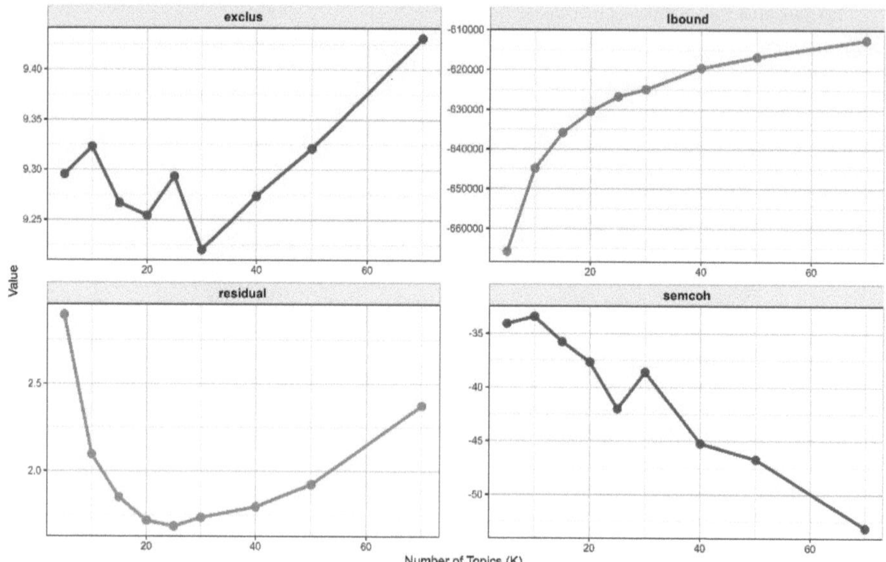

Fig. 3. Model selection diagnostics for Structural Topic Modeling.

reactions, predominantly associated with high-complexity therapies including cardiovascular and analgesic agents. *Bone-Marrow Suppression* and *Cytopenias* and *Oncologic Mucosal* and *Neuropathic Toxicities* captured hematologic and peripheral alignment with oncology-related embeddings.

The nervous system emerged as another major domain encompassing *Neuropsychiatric* and *Autonomic Adverse Effects, Central Nervous System Symptoms, General Nervous System Disorders, and Neuro-immunological Disorders and Behavioral Changes.* These topics distinguished acute autonomic responses from chronic neurodegenerative or behavioral conditions and were prevalent in psychotropic and opioid narratives, consistent with positive PC3 loadings that denote central pharmacological activity.

Complementary topic clusters in Table 1 described adverse phenomena related to drug administration, hypersensitivity, metabolism, infection, and endocrine function. *Infusion-Site and Administration Errors and Injection-Site and Vaccination Reactions* illustrated language associated with parenteral routes, while *Genitourinary and Lower Digestive Disorders* occupied the local-action pole of the semantic space.

Allergic and immune topics—including *Acute Hypersensitivity* and *Gastrointestinal Intolerance, Drug Eruptions* and *Allergic Dermatitis, Urticaria* and *Angioedema of the Face and Eyes,* and *Immediate Hypersensitivity with Autonomic Dysfunction*—demonstrated how cutaneous and visceral reactions frequently overlap in spontaneous reporting. Metabolic and vascular themes such as *Coagulation Control* and *Peripheral Vascular Complications, Diabetes-Related Multiorgan Complications,* and *Metabolic* and *Lipid Disorders* were dominant among antithrombotic and antidiabetic drugs, whereas endocrine and gynecologic events centered on Hormone-Related Uterine Bleeding. Finally, infectious and dermatologic topics, including Dermatologic and Musculoskeletal Infections, Respiratory Tract Infections, and Chronic Cutaneous and Chronic Skin

Table 1. Structural Topic Model–Derived Themes Adverse Drug Reaction Narratives

Topic	Top FREX Terms (8)	Topic Label	Definition (English)
1	cardiaca, infarto, miocardio, paro, cardíaco, miocárdico, corazón, pulmonar	Acute Myocardial Ischemia and Cardiac Failure	Describes ischemic cardiac events and heart function deterioration, including myocardial infarction and acute cardiac failure
2	intermenstrual, menstruación, sangrado, embarazo, implantación, mifepristona, menorragia, anticonceptivo	Hormone-Related Uterine Bleeding	Encompasses menstrual irregularities and bleeding linked to hormonal contraceptives or pregnancy
3	médula, aplasia, ósea, neutropenia, granulocitopenia, febril, trombocitopenia, granulocitosis	Bone-Marrow Suppression and Cytopenias	This condition represents hematological toxicity such as marrow aplasia, neutropenia, and thrombocytopenia
4	artritis, acosa, reumatoide, herpes, otitis, deformidad, pustular, piel	Dermatologic and Musculoskeletal Infections	This condition combines skin infection and rheumatologic inflammatory processes, including herpes and pustular lesions
5	transaminasas, glutámico-oxalacética, hepatitis, aumentadas, fémico, enzimas, glutámico-pirúvica, hepática	Hepatocellular Injury and Elevated Liver Enzymes	This includes hepatic enzyme elevation and toxic hepatitis phenomena
6	sedación, extrapiramidal, anestesia, excesiva, neuroléptico, agitación, distonía, hipotensión	Neuropsychiatric and Autonomic Adverse Effects	The condition reflects sedation, extrapyramidal reactions, anxiety, and cardiovascular autonomic instability
7	brazo, administración, incorrecta, medicamento, flebitis, extravasaciones, vasculares, infusión	Infusion-Site and Administration Errors	This refers to incorrect administration routes, vascular inflammation, and infusion-site reactions

(continued)

Table 1. (*continued*)

Topic	Top FREX Terms (8)	Topic Label	Definition (English)
8	heces, frecuentes, orina, inflamada, prostático, micciones, específica, próstata	Genitourinary and Lower Digestive Disorders	Aggregates urinary, prostate, and stool-related complaints
9	amputación, INR, internacional, normalizada, razón, pie, congestiva, aórtica	Coagulation Control and Peripheral Vascular Complications	Captures anticoagulation issues and peripheral vascular disorders, often of cardiac origin
10	mucositis, neuropatía, estomatitis, metastásis, ulcerativa, linfopenia, cáncer, palmo-plantar	Oncologic Mucosal and Neuropathic Toxicities	This condition represents chemotherapy-type adverse reactions, including mucositis, neuropathy, and ulcerations
11	glaucoma, somnolencia, cerrada, suicida, cabeza, nerviosismo, vértigo, convulsiones	Central Nervous System Symptoms and Seizure-Related Events	Describes CNS-related events such as somnolence, vertigo, seizures, and suicidal ideation
12	sistema, trastornos, nervioso, periférico, del, central, generales, músculo-esquelético	General Nervous System Disorders	Includes central and peripheral neurological symptoms
13	varicela, rinitis, nasal, congestión, neumonía, respiratorio, obstrucción, tos	Respiratory Tract Infections	The groupings include upper and lower respiratory infections, such as pneumonia and rhinitis
14	urticaria, vómitos, rash, diarrea, prurito, náuseas, organismos, generales	Acute Hypersensitivity and Gastrointestinal Intolerance	Combines allergic skin rash with gastrointestinal intolerance manifestations
15	esclerosis múltiple, incontinencia, agravado, recaída, comportamiento, suicidio, manía	Neuroimmunological Disorders and Behavioral Changes	Covers multiple sclerosis relapse and behavioral changes including mood alterations

(continued)

Table 1. (*continued*)

Topic	Top FREX Terms (8)	Topic Label	Definition (English)
16	varicosas, herpes, tuberculosis, otitis, rash, artritis, sinusitis, piel	Chronic Cutaneous and ENT Infections	Describes chronic or recurrent infections of skin, ear, and sinus regions
17	diabética, pie, coma, coronaria, carcinoma, pulmonar, cerebrovascular, retinopatía	Diabetes-Related Vascular and Multiorgan Complications	Encompasses diabetic foot, coma, and systemic vascular pathologies
18	zona, aplicación, inyección, reacción, inflamación, eritema, induración, celulitis	Injection-Site and Vaccination Reactions	Captures local erythema, induration, and inflammation at injection sites
19	erupción, eritematosa, medicamentosas, alergia, eritematoso, eritema, rash, prurito	Drug Eruptions and Allergic Dermatitis	Describes drug-induced rashes and allergic cutaneous eruptions
20	pérdida, cansancio, astenias, dolores, oral, plaquetas, respiratoria, apetito	Constitutional and Hematologic Symptoms	Reflects fatigue, appetite loss, and platelet or respiratory disturbances
21	electrolitos, neoplasia, azúcares, calcio, neoplasma, carcinoma, metabolismo, hiponatremia	Metabolic and Electrolyte Disorders in Neoplastic Contexts	It indicates metabolic and electrolyte changes in neoplastic or post-surgery contexts
22	habón, urticaria, eritema, palpebral, labios, ojo, párpados, edema	Urticaria and Angioedema of the Face and Eyes	The condition is characterized by swelling and tightness around eyelids and lips, typical of immediate hypersensitivity
23	abdominal, abdomen, epigástrico, dispepsia, gastritis, estómago, dolor, mareo	Upper Gastrointestinal Pain and Dyspepsia	This refers to gastric discomfort, epigastric pain, and digestive irritation

(*continued*)

Table 1. (*continued*)

Topic	Top FREX Terms (8)	Topic Label	Definition (English)
24	lípidos, lipodistrofia, hiperlipidemia, hiperglicemia, metabólica, masa, concentración	Metabolic and Lipid Disorders	Includes lipid metabolism disturbances and drug-induced lipodystrophy
25	hipersalivación, hipersensibilidad, inmediata, esfínter, atonía, tónica, rubefacción, disfunción	Immediate Hypersensitivity with Autonomic Dysfunction	Captures acute hypersensitivity accompanied by salivation, sphincter dysfunction, and muscular tone changes

and Ear, Nose, and Throat (ENT) Infections, often shared terminology with allergic narratives, indicating that reporters in Spanish tend to describe inflammatory and infectious events through overlapping linguistic patterns [23, 26].

An important observation from the qualitative review is that several topics encapsulated narrative patterns that are not straightforwardly represented in structured MedDRA fields [10, 27]. For instance, topics combining cutaneous hypersensitivity with gastrointestinal discomfort (e.g. Acute Hypersensitivity and Gastrointestinal Intolerance, Drug Eruptions and Allergic Dermatitis) grouped expressions that are often dispersed across multiple System Organ Classes or recorded inconsistently in coded data. Similarly, Infusion-Site and Administration Errors and Injection-Site and Vaccination Reactions captured descriptions of incorrect routes, local mishandling, and contextual details that are typically underrepresented or collapsed in structured variables.

When topic prevalence was projected onto the PCA coordinates, coherent patterns emerged. Cardiovascular and hepatic toxicities are concentrated among drugs with high therapeutic complexity. Local administration and dermatologic topics mapped toward the negative end of Dim. 2 (localized action). Neuropsychiatric and autonomic reactions spanned both dimensions, bridging systemic and central axes.

The clustering analysis indicated three broad pharmacological groupings:

- Cluster 1, dominated by systemic therapies (antineoplastics, antithrombotic), where Topics 1, 3, 5, 9, 10, and 17 prevailed.
- Cluster 2, comprising neurologically active drugs (opioids and antidepressants), marked by Topics 6, 11, 12, 15, and 25.
- Cluster 3, composed of topical and anti-infective agents, enriched in Topics 4, 13, 16, 18, and 19.

It is worth noting that these topics are not fixed "disease entities" but representations of how Spanish-speaking reporters describe drug effects.

Terms like "mareo," "ansiedad," or "ardor" carry cultural and experiential connotations that slightly diverge from standardized MedDRA terminology. The STM framework allows such expressions to persist rather than forcing them into predefined taxonomies, yielding a more faithful reflection of patient and clinician language.

From a regulatory perspective, these findings expand what can be extracted from the narrative field of pharmacovigilance databases [1, 2]. By coupling embeddings with STM, we move from isolated symptom coding to semantic communities that can be monitored over time. For example, a future temporal STM could reveal whether hypersensitivity-related topics intensify after certain product reformulations or whether neurological complaints cluster seasonally with drug shortages.

The Colombian corpus analyzed here demonstrates that the computational integration of text mining and statistical modeling is feasible even in resource-constrained settings, provided that linguistic normalization and data stewardship are handled carefully. The method does not replace expert review but scaffolds it, offering regulators a way to triage attention toward clusters of semantically coherent concern.

Limitations

This study acknowledges several limitations: the dependence on spontaneous ADR reports from Colombia's pharmacovigilance system influences the transferability of the findings due to underreporting and regional disparities. Combining stories at the drug level makes them more stable, but it hides differences between drugs. The multilingual embedding model employed is not explicitly trained on Spanish pharmacovigilance texts, which may influence domain representation [20, 21, 28]. Standard diagnostics dealt with how sensitive the results were to choices made during preprocessing and the number of topics specified, but other methods might give different results. Lastly, we need to test the results on other datasets and get feedback from experts to see how strong and useful they are in regulatory settings.

Implications for Regulators and Future Applications

Our findings indicate various specific methods through which regulators could implement topic and embedding outputs. Topic-drug prevalence matrices serve as triage instruments to pinpoint medications characterized by narratives dominated by hypersensitivity, neurological disturbances, or administration errors, thereby necessitating focused clinical review. Temporal extensions of STM could track changes in topic prevalence after regulatory actions, new products hit the market, or existing products are reformulated. This would give a different view on disproportionality analysis. Embedding-based similarity can help put together pharmacologically related products that use similar experiential language. This can help signal examiners move from looking at one product at a time to recognizing patterns at the class level. In a more general sense, adding these semantic representations to current dashboards would help pharmacovigilance professionals move through large amounts of free text without losing track of the patient and clinician stories that make up each aggregated profile.

4 Conclusions

We present a text mining framework for the analysis of narratives concerning suspected adverse drug reactions (ADRs) from a national pharmacovigilance database. Our research employs structural topic models, sentence-level embeddings, and clustering techniques to diminish dimensionality and produce coherent semantic representations of spontaneous reports. This facilitates the identification of hints and the generation of novel ideas. Our contributions include the efficient use of STM and embeddings in Spanish pharmacovigilance data, the integration of embedding spaces into cohesive pharmaceutical communities, and the establishment of an open-source repository for replication and methodological enhancement. The findings indicate that language can serve as biomedical evidence, as semantic proximity reveals pharmaceutical associations and various themes of adverse events experienced. Subsequent research ought to broaden this paradigm to encompass multilingual accounts of adverse drug reactions, thereby enhancing transparency and reproducibility in pharmacovigilance through the utilization of open-source tools and public datasets.

References

1. WHO. The importance of pharmacovigilance. World Health Organization 48 (2002)
2. Pan American Health Organization (PAHO). Good Pharmacovigilance Practices for the Americas. Washington D.C (2011)
3. Beninger, P.: Pharmacovigilance: an overview. Clin. Ther. **40**, 1991–2004 (2018). https://doi.org/10.1016/J.CLINTHERA.2018.07.012
4. Pro Pharma Research Organization Pharmacovigilance and Drug Safety in LATAM. https://propharmaresearch.com/en/resources/diffusion/pharmacovigilance-emerging-markets-opportunities-and-challenges. Accessed 7 Dec 2025
5. Pilipiec, P., Liwicki, M., Bota, A.: Using machine learning for pharmacovigilance: a systematic review. Pharmaceutics **14**, 266 (2022). https://doi.org/10.3390/PHARMACEUTICS14020266
6. Mikolov, T., Chen, K., Corrado, G., Dean, J.: Efficient estimation of word representations in vector space. In: 1st International Conference on Learning Representations, ICLR 2013 - Workshop Track Proceedings (2013)
7. Santiso, S., Casillas, A., Pérez, A., Oronoz, M.: Word embeddings for negation detection in health records written in Spanish. Soft. Comput. **23**(21), 10969–10975 (2018). https://doi.org/10.1007/S00500-018-3650-7
8. Schmider, J., Kumar, K., Laforest, C., et al.: Innovation in pharmacovigilance: use of artificial intelligence in adverse event case processing. Clin. Pharmacol. Ther. **105**, 954–961 (2019). https://doi.org/10.1002/CPT.1255
9. Wong, A., Plasek, J.M., Montecalvo, S.P., Zhou, L.: Natural language processing and its implications for the future of medication safety: a narrative review of recent advances and challenges. Pharmacotherapy **38**, 822–841 (2018). https://doi.org/10.1002/PHAR.2151
10. Harpaz, R., Dumouchel, W., Shah, N.H., et al.: Novel data-mining methodologies for adverse drug event discovery and analysis. Clin. Pharmacol. Ther. **91**, 1010–1021 (2012). https://doi.org/10.1038/CLPT.2012.50
11. Aranaz-Andrés, J.M., Aibar-Remón, C., Limón-Ramírez, R., et al.: IBEAS design: adverse events prevalence in Latin American hospitals. Rev. Calid. Asist. **26**, 194–200 (2011). https://doi.org/10.1016/j.cali.2010.12.001
12. Gutiérrez-Fandiño, A., Armengol-Estapé, J., Carrino, C.P., et al.: Spanish biomedical and clinical language embeddings (2021)
13. Oronoz, M., Gojenola, K., Pérez, A., et al.: On the creation of a clinical gold standard corpus in Spanish: Mining adverse drug reactions. J. Biomed. Inform. **56**, 318–332 (2015). https://doi.org/10.1016/J.JBI.2015.06.016

14. Aranaz-Andrés, J.M., Aibar-Remón, C., Vitaller-Burillo, J., et al.: Impact and preventability of adverse events in Spanish public hospitals: Results of the Spanish national study of adverse events (ENEAS). Int. J. Qual. Health Care **21** (2009).https://doi.org/10.1093/INTQHC/MZP047

15. Segura-Bedmar, I., Martínez, P., Revert, R., Moreno-Schneider, J.: Exploring Spanish health social media for detecting drug effects. BMC Med. Inform. Dec. Mak. **15**(2), S6- (2015). https://doi.org/10.1186/1472-6947-15-S2-S6

16. Casillas, A., Pérez, A., Oronoz, M., et al.: Learning to extract adverse drug reaction events from electronic health records in Spanish. Expert Syst. Appl. **61**, 235–245 (2016). https://doi.org/10.1016/J.ESWA.2016.05.034

17. Pérez, A., Weegar, R., Casillas, A., et al.: Semi-supervised medical entity recognition: a study on Spanish and Swedish clinical corpora. J. Biomed. Inform. **71**, 16–30 (2017). https://doi.org/10.1016/J.JBI.2017.05.009

18. Weegar, R., Pérez, A., Casillas, A., Oronoz, M.: Deep medical entity recognition for Swedish and Spanish. In: Proceedings - 2018 IEEE International Conference on Bioinformatics and Biomedicine, BIBM, pp. 1595–1601 (2018). https://doi.org/10.1109/BIBM.2018.8621282

19. Campillos-Llanos, L.: MedLexSp – a medical lexicon for Spanish medical natural language processing. J. Biomed. Semant. **14**(1), 2- (2023). https://doi.org/10.1186/S13326-022-00281-5

20. He, L., Han, D., Zhou, X., Qu, Z.: The voice of drug consumers: online textual review analysis using structural topic model. Int. J. Environ. Res. Publ. Health **17**, 3648 (2020). https://doi.org/10.3390/IJERPH17103648

21. Lösch, L., Brown, P., van Hunsel, F.: Using structural topic modelling to reveal patterns in reports on opioid drugs in a pharmacovigilance database. Pharmacoepidemiol. Drug Saf. **31**, 1003–1006 (2022). https://doi.org/10.1002/PDS.5502;PAGEGROUP:STRING:PUBLICATION

22. Reimers N, Gurevych I Sentence-BERT: Sentence Embeddings using Siamese BERT-Networks Ley 1712 de 2014 - Gestor Normativo - Función Pública. https://www.funcionpublica.gov.co/eva/gestornormativo/norma.php?i=56882. Accessed 7 Dec 2025

23. Kjell, O., Giorgi, S., Schwartz, H.A.: The text-package: an R-Package for analyzing and visualizing human language using natural language processing and transformers. Psychol. Methods **28**, 1478–1498 (2023). https://doi.org/10.1037/MET0000542

24. Lê, S., Josse, J., Husson, F.: FactoMineR: an R package for multivariate analysis. J. Stat. Softw. **25**, 1–18 (2008). https://doi.org/10.18637/JSS.V025.I01

25. Roberts, M.E., Stewart, B.M., Tingley, D.: stm: an R package for structural topic models. J. Stat. Softw. **91**, 1–40 (2019). https://doi.org/10.18637/JSS.V091.I02

26. Chopard, D., Treder, M.S., Corcoran, P., et al.: Text mining of adverse events in clinical trials: deep learning approach. JMIR Med. Inform. **9**, e28632 (2021). https://doi.org/10.2196/28632

27. Airoldi, E.M., Bischof, J.M.: A poisson convolution model for characterizing topical content with word frequency and exclusivity (2012)

28. Botsis, T., Nguyen, M.D., Woo, E.J., et al.: Text mining for the vaccine adverse event reporting system: medical text classification using informative feature selection. J. Am. Med. Inform. Assoc. **18**, 631–638 (2011). https://doi.org/10.1136/AMIAJNL-2010-000022

29. Jacobi, C., Van Atteveldt, W., Welbers, K.: Quantitative analysis of large amounts of journalistic texts using topic modelling. Digit. J. **4**, 89–106 (2016). https://doi.org/10.1080/21670811.2015.1093271

From SAA to Distributionally Robust Portfolio Optimization in R: Markowitz and CVaR in Practice

Diego Fonseca[1(✉)] [iD] and Mauricio Junca[2] [iD]

[1] Universidad EAFIT, Medellín, Colombia
dffonsecav@eafit.edu.co
[2] Universidad de los Andes, Bogotá, Colombia
mj.junca20@uniandes.edu.co

Abstract. This paper shows how classic (Markowitz) and modern (CVaR) portfolio optimization models can be implemented efficiently in R. We compare two paradigms: the standard Sample Average Approximation (SAA) and distributionally robust optimization (DRO) using Wasserstein ambiguity sets. We first show that these advanced DRO models admit tractable reformulations as Second-Order Cone Programs (SOCPs). Then, we present the core contribution: a complete R pipeline that implements these exact SOCPs using the `CVXR` modeling package and the `MOSEK` solver. We provide code listings and discuss practical integration details. A numerical study on S&P 500 data illustrates SAA's fragility, showing its consistent failure to meet out-of-sample return constraints. In contrast, the WDRO models successfully achieve this target, demonstrating a superior trade-off between performance and reliability. The result is a practical, reproducible toolkit for `R` users to build and solve advanced robust optimization problems.

Keywords: Portfolio Optimization · Distributionally Robust Optimization · R Language · CVXR · Conic Optimization · MOSEK

1 Introduction

Stochastic optimization problems are central to decision-making under uncertainty across diverse fields, from finance [6,18] and engineering to operations research [11,15] and machine learning [10,13]. These problems are commonly formulated as

$$\min_{x \in \mathcal{X}} \mathbb{E}_{\xi \sim \mathbb{P}}[F(x, \xi)], \tag{1}$$

where x is the decision vector chosen from a feasible set $\mathcal{X}$, and ξ is a random vector supported on $\Xi \subseteq \mathbb{R}^d$ with a probability distribution $\mathbb{P}$. In practice, the true distribution $\mathbb{P}$ is rarely known. This lack of knowledge is typically addressed by leveraging an available dataset of n independent and identically distributed samples $\hat{\Xi}_n = \{\hat{\xi}_1, \hat{\xi}_2, \ldots, \hat{\xi}_n\}$ drawn from $\mathbb{P}$.

© The Author(s), under exclusive license to Springer Nature Switzerland AG 2026
B. M. Suárez et al. (Eds.): R Day 2025, CCIS 2824, pp. 206–222, 2026.
https://doi.org/10.1007/978-3-032-18455-9_11

A common approach to address (1) is the Sample Average Approximation (SAA) [16]. SAA replaces the true, unknown distribution $\mathbb{P}$ with the empirical distribution $\widehat{\mathbb{P}}_N$ derived from the samples $\hat{\Xi}_n$, effectively replacing the expected value with a sample mean. Despite its asymptotic properties, the SAA method exhibits critical structural weaknesses when applied to finite-sample regimes typical of real-world portfolio optimization. First, SAA is susceptible to the 'Optimizer's Curse,' a phenomenon where the optimization process overfits to idiosyncratic noise in the historical data, leading to an optimistic bias in-sample and subsequent disappointment out-of-sample [17]. Second, SAA solutions are notoriously unstable; small perturbations in the input data can result in drastic changes to the optimal portfolio weights, undermining practical implementation [2]. Furthermore, by assuming the empirical distribution is the ground truth, SAA neglects the statistical ambiguity of the underlying data-generating process, failing to provide reliable performance guarantees [12].

To address this fragility, Distributionally Robust Optimization (DRO) has emerged as a powerful framework. Instead of optimizing against a single empirical distribution, DRO seeks a solution that performs well under the most adverse distribution within a so-called *ambiguity set* $\mathcal{D}$. The DRO problem is formulated as:

$$\min_{x \in X} \sup_{Q \in \mathcal{D}} \mathbb{E}_{\xi \sim Q}[F(x, \xi)], \tag{2}$$

where $\mathcal{D}$ is a set of probability distributions constructed around the empirical data, typically containing $\widehat{\mathbb{P}}_N$. The construction of $\mathcal{D}$ is crucial and influences both the robustness of the solution and the tractability of the problem.

While ambiguity sets can be defined using moment conditions [5] or divergence measures like Kullback-Leibler [9], a popular and effective approach is to define $\mathcal{D}$ as a ball centered at $\widehat{\mathbb{P}}_N$ using the Wasserstein distance [4, 8, 12].

Definition 1 (Wasserstein distance). *The Wasserstein distance $W_p(\mu, \nu)$ of order $p \in [1, \infty)$ between two probability measures μ and ν is defined as:*

$$W_p(\mu, \nu) := \left(\inf_{\Pi \in \mathcal{P}(\Xi \times \Xi)} \left\{ \int_{\Xi \times \Xi} \mathbf{d}^p(\xi, \zeta) \Pi(d\xi, d\zeta) \; : \; \begin{array}{l} \Pi(\cdot \times \Xi) = \mu(\cdot), \\ \Pi(\Xi \times \cdot) = \nu(\cdot) \end{array} \right\} \right)^{1/p} \tag{3}$$

where $\mathbf{d}(\cdot, \cdot)$ is a ground metric on Ξ and Π is a joint distribution with marginals μ and ν.

Using this metric, we define the Wasserstein ambiguity set (or ball) with radius $\varepsilon > 0$ as $\mathcal{B}_\varepsilon\left(\widehat{\mathbb{P}}_N\right) := \left\{ \nu \in \mathcal{P}(\Xi) \mid W_p(\nu, \widehat{\mathbb{P}}_N) \leq \varepsilon \right\}$. The resulting p-Wasserstein DRO (p-WDRO) problem is:

$$\min_{x \in X} \sup_{Q \in \mathcal{B}_\varepsilon(\widehat{\mathbb{P}}_N)} \mathbb{E}_{\xi \sim Q}[F(x, \xi)]. \tag{4}$$

Note that when the radius $\varepsilon = 0$, the ambiguity set contains only $\widehat{\mathbb{P}}_N$, and the p-WDRO problem (4) reduces to the SAA problem.

A key advantage of the Wasserstein distance is its powerful dual representation. As shown in [4,8], the p-WDRO problem (4) has an equivalent reformulation.

Theorem 1. *Assume that $F(x,\xi)$ is upper semicontinuous in ξ. Then the p-WDRO problem (4) is equivalent to the finite-dimensional optimization problem:*

$$\begin{cases} \displaystyle\inf_{x\in\mathbb{X},\lambda\geq 0,s\in\mathbb{R}^N} \lambda\varepsilon^p + \frac{1}{N}\sum_{i=1}^N s_i \\ \text{subject to} \quad \sup_{\xi\in\Xi}\left(F(x,\xi) - \lambda\mathbf{d}^p(\xi,\widehat{\xi}_i)\right) \leq s_i \ \forall i = 1,\ldots,N. \end{cases} \tag{5}$$

However, the reformulation in (5) presents a new challenge. The constraints involve an inner supremum problem over the (potentially continuous) support set Ξ. This makes (5) a semi-infinite optimization problem, which is generally computationally demanding to solve directly.

Fortunately, for specific cases of p and F, this semi-infinite problem can be simplified into a tractable form. A celebrated example is the case for $p = 1$, which is closely related to the Kantorovich-Rubinstein duality.

Corollary 1 (1-WDRO Reformulation for Lipschitz Functions). *For the $p = 1$ case, if the function $F(x,\cdot)$ is L_x-Lipschitz with respect to ξ under the metric $\mathbf{d}$ (i.e., $|F(x,\xi) - F(x,\xi')| \leq L_x\,\mathbf{d}(\xi,\xi')$), the inner supremum in (4) has an exact, closed-form solution:*

$$\sup_{\mathbb{Q}:W_1(\mathbb{Q},\widehat{\mathbb{P}}_N)\leq\varepsilon} \mathbb{E}_{\xi\sim\mathbb{Q}}[F(x,\xi)] = \frac{1}{N}\sum_{i=1}^N F(x,\widehat{\xi}_i) + \varepsilon\cdot L_x.$$

This identity provides a crucial insight: for 1-WDRO, the robust objective is equivalent to the SAA objective plus a regularization term $\varepsilon \cdot L_x$, which penalizes the model's sensitivity (its Lipschitz constant) to perturbations in the data. This reformulation is no longer semi-infinite and becomes tractable if the Lipschitz constant L_x can be computed or bounded efficiently, as is the case for the portfolio optimization models we study.

This tractability is the central theme of our paper. For the portfolio optimization problems of interest (both mean-variance and mean-CVaR), the seemingly complex p-WDRO formulations (for both $p = 1$ and $p = 2$) can be exactly reformulated as standard, solvable convex optimization problems, specifically **Second-Order Cone Programs (SOCPs)**, which is possible thanks to the theoretical foundations of WDRO established in the literature [4,12]. However, a significant gap remains in the availability of accessible, reproducible computational tools for practitioners, particularly in the R community. Most existing implementations are proprietary or dispersed across custom scripts in other languages (e.g., Python or MATLAB).

The main contribution of this paper is to bridge this gap by providing a complete R-based pipeline. We demonstrate that the advanced WDRO theory can be seamlessly translated into practice using the `CVXR` package [7] and the `MOSEK` solver. We explicitly move beyond the theoretical derivations to focus on the algorithmic implementation and empirical validation within the R ecosystem.

The remainder of this paper is structured to deliver this practical toolkit. Section 2 first establishes the theoretical foundation, deriving the tractable Second-Order Cone Program (SOCP) reformulations for both the mean-CVaR and mean-variance WDRO models. This provides the specific mathematical blueprints required for implementation. Section 3 presents the core technical contribution: the practical implementation of these SOCPs in R. We demonstrate how to build these models using `CVXR`, provide code listings for each formulation, and discuss the practical considerations for integrating the high-performance `MOSEK` solver. Finally, Sect. 4 validates this R-based pipeline in a comprehensive numerical study on S&P 500 data, comparing the out-of-sample performance of our WDRO strategies against SAA and other standard industry benchmarks.

2 From WDRO Models to Tractable SOCP Reformulations

In this section, we apply the WDRO framework from Sect. 1 to two canonical portfolio optimization problems: Mean-Risk using Conditional Value at Risk (CVaR) and Mean-Variance (Markowitz). For each model, we present the standard stochastic formulation, its p-WDRO counterpart, and the resulting tractable reformulation that serves as the basis for our R implementation. Throughout this section, we assume the ground metric $\mathbf{d}(\cdot, \cdot)$ is the standard Euclidean norm $\| \cdot \|_2$.

2.1 Mean-Risk (CVaR) Portfolio Optimization

In this subsection, we specialize the generic problem (1) to the portfolio context. Our decision vector x is now composed of the portfolio **weights** vector, which we denote by $w \in \mathbb{R}^m$, and an auxiliary scalar variable $\tau \in \mathbb{R}$ required for the the Conditional Value at Risk formulation.

We first consider the problem of minimizing the **Conditional Value at Risk (CVaR)** of the portfolio loss $L(w, \xi) = -\langle w, \xi \rangle$, subject to a constraint on the minimum expected return μ. For a given confidence level $\alpha \in (0, 1)$, the CVaR_α measures the expected loss in the $(1 - \alpha)\%$ worst-case scenarios. It is formally defined as $\mathrm{CVaR}_\alpha(w) = \mathbb{E}[L(w, \xi) \,|\, L(w, \xi) \geq \mathrm{VaR}_\alpha(w)]$, where $\mathrm{VaR}_\alpha(w)$ is the α-quantile (or Value at Risk) of the loss. CVaR is widely used because it is a **coherent risk measure** [1] and, critically for optimization, it is **convex**. While its definition involves the nested VaR_α term, [14] provide an equivalent and simpler formulation by introducing the auxiliary variable $\tau \approx \mathrm{VaR}_\alpha(w)$. Following their approach, the stochastic problem is formulated as:

$$
\begin{cases}
\displaystyle \min_{w \in \mathbb{R}^m, \tau \in \mathbb{R}} \quad \mathbb{E}_{\xi \sim \mathbb{P}}\left[F(w, \tau, \xi) \right] \\
\text{subject to } \mathbb{E}_{\xi \sim \mathbb{P}}\left[\langle w, \xi \rangle \right] \geq \mu, \\
\qquad\qquad w \in \mathcal{W},
\end{cases}
\tag{6}
$$

where $\mathcal{W}$ is the feasible set for the portfolio weights w (e.g., $w \geq 0$, $\sum w_i = 1$), and the objective function $F(w, \tau, \xi)$ is given by:

$$F(w, \tau, \xi) = \max \left\{ -\frac{\langle w, \xi \rangle}{\alpha} + \left(1 - \frac{1}{\alpha}\right)\tau, \quad \tau \right\} \tag{7}$$

We replace the unknown distribution $\mathbb{P}$ with a p-Wasserstein ambiguity set $\mathcal{B}_\varepsilon\left(\widehat{\mathbb{P}}_N\right)$ centered at the empirical distribution from N samples. This yields the following p-WDRO problem:

$$\begin{cases} \min\limits_{w \in \mathcal{W}, \tau \in \mathbb{R}} \quad \sup_{\mathbb{Q} \in \mathcal{B}_\varepsilon(\widehat{\mathbb{P}}_N)} \mathbb{E}_{\xi \sim \mathbb{Q}}\left[F(w, \tau, \xi)\right] \\ \text{subject to } \inf_{\mathbb{Q} \in \mathcal{B}_\varepsilon(\widehat{\mathbb{P}}_N)} \mathbb{E}_{\xi \sim \mathbb{Q}}\left[\langle w, \xi \rangle\right] \geq \mu. \end{cases} \tag{8}$$

This problem appears challenging, as it involves optimizing over a space of distributions. However, by applying the duality results from Theorem 1 (for $p = 2$) and Corollary 1 (for $p = 1$), this problem can be cast as a Second-Order Cone Program (SOCP). The following lemma presents these key reformulations, which are the basis for our implementation.

Lemma 1 (SOCP Reformulations for WDRO-CVaR). *Assuming $\Xi = \mathbb{R}^m$, the 1-WDRO approach ($p = 1$) for problem (8) is equivalent to the following SOCP:*

$$\begin{cases} \inf\limits_{w, \tau, \lambda, s} \lambda\varepsilon + \frac{1}{N}\sum_{i=1}^{N} s_i \\ \text{s.t.} \quad a_k \langle w, \widehat{\xi}_i \rangle + b_k \tau \leq s_i \qquad \forall i \in [N], k \in \{1, 2\} \\ \qquad \|w\|_2 |a_k| \leq \lambda \qquad\qquad\quad \forall k \in \{1, 2\} \\ \qquad \frac{1}{N}\sum_{i=1}^{N} \langle w, \widehat{\xi}_i \rangle - \varepsilon\|w\|_2 \geq \mu, \\ \qquad w \in \mathcal{W}, \ \tau \in \mathbb{R}, \ \lambda \geq 0, \ s \in \mathbb{R}^N. \end{cases} \tag{9}$$

Furthermore, the 2-WDRO approach ($p = 2$) for problem (8) is equivalent to the following SOCP:

$$\begin{cases} \inf\limits_{w, \tau, \lambda, s, z} \lambda\varepsilon^2 + \frac{1}{N}\sum_{i=1}^{N} s_i \\ \text{s.t.} \quad \frac{a_k^2}{4}z + a_k \langle w, \widehat{\xi}_i \rangle + b_k \tau \leq s_i \qquad \forall i \in [N], k \in \{1, 2\} \\ \qquad \left\| \begin{pmatrix} 2w \\ \lambda - z \end{pmatrix} \right\|_2 \leq \lambda + z \\ \qquad \frac{1}{N}\sum_{i=1}^{N} \langle w, \widehat{\xi}_i \rangle - \varepsilon\|w\|_2 \geq \mu, \\ \qquad w \in \mathcal{W}, \ \tau \in \mathbb{R}, \ \lambda \geq 0, \ s \in \mathbb{R}^N, \ z \geq 0. \end{cases} \tag{10}$$

In both formulations, the constants a_k, b_k are derived from (7) as $a_1 = -1/\alpha$, $b_1 = 1 - 1/\alpha$, $a_2 = 0$, and $b_2 = 1$.

2.2 Mean-Variance (Markowitz) Portfolio Optimization

We now examine the classic Markowitz mean-variance problem. In this case, the generic decision vector x simplifies to *only* the portfolio weights. To maintain notational consistency with the CVaR model, we continue to denote this vector as $w \in \mathbb{R}^m$. The objective is to minimize the portfolio variance, subject to the same constraint on the minimum expected return μ.

$$J^{\mathrm{Var_{cst}}} := \begin{cases} \min_{w \in \mathbb{R}^m} & \mathbb{V}\mathrm{ar}_{\xi \sim \mathbb{P}}\left[\langle w, \xi \rangle\right] \\ \text{subject to } \mathbb{E}_{\xi \sim \mathbb{P}}\left[\langle w, \xi \rangle\right] \geq \mu, \\ \qquad\qquad w \in \mathcal{W}. \end{cases} \tag{11}$$

For this problem, a tractable 2-WDRO reformulation is available [3]. The robust counterpart problem is:

$$\begin{cases} \min_{w \in \mathbb{R}^m} & \sup_{\mathbb{Q} \in \mathcal{B}_\epsilon(\widehat{\mathbb{P}}_N)} \mathbb{V}\mathrm{ar}_{\xi \sim \mathbb{Q}}\left[\langle w, \xi \rangle\right] \\ \text{subject to } \inf_{\mathbb{Q} \in \mathcal{B}_\epsilon(\widehat{\mathbb{P}}_N)} \mathbb{E}_{\xi \sim \mathbb{Q}}\left[\langle w, \xi \rangle\right] \geq \mu, \\ \qquad\qquad w \in \mathcal{W}. \end{cases} \tag{12}$$

As derived in [3], this problem is equivalent to the following formulation:

$$\begin{cases} \inf_{w \in \mathcal{W}} & \left(\sqrt{\langle w, \widehat{\Sigma}_N w \rangle} + \varepsilon \|w\|_2\right)^2 \\ \text{subject to } \langle \widehat{\mathbf{m}}_N, w \rangle - \varepsilon \|w\|_2 \geq \mu, \end{cases} \tag{13}$$

where $\widehat{\mathbf{m}}_N$ is the sample mean vector and $\widehat{\Sigma}_N$ is the (biased) sample covariance matrix, both computed from the sample $\widehat{\xi}_1, \ldots, \widehat{\xi}_N$.

Since the square root function is monotonic, minimizing the squared term in the objective is equivalent to minimizing its base. Therefore, the problem can be classified as an SOCP. This convex optimization problem is computationally tractable. This formulation provides a clear interpretation: the 2-WDRO Markowitz problem is equivalent to the SAA problem (which would only minimize $\langle w, \widehat{\Sigma}_N w \rangle$ subject to $\langle \widehat{\mathbf{m}}_N, w \rangle \geq \mu$) but with an added regularization term $\varepsilon \|w\|_2$. This term, added to the standard deviation in the objective and subtracted from the mean in the constraint, explicitly penalizes portfolio sensitivity and enforces robustness.

The key insight from this section is that both advanced robust portfolio models (WDRO-CVaR and WDRO-Markowitz) can be solved using standard SOCP solvers. The challenge now shifts from theoretical derivation to practical implementation. In the next section, we demonstrate how to build and solve these SOCPs directly in R.

3 Implementation in R with CVXR and MOSEK

The key insight from Sect. 2 is that our advanced WDRO portfolio models are equivalent to tractable Second-Order Cone Programs (SOCPs). This section demonstrates how to build and solve these models efficiently in R, bridging the gap between theory and practice.

3.1 The CVXR Modeling Framework

Our implementation relies on the CVXR package [7]. CVXR is an R package that provides an object-oriented modeling language for convex optimization, inspired by CVX and CVXPY. It is built on the principles of *Disciplined Convex Programming (DCP)*, a ruleset that allows it to verify the convexity of a problem.

The primary advantage of CVXR is that it allows the user to formulate an optimization problem using a natural, mathematical syntax that mirrors the equations on paper. CVXR automatically parses this problem, verifies its convexity, and translates it into the standardized conic form required by a backend solver. This abstracts away the complex and error-prone process of manually constructing the matrices for the solver, making it an ideal tool for rapid, reliable prototyping of complex models like our WDRO formulations.

3.2 Building the 2-WDRO-CVaR SOCP in CVXR

We will use the 2-WDRO-CVaR model (Eq. 10) as our primary example, as it is the most complex. Listing 1.1 shows the R snippet that builds and solves this problem, mapping directly to the mathematical components of the SOCP.

```r
library(CVXR)
library(Rmosek) # Required for the MOSEK solver
# --- 1. Define Problem Data ---
# Assume n (samples), m (assets) are defined
# Assume X_hat (n x m return matrix), eps, mu, alpha are
    defined
n <- nrow(X_hat)
m <- ncol(X_hat)
m_hat <- colMeans(X_hat) # Sample mean vector (m x 1)

# --- 2. Define Constants (from Lemma 1) ---
a1 <- -1.0 / alpha
b1 <- 1.0 - (1.0 / alpha)
a2 <- 0.0
b2 <- 1.0

# --- 3. Define CVXR Variables ---
w <- Variable(m)                    # Portfolio weights
tau <- Variable(1)                  # CVaR auxiliary variable
lambda <- Variable(1, nonneg = TRUE) # Dual variable for
    WDRO
s <- Variable(n)                    # Per-sample epigraph
    variable
z <- Variable(1, nonneg = TRUE) # SOCP auxiliary variable

# --- 4. Define Objective Function ---
# lambda * epsilon^2 + (1/n) * sum(s_i)
objective <- Minimize(lambda * eps^2 + (1/n) * sum(s))
```

```
27 # --- 5. Define Constraints ---
28 constraints <- list()
29 constraints[[1]] <- sum(w) == 1 # Fully invested
30 constraints[[2]] <- w >= 0       # Long-only
31 # Robust mean constraint (from Eq. \ref{
      StochsticProgWithExpectConstVariance})
32 constraints[[3]] <- t(m_hat) %*% w - eps * norm2(w) >= mu
33 # Rotated SOCP constraint (from Eq. \ref{
      StochsticProgWithExpectConstVariance})
34 soc_vector <- vstack(2 * w, lambda - z)
35 constraints[[4]] <- norm2(soc_vector) <= (lambda + z)
36 # Per-sample epigraph constraints (from Eq. \ref{
      StochsticProgWithExpectConstVariance})
37 s_i_lhs1 <- (a1^2 / 4) * z + (X_hat %*% w) * a1 + b1 * tau
38 s_i_lhs2 <- (a2^2 / 4) * z + (X_hat %*% w) * a2 + b2 * tau
39 constraints[[5]] <- s_i_lhs1 <= s
40 constraints[[6]] <- s_i_lhs2 <= s
41
42 # --- 6. Solve the Problem ---
43 problem <- Problem(objective, constraints)
44 solution <- solve(problem, solver = "MOSEK")
45 w_optimal <- solution$getValue(w)
```

Listing 1.1. R code for building the 2-WDRO-CVaR SOCP (Eq. 10) using CVXR.

The CVXR syntax allows for a natural translation of the model's components. For example, the L_2-norm $\|w\|_2$ is implemented as norm2(w). The main rotated cone constraint is built using vstack to create the vector $\begin{pmatrix} 2w \\ \lambda - z \end{pmatrix}$ and then applying norm2(). The per-sample constraints are *vectorized* using matrix multiplication X_hat %*% w for efficiency.

3.3   Adapting the Code for Other WDRO Models

The code in Listing 1.1 serves as a template. The other two SOCP formulations from Sect. 2 can be implemented by modifying this base.

Modification for 1-WDRO-CVaR. To implement the 1-WDRO-CVaR model (Eq. 9), we make two primary changes to Listing 1.1:

1. We remove the variable z and the rotated SOCP constraint (line 38).
2. The objective function (line 26) and the per-sample constraints (lines 41–44) are simplified, as shown in Listing 1.2.

```
1 # --- 4. Define Objective Function (1-WDRO) ---
2 # lambda * epsilon + (1/n) * sum(s_i)
3 objective <- Minimize(lambda * eps + (1/n) * sum(s))
4 # ... (Constraints 1, 2, 3 remain the same) ...
5 # ... (Constraint 4 is REMOVED) ...
```

```
6
7  # --- 5b. Per-sample and SOCP constraints (1-WDRO) ---
8  # Replaces lines 42-45
9  s_i_lhs1 <- (X_hat %*% w) * a1 + b1 * tau
10 s_i_lhs2 <- (X_hat %*% w) * a2 + b2 * tau
11 constraints[[4]] <- s_i_lhs1 <= s
12 constraints[[5]] <- s_i_lhs2 <= s
13
14 # Add the two simple SOCP constraints (from Eq. \ref{eqn:
       ReformulCVaRCasoP2})
15 constraints[[6]] <- abs(a1) * norm2(w) <= lambda
16 constraints[[7]] <- abs(a2) * norm2(w) <= lambda
```

Listing 1.2. Modifications for 1-WDRO-CVaR (replaces lines 27, 42-45 from Listing 2)

Modification for 2-WDRO-Markowitz. The 2-WDRO-Markowitz model (Eq. 13) is implemented using the same CVXR concepts but with a different set of variables. We replace tau, lambda, s, and z with two new non-negative variables, t and r. The core logic is shown in Listing 1.3.

```
1  # --- 1. Define Problem Data ---
2  S_hat <- cov(X_hat) * (n-1)/n # Biased covariance matrix
3  S_half <- base::chol(S_hat)   # Cholesky decomposition S_
       hat = t(S_half) %*% S_half
4  # --- 3. Define CVXR Variables ---
5  w <- Variable(m)
6  t <- Variable(1, nonneg = TRUE) # Epigraph for variance: t
       >= || S_half %*% w ||_2
7  r <- Variable(1, nonneg = TRUE) # Epigraph for norm:   r
       >= || w ||_2
8
9  # --- 4. Define Objective Function ---
10 # Equivalent objective: Minimize t + eps * r
11 objective <- Minimize(t + eps * r)
12
13 # --- 5. Define Constraints ---
14 constraints <- list()
15 constraints[[1]] <- sum(w) == 1 # Fully invested
16 constraints[[2]] <- w >= 0       # Long-only
17 # Robust mean constraint (from Eq. 14)
18 constraints[[3]] <- t(m_hat) %*% w - eps * r >= mu
19 # SOCP constraints for t and r (from Eq. 14)
20 constraints[[4]] <- norm2(t(S_half) %*% w) <= t
21 constraints[[5]] <- norm2(w) <= r
```

Listing 1.3. Core implementation for 2-WDRO-Markowitz (Eq. 14)

3.4   Solver Integration: Practical Setup for MOSEK

While CVXR supports open-source solvers, high-performance, commercial-grade solvers like **MOSEK** are essential for solving complex SOCPs reliably and at scale.

Installation and Configuration A successful integration requires two components: the MOSEK software suite and the R package Rmosek.

1. MOSEK **Software:** The platform-specific binaries and libraries must be downloaded from the MOSEK website and installed. This process includes placing a valid license file (a free academic license is available) in the correct directory.
2. Rmosek **R Package:** This package provides the R-to-MOSEK interface. It is crucial to ensure version compatibility. For example, CVXR versions 1.0 and later work well with Rmosek v9.x or v10.x. A mismatch in versions is a common source of installation failure.

This setup process, which often involves manually setting environment variables, can be non-trivial. However, we found this one-time investment to be critical for stable and fast performance.

Setting MOSEK as the Default By default, CVXR may select an open-source solver. To ensure MOSEK is always used, you can set it as the default solver for your R session by running the following command after loading the library:

```
1 # Set MOSEK as the default solver for all CVXR problems
2 CVXR::set_solver("MOSEK")
```

Alternatively, one can pass the `solver = "MOSEK"` argument in every `solve()` call, as shown in Listing 1.1. All experiments in the following section were conducted using this CVXR + MOSEK stack.

3.5   Implementation of Benchmark Strategies

To ensure a fair comparison and reproducibility, the benchmark strategies (SAA, MinCVaR, MinVar, MaxSR, the last three are described in the following section) were also implemented using the CVXR framework. For instance, the standard SAA Mean-Variance problem is solved using the same structure as Listing 1.3, but setting $\varepsilon = 0$ (removing the regularization terms). The Maximum Sharpe Ratio (MaxSR) portfolio was computed using the standard analytical solution $w \propto \widehat{\Sigma}^{-1}\widehat{\mu}$ for unconstrained problems, or via CVXR for constrained versions. This unified modeling approach minimizes implementation discrepancies between the robust and nominal models.

4   Numerical Experiments

To validate the proposed framework, we conduct a comprehensive numerical study using the R pipeline developed in Sect. 3. All optimization models were solved using the CVXR package with the MOSEK solver. The resulting data and performance metrics were processed and visualized using R, leveraging packages from the `tidyverse`, such as `ggplot2` for plotting.

Table 1. Asset universe selected from the S&P 500 index.

AAPL - Apple	INTC - Intel	PG - P&G
AMZN - Amazon	JNJ - Johnson & Johnson	T - AT&T
BAC - Bank of America	JPM - J.P Morgan	UNH -UnitedHealth Group
BRKA - Berkshire Hathaway	KO - Coca Cola	VZ - Verizon
CVX - Chevron	MA - Mastercard	WFC - Wells Fargo
DIS - Disney	MRK - Merck & Co	WMT - Walmart
HD - The Home Depot	XOM - Exxom Mobil	MSFT - Microsoft
PFE - Pfizer	GOOG - Alphabet Google	

4.1   Data and Experimental Setup

We test our strategies on real market data, specifically the daily returns of 23 companies selected from the S&P 500 index, listed in Table 1.

The data encompass the time window from January 1, 2008, to June 30, 2021. We analyze the cumulative wealth over time by employing a rolling horizon procedure with daily rebalancing. For example, we use data from January 1, 2008, to February 13, 2018, to estimate the portfolio for February 14, 2018. Subsequently, we use data from January 2, 2008, to February 14, 2018, to estimate the portfolio for February 15, 2018, and so on. This process continues by removing the earliest return and adding the next period's return until we reach the end of the dataset.

We compare our WDRO approaches against several standard portfolio optimization techniques: CVaR SAA, Var SAA, EW, MinCVaR, MinVar, and MaxSR. The SAA techniques involve solving the non-robust stochastic programs (6) and (11) using Sample Average Approximation. The other four benchmarks are:

- Equal Weight (EW): Assigns equal weight to all assets.
- Minimum CVaR (MinCVaR): Minimizes the empirical CVaR($\alpha = 0.05$) without a return constraint.
- Minimum Variance (MinVar): Minimizes the empirical variance $\langle w, \widehat{\Sigma}_N w \rangle$ without a return constraint.
- Maximum Sharpe Ratio (MaxSR): Maximizes the empirical Sharpe Ratio $\frac{\langle \widehat{\mathbf{m}}_N, w \rangle}{\langle w, \widehat{\Sigma}_N w \rangle}$.

For a given μ, we focus on analyzing the WDRO strategies. Specifically, for the CVaR model, we implement the **1-WDRO** reformulation (Eq. 9), which we refer to as "CVaR Wass". For the Markowitz model, we implement the **2-WDRO** reformulation (Eq. 13), referred to as "Var Wass". The choice of parameters reflects standard practice in portfolio management and robust optimization. The target return $\mu = 0.001$ represents a hypothetical investor's requirement; while arbitrary, it serves as a rigorous test for feasibility constraints.

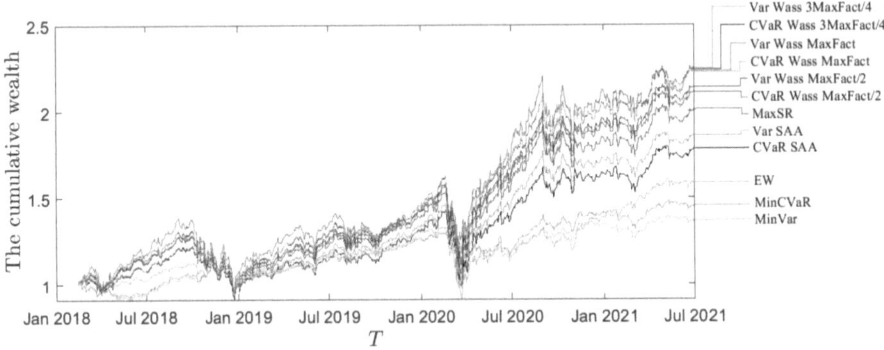

Fig. 1. The cumulative wealth of the trading strategies with $\mu = 0.001$.

Regarding the Wasserstein radius ε, its selection is critical. Rather than relying on asymptotic bounds which can be overly conservative [4], we adopt a data-driven heuristic approach often used in practice. We define $\widehat{\varepsilon}_N^{\max}(\mu)$ as the maximum radius that keeps the problem feasible for a given μ. We then explore a sensitivity range defined by fractions of this feasibility limit ($\varepsilon \in \{1.0, 0.75, 0.5\} \times \widehat{\varepsilon}_N^{\max}(\mu)$). We append "MaxFact", "3MaxFact/4", and "Max-Fact/2" to the strategy names to denote the radius used, respectively. This approach avoids the need for computationally expensive cross-validation while ensuring the ambiguity set is large enough to provide robustness but small enough to maintain financial meaning.

4.2 Performance Analysis and Results

Figure 1 displays the cumulative wealth generated by the portfolios from the described strategies. The WDRO approach leads to the highest cumulative wealth, surpassing all traditional strategies. Notably, the evaluation period includes the onset of the COVID-19 pandemic, which impacted all investment strategies. Nevertheless, our proposed WDRO strategies appear to mitigate the long-term negative effect on portfolio value.

Table 2 presents various out-of-sample performance indicators. Notably, the mean return of all WDRO strategies surpasses the target $\mu = 0.001$. This is a significant finding, as none of the other strategies, including their SAA counterparts, achieve this. While SAA is expected to converge to the true optimal solution with enough data, this data period contains structural shifts (like the pandemic), suggesting the underlying distribution is not stationary. This implies that the SAA strategies fail to meet the out-of-sample return target. In contrast, the WDRO strategies successfully overcome this challenge.

From an economic perspective, the superior performance of WDRO can be attributed to its implicit regularization. As shown in the reformulations, the Wasserstein constraint acts as a penalty on the norm of the weights (for Markowitz) or the Lipschitz constant (for CVaR). This prevents the optimizer

Table 2. Performances of different portfolio strategies.

	Mean	Standard deviation	Sharpe Ratio	Turnover	Avg. Portfolio Assets	$\mathrm{CVaR}_{\alpha=0.05}$
CVaR Wass MaxFact	0.0010932	0.01719	0.063594	0.045383	4.3176	0.040446
CVaR Wass 3MaxFact/4	0.0010825	0.01599	0.067698	0.035326	6.2682	0.037618
CVaR Wass MaxFact/2	0.0010047	0.015693	0.064023	0.040298	7.3953	0.036657
Var Wass MaxFact	0.0010947	0.01719	0.063679	0.045387	4.2953	0.040437
Var Wass 3MaxFact/4	0.0010793	0.015938	0.06772	0.033119	5.9965	0.037457
Var Wass MaxFact/2	0.0010202	0.01564	0.065232	0.032776	7.2988	0.036351
CVaR SAA	0.00080191	0.0153	0.052413	0.065448	6.8388	0.03512
Var SAA	0.00085056	0.01522	0.055883	0.040939	8.0188	0.034432
MinCVaR	0.00051007	0.011487	0.044403	0.015437	13.711	0.026447
MinVar	0.00043373	0.011263	0.03851	0.0096731	11.331	0.026286
MaxSR	0.0009563	0.016398	0.058319	0.026923	5.3859	0.038047
EW	0.00063778	0.01362	0.046828	0.97409	23	0.031383

from "chasing" noise in the data, leading to more diversified portfolios (lower concentration) as observed in the 'Avg. Portfolio Assets' column. While this robustness comes at the cost of slightly higher tail risk (higher out-of-sample CVaR) compared to the most conservative benchmarks (MinVar), it achieves the primary goal: satisfying the return constraint μ where SAA fails, offering a more reliable trade-off for risk-averse investors.

Another crucial indicator is standard deviation. Here, the WDRO approaches are among the highest. However, the difference relative to the SAA approach is not particularly significant, especially for the MaxFact/2 approaches. Furthermore, the Sharpe Ratios of the WDRO strategies are the highest; all exceed 0.063, while the Sharpe Ratios of the other strategies do not surpass 0.059. This suggests the risk-return trade-off is more favorable for the WDRO strategies.

Regarding other indicators, turnover (a metric for trading costs) increases with ε, though in some instances, it remains lower than that of SAA. For CVaR, all WDRO strategies yield lower turnover than their SAA counterpart. For variance, the 3MaxFact/4 and MaxFact/2 WDRO strategies provide lower turnover than their SAA version.

Table 2 also presents the average number of assets in the portfolio. This metric shows that portfolio concentration (fewer assets) increases as the robustness

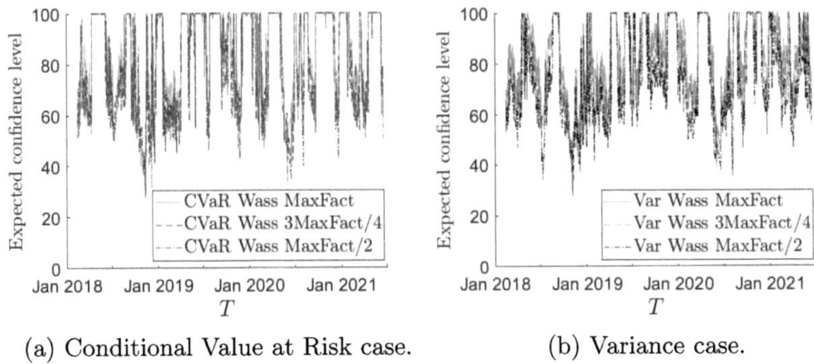

(a) Conditional Value at Risk case.   (b) Variance case.

Fig. 2. Daily Expected Confidence level for $\mu = 0.001$.

radius ε increases. For variance-focused WDRO strategies, this value is lower than that of the SAA version. The same holds true for WDRO-CVaR, except for the MaxFact/2 case. The more conservative benchmark strategies (MinVar, MinCVaR) seek greater diversification, resulting in a higher average number of assets.

The final indicator is the out-of-sample CVaR at $\alpha = 0.05$. The WDRO strategies exhibit the highest (worst) CVaR values, with the exception of MaxSR. However, the difference relative to SAA is not significant. This observation is justifiable: the WDRO strategies deliver a higher average return while maintaining a favorable balance with respect to variance (Sharpe Ratio), and this performance may come at the cost of higher tail risk.

Another aspect warranting discussion is the concept of expected confidence level, which can be interpreted as a lower estimate of the probability that the return constraint will be satisfied out-of-sample. Figure 2 displays the daily expected confidence level induced by the WDRO strategies, which remains above 40% almost always.

When examining the averages, the expected confidence levels for the CVaR Wass strategies were 84.9% (MaxFact), 83.3% (3MaxFac/4), and 81.2% (MaxFact/2). For the Var Wass strategies, the averages were 83.8%, 80.3%, and 75.3%, respectively. This theoretical estimate is reflected in reality: as seen in Table 2, all WDRO strategies exceeded the minimum required expected return of $\mu = 0.001$ on average, confirming that the expected confidence level provides useful insight into out-of-sample performance.

Finally, we consider $\widehat{\mu}_N^{\max}$, which represents the maximum μ level that can be targeted for the robust problem to remain feasible. Figure 3 presents the daily $\widehat{\mu}_N^{\max}$ values, suggesting that the value of μ could have been increased from our test value of 0.001.

However, elevating μ also impacts the set of feasible ε values and, in turn, affects the expected confidence level. Reducing the proximity between μ and $\widehat{\mu}_N^{\max}$ implies that the feasible ε values will have a lower expected confidence level.

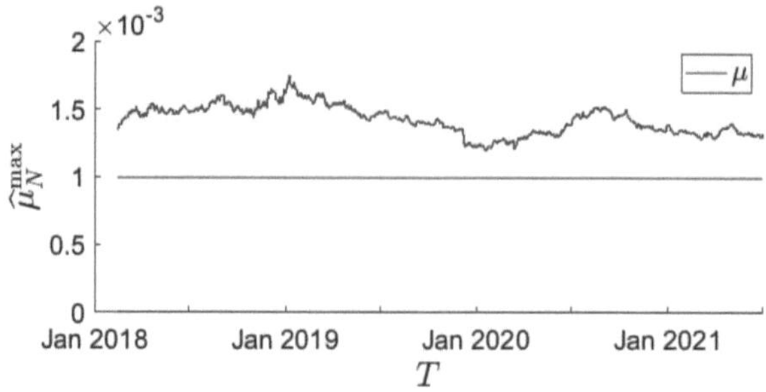

Fig. 3. Daily maximum expected return level that guarantees feasibility. In this case, $\mu = 0.001$.

Consequently, it becomes more challenging to ensure that the portfolios satisfy the constraint with high probability. Thus, $\widehat{\mu}_N^{\max}$ and the expected confidence level can be considered practical tools that allow investors in R to establish a trade-off between the minimum expected return (μ) and the feasibility of the robust strategies before making decisions.

5   Conclusion

This paper addressed the practical implementation of modern robust portfolio optimization techniques within the R ecosystem. We began by revisiting the Distributionally Robust Optimization (DRO) framework using the Wasserstein metric, which provides a principled method for creating data-driven ambiguity sets to hedge against estimation error. We confirmed that for both classic mean-variance (Markowitz) and modern mean-risk (CVaR) models, the resulting WDRO problems admit equivalent and tractable reformulations as Second-Order Cone Programs (SOCPs).

The primary contribution of this work is a practical demonstration that these advanced models are not merely theoretical. We provided a complete, reproducible toolkit in R, leveraging the CVXR package for its intuitive, mathematical syntax and the high-performance MOSEK solver. We detailed the code structure for these models and highlighted practical considerations for adapting the code and successfully integrating the CVXR + MOSEK stack.

Our numerical experiments on S&P 500 data validated this pipeline. The WDRO strategies, implemented in R, consistently outperformed their standard Sample Average Approximation (SAA) counterparts in a volatile, real-world setting. Most notably, the robust models successfully met the minimum return constraint out-of-sample, a key objective that the SAA models failed to achieve. This work provides a practical and validated toolkit for R users to move from fragile SAA models to more reliable robust optimization, achieving a superior trade-off between performance and reliability.

Finally, we acknowledge certain limitations. First, the reliance on SOCP formulations, while tractable, can become computationally intensive for very large-scale portfolios (thousands of assets) compared to simple quadratic programming. Second, our study focuses on the Wasserstein distance; other ambiguity sets (e.g., Kullback-Leibler) might offer different insights. Future work could explore the integration of multi-period WDRO models in R and the application of this pipeline to non-convex risk measures using appropriate convex approximations.

Data Availibility Statement. The complete R code for data preparation (cleaning, return calculation, rolling-window setup), along with the optimization scripts used in this study, is available in the following public repository: https://anonymous.4open.science/r/WDRO-Portfolio-R-AF31/README.md.

References

1. Artzner, P., Delbaen, F., Eber, J.M., Heath, D.: Coherent measures of risk. Math. Financ. **9**(3), 203–228 (1999)
2. Bertsimas, D., Gupta, V., Kallus, N.: Robust sample average approximation. Math. Program. **171**(1–2), 217–282 (2018)
3. Blanchet, J., L., C., Zhou, X.Y.: Distributionally robust mean-variance portfolio selection with Wasserstein distances. Manag. Sci. **68**(9), 6382–6410 (2022)
4. Blanchet, J., Murthy, K.: Quantifying distributional model risk via optimal transport. Math. Oper. Res. **44**(2), 565–600 (2019)
5. Delage, E., Ye, Y.: Distributionally robust optimizatión under moment uncertainty with application to data-driven problems. Oper. Res. **58**(3), 595–612 (2010)
6. Dentcheva, D., Ruszczyńsk, A.: Optimization with stochastic dominance constraints. SIAM J. Opti. **14**(2), 548–566 (2003)
7. Fu, A., Narasimhan, B., Boyd, S.: CVXR: An R package for disciplined convex optimization. J. Stat. Softw. **94**(14), 1–34 (2020)
8. Gao, R., Kleywegt, A.: Distributionally robust stochastic optimization with Wasserstein distance. Math. Oper. Res. **48**(2), 603–655 (2022)
9. Jiang, R., Guan, Y.: Data-driven chance constrained stochastic program. Math. Program. **158**, 291–327 (2016)
10. Lan, G.: First-order and Stochastic Optimization Methods for Machine Learning. Springer Series in the Data Sciences (2020)
11. Miller, B.L., Wagner, H.M.: Chance constrained programming with joint constraints. Oper. Res. **13**(6), 930–945 (1965)
12. Esfahani, M.P., Kuhn, D.: Data-driven distributionally robust optimization using the Wasserstein metric: performance guarantees and tractable reformulations. Math. Program. **171**(1), 115–166 (2018)
13. Rigollet, P., Tong, X.: Neyman-Pearson classification, convexity and stochastic constraints. J. Mach. Learn. Res. **12**(3), 2831–2855 (2011)
14. Rockafellar, R., Uryasev, S.: Optimization of conditional value-at-risk. J. Risk **2**, 21–42 (2000)
15. Ruszczynski, A., Shapiro, A.: Stochastic programming (handbooks in operations research and management science) (2003)

16. Shapiro, A.: Monte Carlo sampling methods. In: Stochastic Programming, Handbooks in Operations Research and Management Science, vol. 10, pp. 353–425. Elsevier (2003)
17. Smith, J.E., Winkler, R.L.: The optimizer's curse: Skepticism and postdecision surprise in decision analysis. Manage. Sci. **52**(3), 311–322 (2006)
18. Ziemba, W.T., Vickson, R.G.: Stochastic Optimization Models in Finance. WORLD SCIENTIFIC, 2006 edn. (2006)

Quality Control in Beverage Production – Cap Defects Analysis Using an Integration of R-Studio in Power BI

Espitia Rodriguez Arcesio[⊠] [ID] and Carmen E. Patiño-Rodríguez

University of Antioquia, 050010 Medellín, Colombia
`{arcesio.espitia,elena.patino}@udea.edu.co`

Abstract. This case study, developed in the Standardization and Quality Control 2025-1 course, analyzed the capping and sealing process in beverage production under Quality 4.0. Using SIPOC, key variables were defined: defective caps (discrete) and closing turns (continuous). Data collection stage enabled BI control charts, p-charts, and Gage R&R by attributes with risk analysis method for assess measurement reliability. From 1,978 caps across 56 batches with 3 operators, high defect rates linked to raw material variability, poor inspection, and unstandardized procedures were found. Corrective actions included stricter inspections, supplier renegotiation, and process standardization. Integrating R-Studio with Power BI enabled real-time dashboards for defect monitoring, supporting data-driven decisions. This approach demonstrates how statistical methods and BI tools enhance quality control, offering potential for predictive analytics, anomaly detection, and broader industrial applications.

Keywords: Quality 4.0 · Statistical Process Control · Digitalization · Control Charts · Power BI · R-Studio

1 Introduction

Global manufacturing industries are increasingly adopting the principles of Quality 4.0, which emphasize digital integration, advanced analytics, and real-time monitoring. Traditional statistical tools, such as Statistical Process Control (SPC), remain indispensable but require alignment with modern Business Intelligence (BI) solutions to address the current complexity of industrial operations.

The beverage industry is particularly sensitive to packaging and sealing processes, where cap defects can compromise product safety, consumer trust, and regulatory compliance. While control charts, Gage R&R, and measurement system analysis have historically been applied, their integration into real-time BI systems remains underexplored.

This study advances the field of Quality 4.0 by demonstrating a reproducible, statistically rigorous integration of R-based quality control methods within an enterprise business intelligence platform (Power BI). Unlike traditional SPC workflows that are

B. M. Suárez et al. (Eds.): R Day 2025, CCIS 2824, pp. 223–236, 2026.
https://doi.org/10.1007/978-3-032-18455-9_12

executed exclusively in statistical software or custom dashboards, the proposed framework embeds R's classical quality control tools, such as p-charts, operating characteristic curves, and attribute-based R&R, in a real-time, industry-standard BI environment. This fusion enables organizations to adopt advanced statistical techniques without migrating to R-native ecosystems such as Shiny.

From a scientific standpoint, the contribution lies in showing how R's computational capabilities can be operationalized as an analytical engine inside Power BI, creating a hybrid architecture that supports real-time SPC monitoring, reproducible analyses, and traceable statistical computation. The work also formalizes the theoretical underpinnings of the analyses used, linking Statistical Process Control, Measurement System Analysis, and reliability modeling, with the technological enablers of Quality 4.0 (connectivity, interoperability, and digital integration).

This positions the study not merely as an applied case, but as a methodological template for how statistical computing languages can be embedded inside BI ecosystems to expand analytical depth while maintaining industrial usability.

2   Bibliometric Analysis

In this Section, we provide a bibliometric analysis of the reviewed papers. To structure the bibliographic search in Scopus, the query was divided into three main criteria blocks that reflect the conceptual, technological, and industrial dimensions of the research. The first block addresses the classical statistical quality methods that form the theoretical backbone of quality control. The second block emphasizes the integration of business intelligence and digital tools, particularly Power BI and R-Studio, as enablers of Quality 4.0. Finally, the third block narrows the focus to the industrial application context, with special emphasis on the beverage and packaging sector where capping and sealing defects occur. Table 1 summarizes the query strings and scope of each block.

To better understand the scientific landscape surrounding Quality 4.0, SPC, and digital integration (Power BI, R, machine learning), a bibliometric analysis was conducted using VOSviewer. The analysis focused on keyword co-occurrence extracted from Scopus-indexed publications between 2000 and 2024 (Fig. 1).

The bibliometric evidence reinforces that SPC remains the backbone of quality management research; there is a clear trend toward AI, predictive models, and real-time monitoring, aligning with Quality 4.0 principles; the integration of BI tools (Power BI + R-Studio) into manufacturing quality remains underexplored, validating the novelty of this study.

This bibliometric analysis was included to establish the conceptual landscape that frames this study, specifically highlighting the growing convergence between classical SPC, digital quality systems, and analytics platforms. By mapping the evolution of Quality 4.0 literature, the section contextualizes the relevance of integrating R into BI environments and clarifies the research gap that motivates this work.

Table 1. Query string for bibliography search

Criteria block	Query string	Scope of search
Quality & Statistical Methods	"Quality 4.0" OR "digital quality management" OR "statistical process control" OR "SPC" OR "Gage R&R" OR "measurement system analysis" OR "Cohen's Kappa" OR "control charts"	Captures literature related to quality management paradigms, SPC techniques, and measurement system analysis as classical foundations of quality control
Business Intelligence & Digital Tools	"Power BI" OR "business intelligence" OR "dashboard" OR "real-time monitoring" OR "data visualization" OR "predictive quality" OR "machine learning" OR "R-Studio" OR "R language"	Focuses on modern BI and digital analytics tools (Power BI, R, dashboards, ML) enabling Quality 4.0 integration and real-time process monitoring
Industrial Application Context	"beverage industry" OR "food and beverage" OR "manufacturing process" OR "packaging" OR "sealing process" OR "capping" OR "quality defects" OR "supply chain quality"	Narrows down to industrial applications, especially the beverage/food sector, packaging, capping/sealing defects, and supply chain quality issues

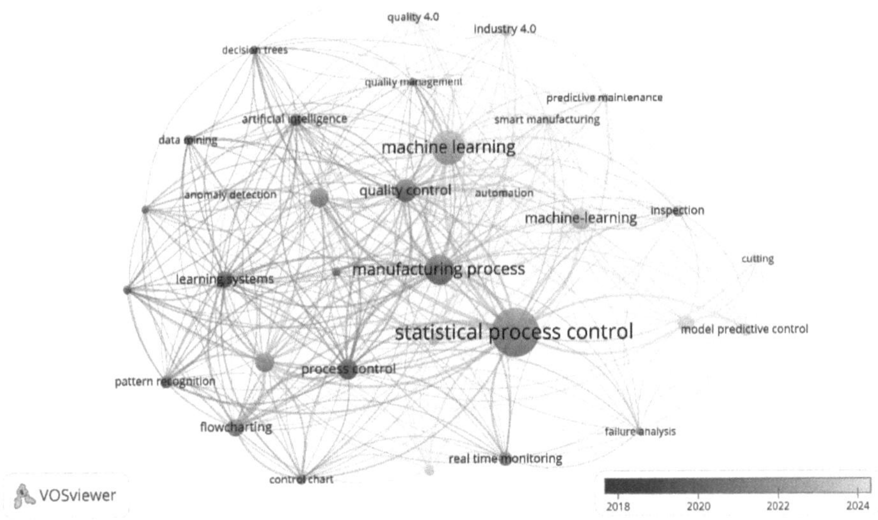

Fig. 1. Bibliography analysis

3 DMAIC Methodology and Data Validation

The statistical procedures implemented in this study build directly on foundational SPC theory from Montgomery (2019) and Wheeler (2010), ensuring that the analytical workflow adheres to established best practices. The p-chart methodology follows the binomial-based control structure for attribute data, using subgroup sizes and defect counts to estimate process stability and inherent capability. The OC curve computation relies on the theoretical relationship between sample size, process proportion nonconforming, and the probability of detecting assignable causes.

Measurement System Analysis follows ISO 22514-7 and AIAG MSA guidelines, using inter-observer agreement (Cohen's Kappa) as the inferential measure of attribute repeatability and reproducibility. The computational execution of these procedures is handled entirely through R, ensuring that the statistical calculations follow open-source standards rather than proprietary BI algorithms.

The project was developed following the DMAIC methodology (Define, Measure, Analyze, Improve, Control) according to the following guidelines: (Park, 2003):

- **Define**. Identification of the process or product that needs improvement.
- **Measure**. Identify those characteristics of the product or process that are critical to the customer's requirements for quality performance, and which contribute to customer satisfaction.
- **Analyze**. Evaluate the current operation of the process to determine the potential sources of variation for critical performance parameters.
- **Improve**. Select those product or process characteristics which must be improved to achieve the goal. Implement improvements.
- **Control**. Ensure that the new process conditions are documented and monitored via statistical process control methods (SPC).

Depending on the outcome it may become necessary to revisit one or more of the preceding phases.

Together, these foundations situate the study within classical SPC and MSA theory while leveraging modern data workflows enabled by Quality 4.0 technologies.

3.1 Define and Measure Phase: Variables and Baseline

Beverage production is divided into four major macro-processes, which are indicated simultaneously in Fig. 2.

Fig. 2. Beverage production macroprocess

The Definition and Measurement phase began with a SIPOC diagnosis, which describes the process carried out in the covering and sealing of the subsequent identification of the quality variables of interest (Fig. 3).

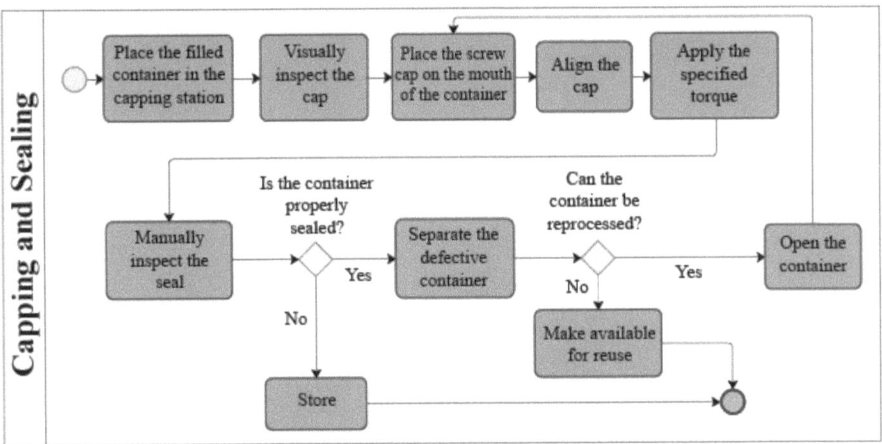

Fig. 3. Capping and sealing process

- **Discrete variable (Attributes):** Cap not properly fitted (Yes/No). This variable affects the Reliability quality dimension.
- **Continuous variable:** Closing torque (Nm) or Number of turns to close (Continuous).

The analysis focuses on the capping and sealing process, illustrated in the flowchart, which details the sequence from placing the filled container to verifying the sealing quality and managing defective units. Two critical variables are monitored to ensure process reliability. The discrete variable, Cap not properly fitted (Yes/No), reflects whether the sealing meets the required reliability standard, serving as an attribute-based indicator of product conformity. Meanwhile, the continuous variable, closing torque (Nm) or Number of turns to close, provides quantitative control over the precision of the capping operation. Together, these variables enable a comprehensive quality evaluation of the process, linking operational consistency with reliability outcomes as shown in the procedural flow.

To establish the baseline and monitor the proportion of defective units, a set of exploratory visualizations was developed to describe the initial defect patterns observed during the preview stage. These graphical tools (boxplot, histogram, and dotplot) were generated using R within the Power BI environment to visualize the dispersion, central tendency, and distribution of cap defect data across the first 16 batches. Together with the descriptive data table, these visualizations provide an initial diagnostic view of process stability, operator consistency, and potential improvement areas before implementing more advanced statistical controls (Fig. 4).

In the first stage of data collection, where an average of 36.13 caps were analyzed for each of the 16 batches, a moderately consistent process with some variability in defect proportions was revealed, centered around a median of approximately 0.32. The histogram indicates a unimodal, slightly right-skewed distribution, suggesting sporadic increases in defects tied to material or operational inconsistencies. Early batches show higher defect ratios (up to 0.43), while later ones exhibit improvement, potentially due to operator adaptation or initial process adjustments.

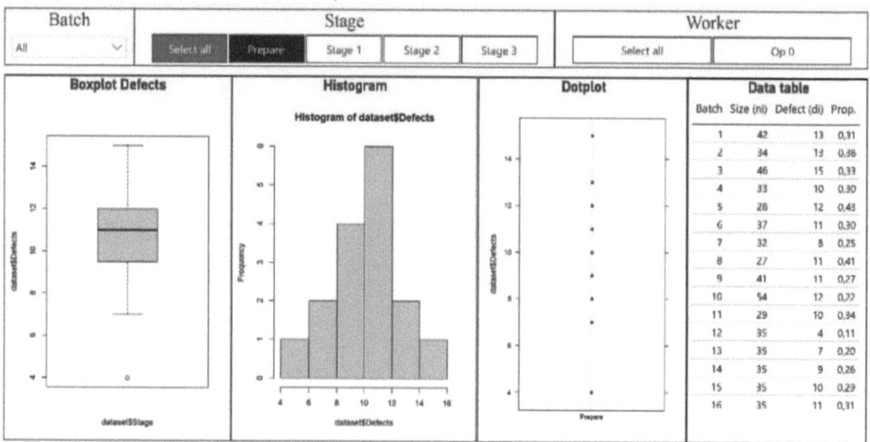

Fig. 4. Distribution chart for the proportion of defective caps in the preview stage

Overall, the process displays emerging stability with indications of learning and quality enhancement, warranting continued monitoring through control charts to confirm sustained improvement.

3.2 Validation of the Measurement System (R&R by Attributes)

The rapid advancement of technology has left a knowledge gap for manufacturing and process industries that do not have state-of-the-art technologies or that, by their nature, do not produce the quantity and variety of data needed to apply models that require high volumes of data.

Validation of the inspection system was crucial to ensure that decisions (accept/reject) were accurate, reducing the subjectivity associated with attribute-based evaluation. A Gage R&R analysis by attributes was performed using the risk analysis method, supported by contingency tables and Cohen's Kappa index (Fig. 5).

Fig. 5. Risk analysis method

The study included two operators, evaluating 45 parts with five measurements (or trials), resulting in a total of 225. Cohen's Kappa index is a measure of agreement that compares the observed agreement with that which could occur by chance. In this case, the results of the R&R analysis were summarized in a contingency table shown in Table 2, the observed agreement was Po = 0.947, and the Kappa agreement was kappa = 0.787.

Table 2. Contingency table

Contingency table		Observer 2		Total
		Satisfied	Not Satisfied	
Observer 1	Satisfied	186	8	194
	Not Satisfied	4	27	31
Total		35	190	225

4   Analysis and Results

The results of the monitoring stage carried out by the two operators in charge provided a preliminary P-chart based on the 21 samples inspected, which did not identify any patterns of special causes (out-of-control points or trends). However, the control chart for the proportion of defective caps indicated a high average percentage of defective caps. This implies that, although the process is stable, it is inherently incapable (consistently poor) due to the variability of the raw material.

At the same phase, the main assignable causes of variation were identified, including shortages in raw material supply, inadequate inspection processes—which justified the application of the R&R study—and the presence of unstandardized procedures.

Data Analytics (DA) is identified as a key enabler and fundamental success factor for the transformation of manufacturing processes, especially in the context of Quality 4.0, for this one in the next session are described the steps to control and stabilize the procedure of described.

5   Improvement, Control, Quality 4.0

In order to improve the performance of the capping process, various interventions were implemented to reduce process variability. These included, but were not limited to, increased inspection procedures prior to sealing, renegotiation with suppliers to ensure higher quality materials, as well as the adoption of waste practices and standardization of procedures to reduce variability.

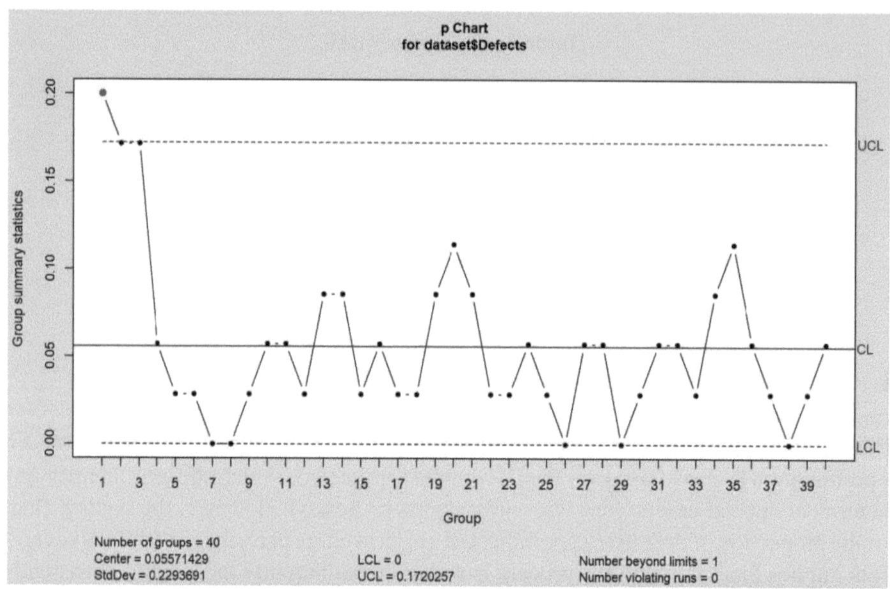

Fig. 6. P-chart proportion of defective caps monitoring stage

During the monitoring stage, 40 groups were sampled with an average batch size of 35 units. The process center is located at 0.055 (5.5%), with significant variability ($\sigma =$ 0.22). One point is observed outside the control limits (group 1, in red), indicating the presence of a special cause of variation. In addition, there is a low number of cases in the proportion of defective items in the groups, which suggests that an increase in the size of these monitoring groups is necessary.

Integration and Continuous Monitoring (Control)
The control phase focused on implementing a continuous monitoring system. To do this, R-Studio and Power BI were integrated under the Quality 4.0 paradigms. The integration of these two tools made it possible to:

1. Dynamic visual representation of process metrics and defect trends.
2. A real-time monitoring dashboard, accessible from multiple devices (mobile-friendly).
3. Facilitate faster, more informed decision-making by combining business intelligence with interconnectivity.

This application aligns with the concept that computer systems specialized in statistics facilitate descriptive analysis and the application of methods. Framework illustrating the integration of classical quality tools such as Statistical Process Control, operations curves and so on, with modern digital platforms (R-Studio & Power BI) for real-time monitoring (Fig. 7).

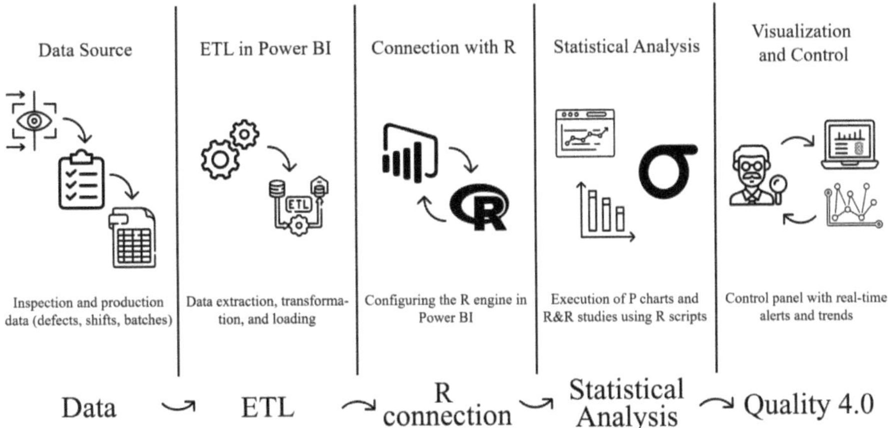

Fig. 7. Power BI–RStudio integration flow for statistical monitoring

Figure 7 illustrates the sequential workflow used for the data integration and statistical control process in the beverage capping study. It begins with the data acquisition and preparation phase, where inspection records are cleaned and structured in Power BI through ETL (Extract, Transform, Load) operations. The process continues with the connection and configuration of R within Power BI, enabling the execution of R scripts for statistical computation. Subsequently, the control chart generation and defect analysis are performed using R code directly embedded in Power BI visuals, allowing dynamic interaction with the underlying data. Finally, results are visualized in real-time through dashboards, enabling ongoing process monitoring and facilitating rapid decision-making.

Data Source and ETL

In Power BI, a data source is the entity from which data is extracted for creating reports and visualizations. It can be a file, webpage, database, or another app/service. The ETL process refers to a series of steps to extract data from numerous sources, transform that data according to business rules or analytical needs, and finally load it into a system where it can be used for analysis and decision-making.

For this case study, we used a flat Excel file containing information about the stage, workers, group labels, and defects, i.e., the number of parts inspected, and the total number of defective parts found in that specific group. During the transformation process, the columns of the file used were configured and a new column was added to calculate the proportion of defective parts in each group. Finally, the final data is loaded to create the visualizations in Power BI.

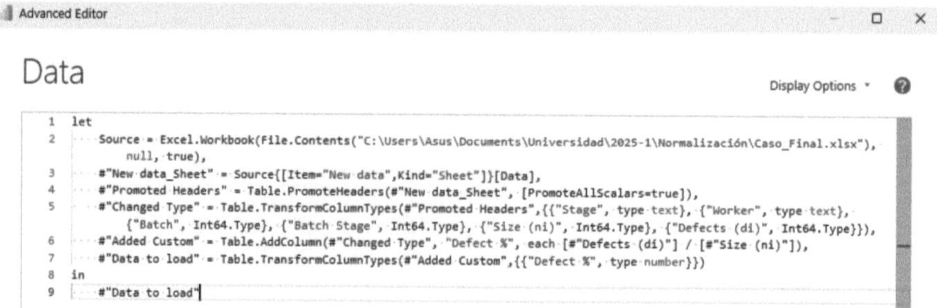


```
1  let
2      Source = Excel.Workbook(File.Contents("C:\Users\Asus\Documents\Universidad\2025-1\Normalización\Caso_Final.xlsx"),
           null, true),
3      #"New data_Sheet" = Source{[Item="New data",Kind="Sheet"]}[Data],
4      #"Promoted Headers" = Table.PromoteHeaders(#"New data_Sheet", [PromoteAllScalars=true]),
5      #"Changed Type" = Table.TransformColumnTypes(#"Promoted Headers",{{"Stage", type text}, {"Worker", type text},
           {"Batch", Int64.Type}, {"Batch Stage", Int64.Type}, {"Size (ni)", Int64.Type}, {"Defects (di)", Int64.Type}}),
6      #"Added Custom" = Table.AddColumn(#"Changed Type", "Defect %", each [#"Defects (di)"] / [#"Size (ni)"]),
7      #"Data to load" = Table.TransformColumnTypes(#"Added Custom",{{"Defect %", type number}})
8  in
9      #"Data to load"
```

Fig. 8. Data extraction, transformation and loading script in Power Query (Power BI)

Figure 8 shows the Power BI Advanced Editor containing an M language script used to perform the ETL (Extract, Transform, Load) process. In this code, data is imported from an Excel file, column headers are promoted, data types are assigned, and a calculated column named "Defect %" is created by dividing the number of defects by the sample size. This script ensures that the dataset is properly structured and ready for analysis within Power BI. The first rows of the data table loaded into the model in Power BI are shown in Fig. 9. The data includes key variables such as stage, worker, batch, sample size (ni), defects (di), and defect percentage (%).

Stage	Worker	Batch	Batch Stage	Size (ni)	Defects (di)	Defect %
Prepare	Op 0	1	1	42	13	0,30952380952381
Prepare	Op 0	2	2	34	13	0,382352941176471
Prepare	Op 0	3	3	46	15	0,326086956521739
Prepare	Op 0	4	4	33	10	0,303030303030303
Prepare	Op 0	5	5	28	12	0,428571428571429
Prepare	Op 0	6	6	37	11	0,297297297297297
Prepare	Op 0	7	7	32	8	0,25
Prepare	Op 0	8	8	27	11	0,407407407407407
Prepare	Op 0	9	9	41	11	0,268292682926829
Prepare	Op 0	10	10	54	12	0,222222222222222
Prepare	Op 0	11	11	29	10	0,344827586206897
Prepare	Op 0	12	12	35	4	0,114285714285714
Prepare	Op 0	13	13	35	7	0,2
Prepare	Op 0	14	14	35	9	0,257142857142857
Prepare	Op 0	15	15	35	10	0,285714285714286
Prepare	Op 0	16	16	35	11	0,314285714285714
Stage 1	Op 1	17	1	35	7	0,2
Stage 1	Op 1	18	2	35	6	0,171428571428571
Stage 1	Op 1	19	3	35	6	0,171428571428571
Stage 1	Op 1	20	4	35	2	0,0571428571428571

Fig. 9. Structured dataset loaded and processed in Power BI

Connection with R

The following configuration window represents the initial setup of the R integration within Power BI, a fundamental step that enables the execution of statistical analyses and the creation of advanced visualizations directly within the BI environment. Through this interface, the R home directory and the preferred R Integrated Development Environment

(IDE)—in this case, RStudio—are defined, ensuring that Power BI correctly identifies the installed R engine (Fig. 10).

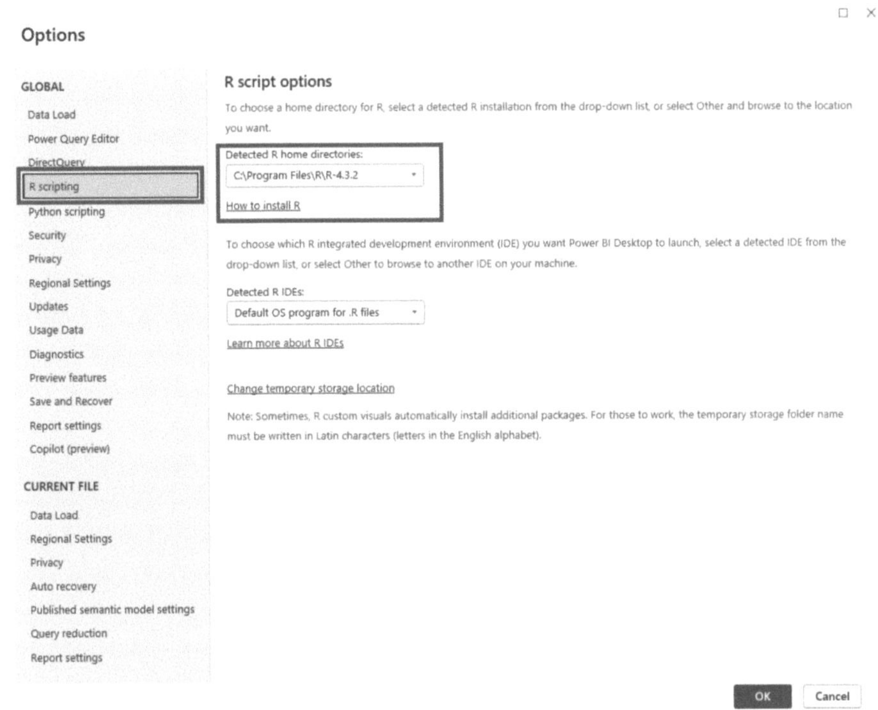

Fig. 10. Configuration of R scripting environment in Power BI

This integration enables the creation of data-driven control charts and quality diagnostics directly within Power BI, enhancing analytical depth, reproducibility, and the visualization of complex statistical models in a clear, dynamic, and research-oriented environment.

Statistical Analysis

Once the data has been loaded into Power BI and the integration with RStudio has been completed, the statistical analysis process begins executing R scripts for statistical computation.

OC Curves Script. Creating statistical graphs in Power BI using R integration. The red boxes highlight key components: on the right, the "R script visual" icon is selected from the visualization pane, and relevant variables (*Batch*, *Size*, and *Defects*) are assigned to the Values section to feed the R script. At the bottom, the R script editor contains the code that loads the qcc library, defines a p-type control chart, and generates the Operating Characteristic (OC) curve using the oc.curves() function. The resulting plot, shown at the top, represents the probability of detecting process variations for a sample size of

35, integrating R's analytical power directly into Power BI's visualization environment. This group size parameter can be made dynamic using DAX language (Fig. 11).

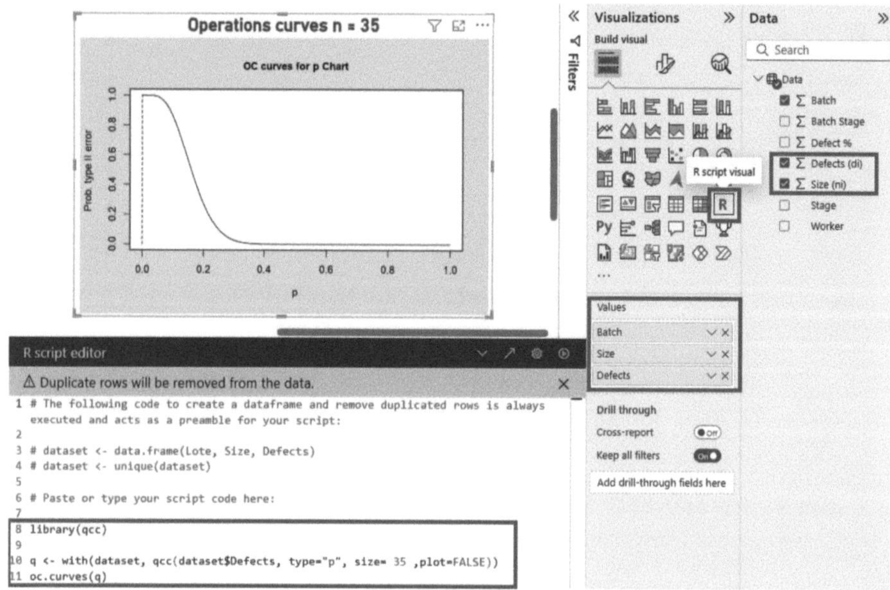

Fig. 11. Generation of OC Curves in Power BI Using R Script

P-Chart Script. The figure shows a screenshot of the R script integrated into Power BI, used to generate a p-type control chart using the qcc library. This chart allows you to monitor the proportion of defective units in the process, taking the number of defects and sample size as variables. Its implementation within Power BI facilitates real-time statistical quality analysis, integrating visualization and process control directly into management dashboards (Fig. 12).

```
R script editor                                    ∨  ↗  ⚙  ⊙

 1  # The following code to create a dataframe and remove duplicated rows is
    always executed and acts as a preamble for your script:

 2

 3  # dataset <- data.frame(Defects, Size)

 4  # dataset <- unique(dataset)

 5

 6  # Paste or type your script code here:

 7

 8  library(qcc)

 9

10  qcc(dataset$Defects, type="p", dataset$Size)
```

Fig. 12. Configuration of R script for p-chart generation in Power BI

Visualization and Control Dashboard

Figure 13 illustrates how statistical monitoring and BI dashboards enable you to obtain predictive information and make decisions more quickly, combining the capabilities of Power BI with the statistical tools of R-Studio.

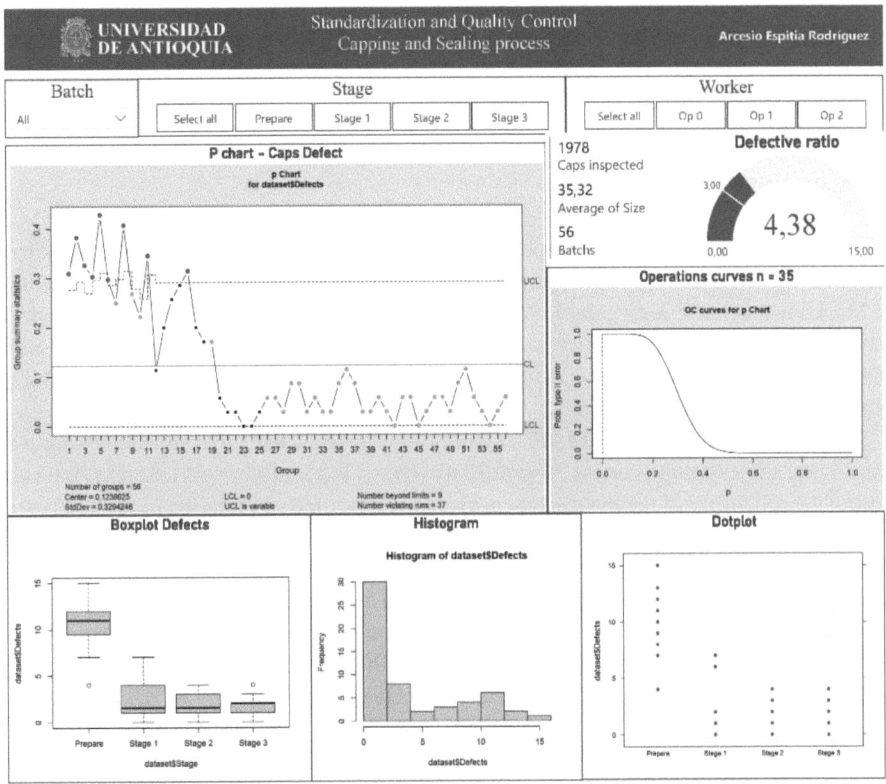

Fig. 13. Power BI dashboard for standardization and quality control in the capping and sealing process

This dashboard integrates Power BI with R scripting to visualize and analyze the capping and sealing process through control charts, descriptive statistics, and defect distribution graphs. It provides a comprehensive overview of process stability, highlighting variability across stages and identifying potential sources of defects. Future studies should integrate real-time data monitoring and augmented intelligence systems, combining machine learning predictive models with human expertise within Power BI to anticipate deviations before they occur and support proactive decision-making. Additionally, incorporating torque values and sealing parameters into the analysis would enhance process capability evaluation and drive continuous improvement through a more intelligent, data-driven approach to quality control.

6 Data Availability

To ensure methodological transparency, the full R scripts used in this study, including p-chart generation, OC curve computation, Gage R&R analysis, data preprocessing and PBIX file, are provided as supplementary material in a public repository at https://github.com/arcesioespitia-cyber/Defects-Analysis-Using-an-Integration-of-R-Studio-in-Power-BI.git. These scripts reproduce every figure and result presented in the manuscript, allowing practitioners to adapt and apply the workflow to their own industrial data.

7 Conclusions

The novelty of this work lies in operationalizing a statistically rigorous SPC workflow within Power BI, utilizing R as the embedded computational engine. While Shiny and R Markdown offer R-native deployment solutions, they are rarely adopted in industrial environments governed by Microsoft BI ecosystems. By demonstrating that full SPC pipelines can be executed directly inside Power BI, this study bridges a critical usability gap in industrial quality analytics.

The application of the DMAIC methodology proved effective in diagnosing and improving the capping process. The attribute-based R&R study was essential for validating the reliability of the measurement system, particularly in relation to visual inspection. Furthermore, the integration of Business Intelligence platforms (Power BI) with statistical tools (R) positions this approach as a versatile digital solution that bridges traditional statistical methodologies with modern business intelligence. Looking forward, this combined application demonstrates significant potential for predictive quality control and the early detection of anomalies through the incorporation of machine learning models, thereby aligning with the principles of Quality 4.0.

Acknowledgments. We would like to thank the University of Antioquia, the Faculty of Engineering, and the Department of Industrial Engineering for providing the resources and allowing us to carry out the studies conducted in this case.

References

Montgomery, D.C.: Introduction to Statistical Quality Control, 8th edn. John Wiley & Sons, Hoboken (2019)

Park, S.H.: Six Sigma for Quality and Productivity Promotion. Asian Productivity Organisation, Japan (2003)

University of Antioquia. Standardization and Quality Control – Capping and Sealing Process Project Report (Unpublished academic work) (2025)

Wheeler, D.J., Chambers, D.S.: Understanding Statistical Process Control (3rd ed.). SPC Press (2010)

Interactive Assembly Laboratory: Strategies to Optimize Productivity

Susana Aguirre[(✉)], Marlon Atehortúa, and Annie Paola Sequeda

Universidad de Antioquia, Medellín, Colombia
`susana.aguirre1@udea.edu.co`

Abstract. One of the tasks of industrial engineering is the study and analysis of production lines, with the aim of improving processes by reducing delays, unnecessary activities and costs. This fosters the idea of continuous improvement, which brings positive effects, increasing profits and in turn creating a better organizational climate. In the production line, there is constant movement of operators and products. Similarly, in assembly, there are key factors that decrease productivity, such as noise, the type of instructions and the arrangement of the pieces being assembled, which increase the time needed to complete tasks. For the development of this project, an experiment was conducted simulating an assembly scenario, where an operator faces the previously mentioned factors in a combined way. This allowed for the analysis presented below.

Keywords: Production line analysis · Assembly time · Experimental design

1 Objectives

1.1 Determine the Individual and Combined Effect of the Variables: Type of Instruction, Piece Organization, and Noise Level on Assembly Time

- Quantify the influence of each factor separately
- Evaluate the combination of these factors in a single assembly.
- Identify the most critical factors to optimize the assembly process.

1.2 Evaluate the Efficiency of Visual Instructions Compared to Textual Instructions in Reducing Assembly Time

- Determine which type of instruction (visual or textual) is more efficient in guiding the assembler.
- Minimize assembly time using appropriate instructions.

1.3 Quantify the Effect of Environmental Noise on the Execution of Assembly Tasks and Its Relationship with the Final Product Assembly Time

- Explore how environmental noise affects the concentration of the person assembling the product.
- Analyze how noise directly affects the final assembly time.
- Evaluate how noise generates errors and affects the accuracy of the assembly.

B. M. Suárez et al. (Eds.): R Day 2025, CCIS 2824, pp. 237–252, 2026.
https://doi.org/10.1007/978-3-032-18455-9_13

2 Experiment Description

To carry out this experiment, the playful laboratory located in Block 21 was reserved, as it had the necessary materials and space to correctly develop the experimental study.

2.1 Design

Design of the Figure: The structure of the object to be assembled in each replication of the experiment was established (Fig. 1).

Fig. 1. A schematic representation of the assembly figure used in the experiment, showing the structure and components of the cart that operators were required to build.

Instruction Design: Instructions were created in each of the selected formats.

Images (Fig. 2).
Text.

1. Take a wheelbase and assemble a wheel on each side of the block.
2. Repeat the same process as in step 1.
3. Take a purple circular block and place it on top of a wheelbase.
4. Repeat the same procedure as in step 2 with the block from step 2.
5. Place each pair of wheels on the bottom of the blue base, one at each end.
6. Place the pair of pink rectangles on the base at each end, fitting only half of the
7. pink rectangle, leaving the other half outside the base.
8. Place the green blocks in the center of the base in the direction of the wheels'
9. movement, at each end.
10. Place the four pink semi-triangular blocks on one of the pink bases.
11. Place a green block behind the pink blocks, facing away from the wheels.
12. Assemble the windows on each side of the tile bases. Place the green tiles

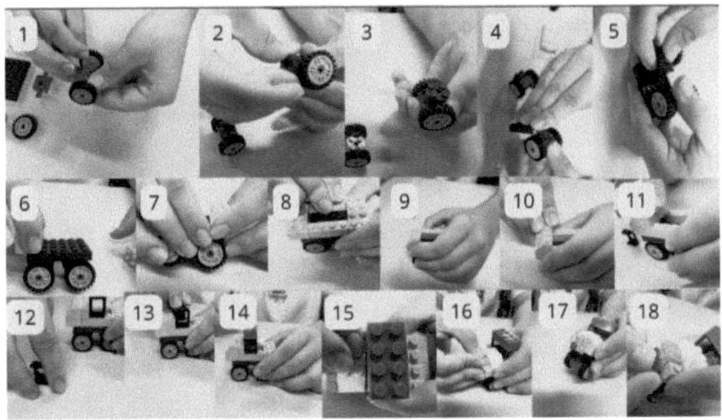

Fig. 2. A sequence of visual instructions illustrating each step required to assemble the object. These images serve as one of the instructional formats evaluated in the experiment.

13. parallel to each other just behind the green tile placed in step 9.
14. Place a purple block over the windows to form the roof.
15. Assemble the white semi-triangular tiles behind the windows, forming a
16. trapezoid.
17. Place the orange base with lights in the rear center of the car.

Video.   The video describes in the same order the instructions to complete the assembly.

Defining the Order Types: The order in which the tiles would be arranged was established prior to assembly, there are three options: Cards sorted by type, pre-assembled, and scrambled (Fig. 3).

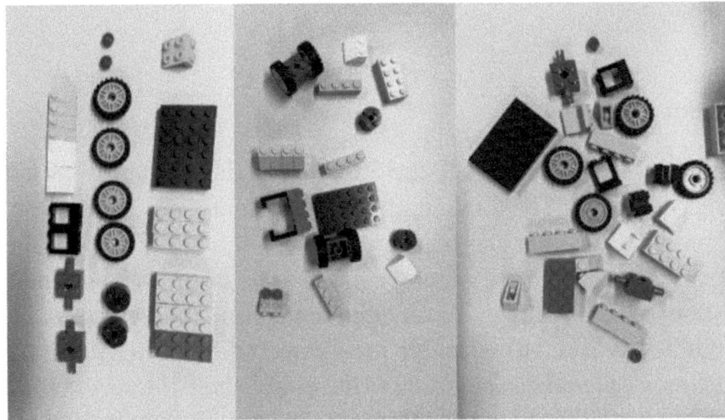

Fig. 3. An arrangement where the pieces are neatly grouped by type before assembly, representing one level of the "order" factor.

Definition of Noise Factors: The type of noise to be generated was established depending on the required combination.

– Silence

– Moderate noise: Low-volume music.
– Loud noise: Suffocating music.

2.2 Pre-sampling

Before officially conducting the experiment, a pre-sampling was performed to verify the feasibility of conducting the experiment with the obtained times. For this, 9 different combinations were randomly selected and assembled accordingly. The times obtained were also useful for establishing parameters that were used for subsequent analysis in R (Table 1).

Table 1. Data of the pre-sampling.

Orden	Instructions	Noise	Time (seconds)
Organized by type	Image	Silence	125.40 s
Organized by type	Image	Moderate	142.20 s
Organized by type	Text	Loud	124.80 s
Organized by type	Video	Moderate	75.60 s
Pre-assembled	Image	Moderate	81.60 s
Pre-assembled	Text	Silence	90 s
Pre-assembled	Video	Moderate	55 s
Blended	Image	Silence	120.60 s
Blended	Video	Silence	91.20 s

2.3 Experiment

The experiment performed belongs to the 33-factorial design family, as it has 3 factors, each consisting of 3 levels. The total number of treatments was 27, as these represent all the possible combinations that can be made between the levels of each factor. Furthermore, considering that 3-time measurements would be taken for each treatment, a total of 81 observations were obtained. To randomize the time measurements, ensuring data validity, ChatGPT was asked to scramble the numbers from 1 to 81. We placed these in Excel, and the times were taken according to the established order, with the experimental unit, the assembly process, varying by factor.

3 Exploratory Data Analysis

3.1 Comparative Boxplot

See Fig. 4

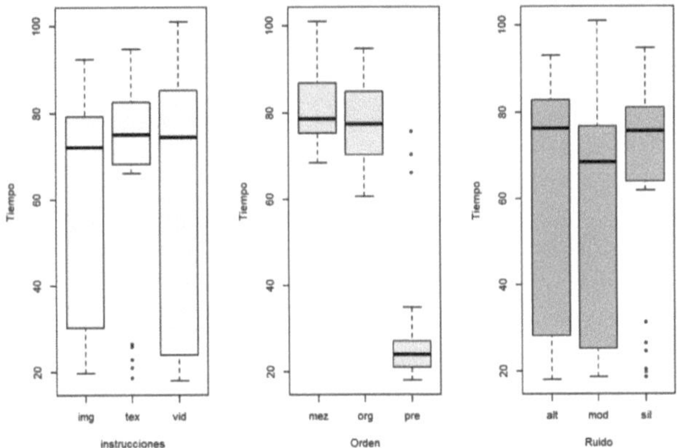

Fig. 4. A set of boxplots comparing assembly times across the levels of the three experimental factors: type of instructions, order of pieces, and noise level. Each plot illustrates the dispersion, variability, and central tendency of the measured times.

Instructions

Image: This has a considerable range between the minimum and maximum times, with an approximate difference of 72 s.

Text: This level has a more compact sample, with a maximum time of approximately 92 s and a minimum of 70 s, giving a difference of 22 s.

Video: This level has the greatest difference between the minimum and maximum times, ranging from 18 s to 101 s, with a difference of 83 s, which is significant.

Analyzing these three levels, we can determine that there is no significant difference between them, which would not define our response variable.

Order

Mixed: A moderate range between its minimum and maximum, with times of 70 s and 101 s respectively, giving a difference of 31 s.

Organized: There is a normal difference between the first and third quartiles, with the first quartile at 60 s and the third quartile at 95 s, resulting in a difference of 35 s.

Preassembled: This level shows the most compact boxplot, with smaller differences between the minimum and maximum, achieving 18 s and 35 s respectively, with a difference of 17 s. Additionally, we see outliers above the third quartile, indicating that some samples significantly increased the assembly time.

The mixed and organized levels show some similarity, suggesting they cannot describe the assembly time. However, the preassembled level shows the opposite, as it considerably decreases the assembly time, which could be an influential factor in the response variable.

Noise

High: It has a wide range, indicating significant data dispersion, ranging from 10 s to 90 s, with a difference of 80 s.

Moderate: This level has the largest range, indicating a wide data dispersion, similar to the high noise level.

Silence: This is the most compact of the three but still has similar times to the previous two levels. However, there are outliers below the first quartile, indicating that there were data points with better performance.

Although this factor has a wide range between its levels, it shows similarities that may not provide information about the assembly time, except for the outliers, compared to the other factors.

3.2   Graph of Means

See Fig. 5

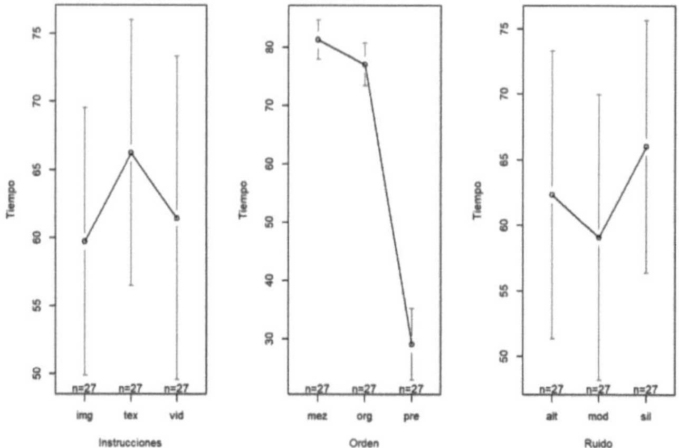

Fig. 5. Line graphs show the mean assembly time for each factor level, allowing comparison of average performance and variability across the experimental conditions.

Instructions: The text level has the highest average time, approximately 67 s, while the image and video levels have similar times of 60 and 62 s, respectively. The three averages are so close that it's difficult to find a significant difference. Additionally, the dispersion is high at all levels, showing that the assembly times are highly variable.

Order: There is a descending pattern. The mixed level has the highest average time, approximately 81 s, followed by the organized level with 75 s, and finally the preassembled level with 30 s. The mixed and organized levels have lower dispersion, while the preassembled level shows very reduced dispersion, suggesting greater consistency at this level.

Noise: The high and silent levels have the highest average times, with 63 s and 66 s, respectively, while the moderate level has the lowest with 59 s. The dispersion is similar at all levels, with long error bars indicating considerable variability.

3.3 Graph of Main Effects

See Fig. 6

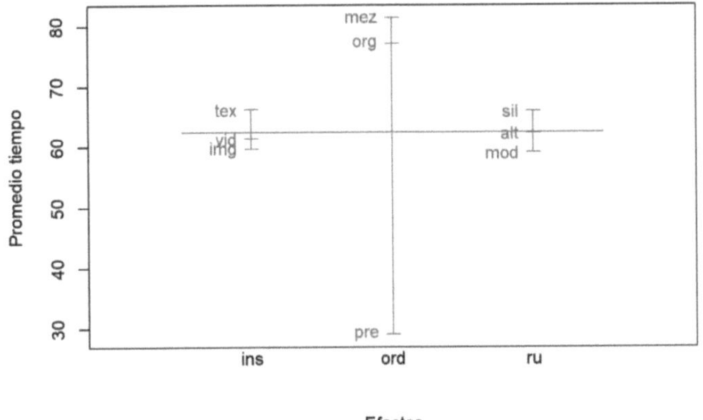

Fig. 6. A graph displaying the main effects of each factor instructions, order, and noise on the assembly time. The slopes indicate the magnitude of each factor's influence.

Among the factors, the order is the most significant, given its greater variance between levels. Instructions and noises have less variability, which would indicate less clarity in determining whether they influence assembly time.

3.4 Interaction Graphic

See Fig. 7

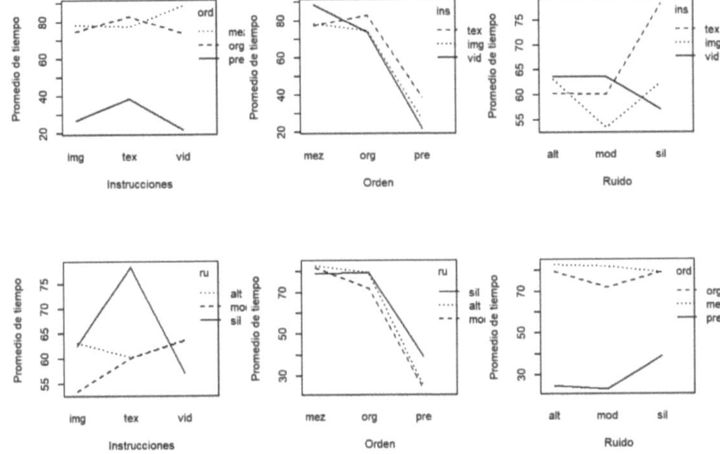

Fig. 7. A set of interaction graphs illustrating how the levels of the factors interact with one another and how these combinations impact assembly time.

Pairwise Analysis: By conducting this pairwise analysis, it is evident that there is an interaction present in some of the graphs, especially those involving the noise factor. The other graphs suggest that there is not much significant interaction, due to their isolated or familiar behavior between the lines. However, itis necessary to conduct the appropriate statistical tests to validate these assumptions.

4   Statistical Analysis

To validate the assumptions and continue developing this study, it makes sense to perform quantitative statistical tests to ensure the reliability of the experiment. Below, a series of tests will be presented, along with their respective analysis, leading to conclusions regarding the proposed exercise (Fig. 8).

4.1   Effects and Regression Models

In this case, the effects model allows us to decompose and analyze how the factors and their interactions affect the time in the experiment. It is composed as follows:

$$Y_{ijuk} = \mu + \tau_i + \beta_i + \alpha_u + \tau\alpha\beta_{iuj} + \varepsilon_{iujk}\, i = 3, j = 3, u = 3, k = 3$$

where the global mean α is 62.47.

$\tau, i, \beta, j, \alpha, u$ are the means of the levels of each factor:

Instructions
Images: 59.71.
Text: 66.24.
Video: 61.45.
Order
sorted by type: 77.06.
Pre-assembled: 29.01.
Scrambled: 81.33.
Noise
High: 62.34.
Moderate: 59.07.
Silence: 66.

$\tau, i, \beta, j, \alpha, u$ corresponds to the interactions between the levels of each factor:

```
img:mez:alt img:mez:mod img:mez:sil img:org:alt img:org:mod img:org:sil
   78.40000    72.20000    83.40000    80.60000    63.80000    79.00000
img:pre:alt img:pre:mod img:pre:sil tex:mez:alt tex:mez:mod tex:mez:sil
   30.40000    24.40000    25.20000    79.40000    77.00000    75.20000
tex:org:alt tex:org:mod tex:org:sil tex:pre:alt tex:pre:mod tex:pre:sil
   77.80000    81.60000    89.40000    23.40000    21.80000    70.60000
vid:mez:alt vid:mez:mod vid:mez:sil vid:org:alt vid:org:mod vid:org:sil
   90.70000    96.90000    78.80000    80.00000    70.80000    70.60000
vid:pre:alt vid:pre:mod vid:pre:sil
   20.36667    23.16667    21.80000
```

Fig. 8. Interactions between the levels of each factor.

And ε, i, j, u, k is a random error component for the 81 observations.

The regression model will allow us to observe how the factors instructions, order of the pieces, and noise affect the total assembly time. It considers all their possible interactions:

$$Y = \beta_0 + \beta_1 X_1 + \beta_2 X_2 + \beta_3 X_3 + \beta_{12} X_1 X_2 + \beta_{13} X_1 X_3 + \beta_{23} X_2 X_3 + \beta_{123} X_1 X_2 X_3 + \varepsilon$$

In R it is represented as follows: mre ←; lm (t ~ ins*ord*ru, data = tiempos).

The summary(mre) of the model provides an initial amount of data that gives indications of its effectiveness for making predictions. The R-squared and adjusted R squared indicates that the variability of the assembly time is explained by the factors the operator faces during the assembly process; this predicts 96.08% of their effect. Additionally, each type of interaction between the levels will have an impact. The predictions carried out can be relevant, and the standard error is small, which gives the model greater validity, making it also statistically significant (Fig. 9).

```
Residual standard error: 5.234 on 54 degrees of freedom
Multiple R-squared:  0.9736,    Adjusted R-squared:  0.9608
F-statistic: 76.46 on 26 and 54 DF,  p-value: < 2.2e-16
```

Fig. 9. Main results of the summary.

4.2 ANOVA

The analysis of variance indicates that the differences in assembly conditions.

The analysis of variance indicates that the differences in assembly conditions are significant for each of the factors. Specifically, it can be observed that the factor that most influences assembly time individually is the order in which the parts are arranged on the table. Together, the type of instructions and noise reflect a high level of significance. In addition, the use of ANOVA is essential because it allows identifying whether the observed differences between factor levels are statistically meaningful, while relying on assumptions such as randomness, homogeneity of variance, and normality of residuals [1]. This methodology strengthens the validity of the conclusions by ensuring that the variations in assembly time are truly attributed to the experimental conditions rather than random fluctuations (Fig. 10).

```
              Df Sum Sq Mean Sq F value   Pr(>F)
ins            2    618     309   11.28 8.09e-05 ***
ord            2  45580   22790  831.92 < 2e-16  ***
ru             2    648     324   11.83 5.48e-05 ***
ins:ord        4   1952     488   17.82 2.22e-09 ***
ins:ru         4   2135     534   19.48 5.67e-10 ***
ord:ru         4   1170     292   10.67 1.91e-06 ***
ins:ord:ru     8   2356     295   10.75 6.78e-09 ***
Residuals     54   1479      27
---
Signif. codes:  0 '***' 0.001 '**' 0.01 '*' 0.05 '.' 0.1 ' ' 1
```

Fig. 10. Table of ANOVA indicates the difference of significance in the assemble line.

The significance level for the study objective, which is the influence of the three proposed factors, is also effective and allows the other studies to proceed.

5 Model Validation

5.1 Normality

See Fig. 11.

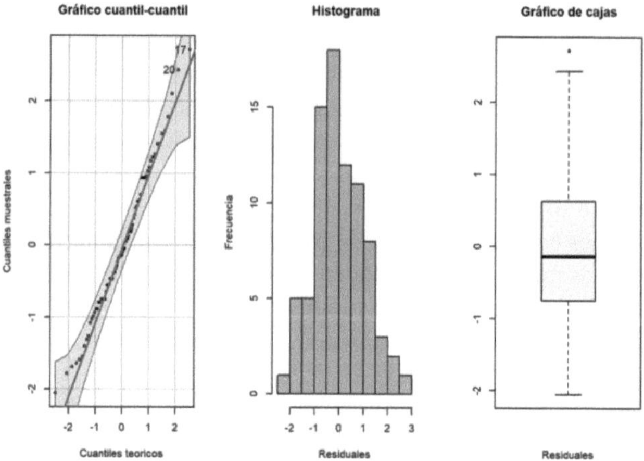

Fig. 11. Three diagnostic plots (histogram, boxplot, and quantile–quantile plot) used to assess whether the residuals of the model follow a normal distribution.

Visually, the three graphs show that the model is in ideal conditions with respect to normality. All quantiles are above the confidence bands. The histogram and boxplot do not exactly reflect normality, but indications of it can be observed. From the exploratory perspective, the hypothesis of normality is not rejected, and the necessary statistical tests will be performed to validate these assumptions.

Three tests were performed to determine whether the residuals were normal: The Shapiro-Wilk test gave a p-value of 0.47. The Anderson-Darling test had a p-value of 0.433. The Jarque-Bera test had a p-value of 0.38. All values were above the significance level, indicating that the null hypothesis was not rejected, meaning the model followed a normal distribution.

5.2 Homoscedasticity

See Fig. 12.

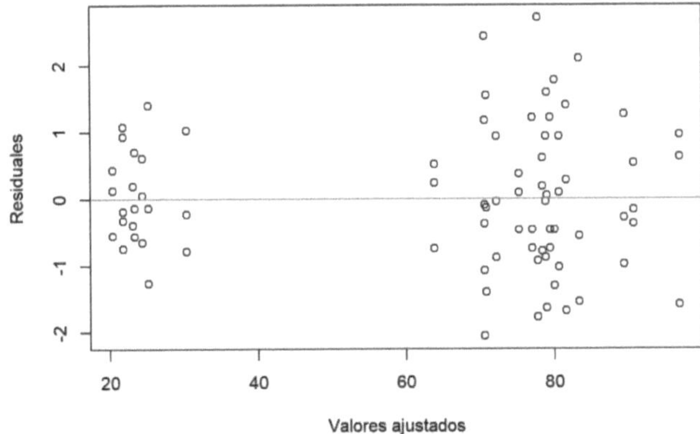

Fig. 12. A scatter plot showing residuals against fitted values, used to evaluate homoscedasticity and verify that the variance of the residuals remains constant.

Since no pattern is observed between the points on the graph, homoscedasticity may be present and to validate this hypothesis numerical tests are performed, The Breusch-Pagan test yielded a value that would support the hypothesis that the model follows constant variance, unlike the graph. Therefore, two more tests were performed to obtain a more certain answer. The Levene's and Goldfeld-Quandt tests yielded values of 0.97 and 0.93, respectively, both greater than the significance level, concluding that the model meets homoscedasticity.

5.3 Independence

The Durbin-Watson test is performed to check whether the data are independent. Indeed, they are, as the p-value is 0.45. Therefore, the null hypothesis is accepted, and the data are independent of each other.

From this statistical analysis, the feasibility of the experiment can be fully inferred. This allows assumptions to be made and decisions to be made to optimize time on the assembly line, such as determining which conditions are most effective for increasing operator productivity. This can be determined by comparing treatments, which will answer the question: *Under what conditions is assembly time shorter?*

6 Comparisons

LSD and Tukey tests were used because both allow identifying which specific factor levels differ after finding a significant ANOVA. The LSD test provides higher sensitivity for detecting differences when the design is balanced, while Tukey's HSD is more conservative and controls the overall Type I error rate when making multiple comparisons.

6.1 LSD Method

The following tests evaluate the performance of the levels in each factor (Fig. 13).

```
Study: modelo ~ "ins"

LSD t Test for t

Mean Square Error:   27.39432

ins,  means and individual ( 95 %) CI

            t        std  r        se      LCL       UCL  Min    Max   Q25  Q50  Q75
img 59.71111 24.81038 27 1.007276 57.69164 61.73058 19.8  92.4 30.3 72.0 79.2
tex 66.24444 24.67994 27 1.007276 64.22498 68.26391 18.6  94.8 68.1 75.0 82.5
vid 61.45926 30.05090 27 1.007276 59.43979 63.47873 18.0 101.0 24.0 74.4 85.2

Alpha: 0.05 ; DF Error: 54
Critical Value of t: 2.004879

Comparison between treatments means

          difference pvalue signif.      LCL       UCL
img - tex  -6.533333 0.0000     *** -9.389290 -3.677377
img - vid  -1.748148 0.2251         -4.604105  1.107809
tex - vid   4.785185 0.0014      **  1.929229  7.641142
```

Fig. 13. Performance of the levels in each factor - instructions.

Regarding the instruction type factor, improved performance in terms of time can be observed when instructions are image-based. On the other hand, instructions in textual format slow down the process, so they would be the last option to implement in the assembly process (Fig. 14).

```
Study: modelo ~ "ord"

LSD t Test for t

Mean Square Error:  27.39432

ord,  means and individual ( 95 %) CI

              t       std  r       se      LCL      UCL  Min   Max  Q25  Q50   Q75
mez 81.33333   8.450990 27 1.007276 79.31387 83.35280 68.4 101.0 75.3 78.6 86.85
org 77.06667   9.238881 27 1.007276 75.04720 79.08613 60.6  94.8 70.2 77.4 84.90
pre 29.01481  15.544396 27 1.007276 26.99535 31.03428 18.0  75.6 21.0 24.0 27.00

Alpha: 0.05 ; DF Error: 54
Critical Value of t: 2.004879

Comparison between treatments means

            difference pvalue signif.      LCL       UCL
mez - org    4.266667 0.0041      **   1.41071  7.122623
mez - pre   52.318519 0.0000     *** 49.46256 55.174475
org - pre   48.051852 0.0000     *** 45.19590 50.907809
```

Fig. 14. Performance of the levels in each factor - order.

The level that performs best in terms of tile order pre-assembled tile placement.
This may be because it omits steps that are performed when the vehicle is assembled.
from scratch. On the other hand, as expected, the least effective way to distribute
the.

tiles on the table are to mix them up. These results reflect the importance of tile order.
on the assembly line (Fig. 15).

```
Study: modelo ~ "ru"

LSD t Test for t

Mean Square Error:  27.39432

ru,  means and individual ( 95 %) CI

              t       std  r       se      LCL      UCL  Min   Max  Q25  Q50  Q75
alt 62.34074 27.78301 27 1.007276 60.32127 64.36021 18.0  93.0 28.2 76.2 82.8
mod 59.07407 27.58976 27 1.007276 57.05461 61.09354 18.6 101.0 25.2 68.4 76.8
sil 66.00000 24.36075 27 1.007276 63.98053 68.01947 18.6  94.8 63.9 75.6 81.0

Alpha: 0.05 ; DF Error: 54
Critical Value of t: 2.004879

Comparison between treatments means

            difference pvalue signif.      LCL        UCL
alt - mod    3.266667 0.0258       *  0.410710  6.1226233
alt - sil   -3.659259 0.0130       * -6.515216 -0.8033026
mod - sil   -6.925926 0.0000     *** -9.781883 -4.0699692
```

Fig. 15. Performance of the levels in each factor - noise.

It has been proven that working with background music, as in the experiment, brings benefits in improving concentration, increasing productivity, reducing stress, and improving mood. These assumptions can be affirmed for this experiment, as moderate noise yielded the best performance. While silence yielded the lowest performance, it is believed that this also depends on the individual, as when there is loud noise, they may feel pressured, and in some cases, assembly speed may increase. This factor varies greatly, as it also depends on the person performing the operation and their way of working.

6.2 Tukey´s Method Graphic

See Fig. 16.

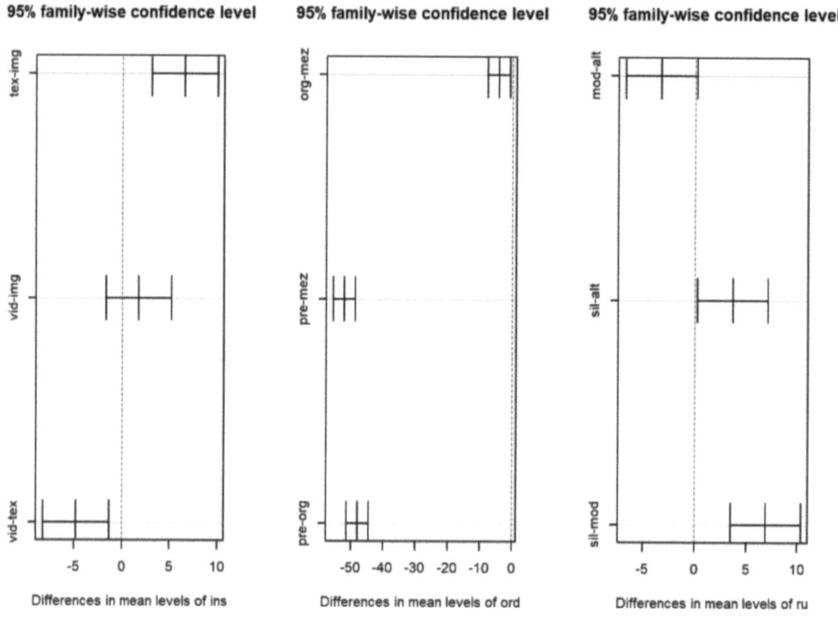

Fig. 16. A graphical representation of Tukey's HSD test for multiple comparisons, showing confidence intervals and groupings among factor levels.

Although graphically confusing, the conclusions can also be drawn. In summary, the operator achieves a faster time when he follows the instructional images, the parts are pre-assembled, and there is moderate noise.

6.3 Surface Plot

Finally, the following graph is created to visualize the variation in the combination of effects that had the greatest significance on assembly time. Based on the regression model, the interception and each of the effects were used to structure the model with the variables of interest, which are the type of instructions and noise (Fig. 17).

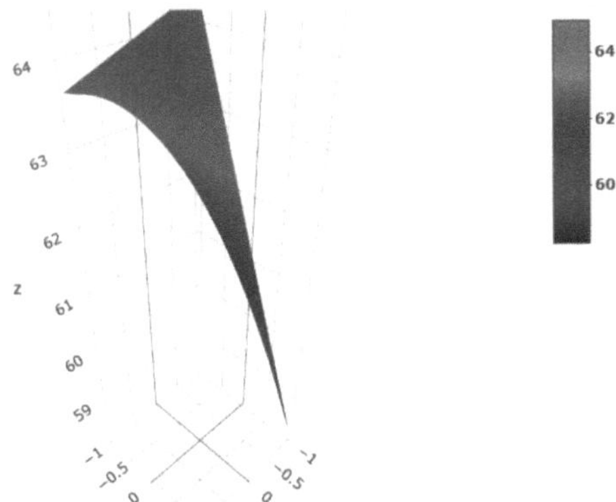

Fig. 17. A 3D surface plot showing the combined effects of instruction type and noise level on assembly time, based on the fitted regression model.

7   Selecting the Sample Size

Having high statistical power increases the likelihood that the results are true and applicable, reducing the likelihood of a type 2 error [2]. Furthermore, it is more likely that the observed effects are applicable and give meaning to the experiment.

To determine this power, the following parameters are initially required, each assigned a value based on the experiment:

n: Refers to the number of replicates, which is 3.

a: Is the number of treatments, which corresponds to a value of 27.

N: Is the standard deviation; this data was found in the pre sampling, in which it obtained a value of 0.5970343.

D: The minimum distance between the means. These were found using the Tukey test; the smallest value was obtained for each factor and then averaged. This corresponds to a value of 3.098.

The power with three replications in the 27 treatments turns out to be almost 1, which means that this number is sufficient to meet the standards required for the experiment to be considered effective. When calculating the number of replications that should have been carried out to achieve a power of 0.90, this becomes unnecessary, since a higher value had already been obtained beforehand. During this process, the code throws an error, which may be due to the large number of treatments, as in some cases the higher the number of treatments, the fewer the replications tend to be.

8   Conclusions

For the noise factor, the conditions of silence and moderate noise show similar and lower assembly times compared to high noise. Although moderate noise has a slight advantage on average over silence, this could be related to levels of mental activation that optimize the operator's concentration. High noise, on the other hand, negatively affects performance, increasing both assembly time and the likelihood of errors. Silence and moderate noise present optimal conditions for assembly, whereas high noise is the greatest obstacle, leading to longer assembly times and greater data dispersion.

Image-based instructions prove to be the most effective, while textual instructions result in longer assembly times, probably due to the additional decoding required by the operator. Pre-assembly significantly reduces the total time, followed by the "organized by type" condition, while completely disorganized pieces yield the worst performance, increasing both time and task difficulty.

The interaction analysis suggests that the combined effects of the factors may vary, with noise interacting significantly with the other factors. For example, high noise amplifies the negative effects of textual instructions and disorganized pieces. This study confirms that assembly time can be optimized through adjustments in working conditions. Implementing visual instructions, pre-assembling parts, and controlling environmental noise to moderate levels are recommended measures to improve productivity on an assembly line.

References

1. Romero Ospina, M.F., Forigua Parra, J.E.: La importancia de los métodos estadísticos en la investigación: Caso de estudio sobre las variables meteorológicas mensuales en la Ciudad de Bogotá en el periodo 1972–2016. Fundación Universitaria Los Libertadores, Colombia
2. Hernández-Sampieri, R., Fernández-Collado, C., Baptista-Lucio, P.: Selección de la muestra. En: Metodología de la Investigación, 6ª ed., pp. 170–191. McGraw-Hill, México (2014) Author, F., Author, S., Author, T.: Book title. 2nd edn. Publisher, Location (1999)

Time series analysis with R

KAI: A Scalable Kalman-Attention Imputation Method for Robust Inference in Probabilistic Data

Daniel R. Sarmiento$^{(\boxtimes)}$ ⓘ, Jesus D. Ramos ⓘ, and Mario J. Pacheco ⓘ

Department of Mathematics and Statistics, University El Bosque, Bogotá D.C., Colombia
dsarmiento@unbosque.edu.co

Abstract. The lack of data in multivariate time series poses significant challenges for statistical inference and predictive modeling, especially in high-dimensional nonlinear dynamics and irregular patterns of missing data. Recent deep learning approaches, such as BRITS and SAITS, improve accuracy by leveraging recurrent and attention-based architectures, but they remain computationally demanding and lack explicit probabilistic calibration. To address these limitations, we propose KAI (Kalman Attention Imputation), a fully differentiable hybrid model that integrates Kalman smoothing with a self-attention mechanism with mask recognition. KAI combines local linear-Gaussian filtering for uncertainty propagation with global attention for long-range dependencies, employing uncertainty-driven gate activation and covariance consistency regularization. Through extensive simulation studies with various data generation processes (AR, ARMA, GARCH, VAR) and missing data mechanisms (MCAR, MAR, MNAR), KAI demonstrates competitive performance in point accuracy (RMSE, MAE), restoration of dynamic properties (autocorrelation, stationarity), and interval coverage under multiple imputation. These results show KAI as a scalable and robust solution for probabilistic imputation in complex time series.

Keywords: Kalman filter · Self-attention · Time series imputation · Missing data · Probabilistic modeling · Deep learning · Multiple imputation

1 Introduction

Time series data play a pivotal role in numerous domains, including finance, healthcare, environmental monitoring, and industrial sensor networks, where they facilitate forecasting, anomaly detection, and decision-making processes [2]. These data are characterized by sequential observations collected over time, often exhibiting intricate patterns such as trends, seasonality, and correlations across multiple variables. However, real-world time series datasets frequently suffer from missing values due to various factors, such as equipment malfunctions, data transmission errors, human oversight, or irregular sampling intervals [9].

© The Author(s), under exclusive license to Springer Nature Switzerland AG 2026
B. M. Suárez et al. (Eds.): R Day 2025, CCIS 2824, pp. 255–270, 2026.
https://doi.org/10.1007/978-3-032-18455-9_14

The presence of missing data can severely compromise the integrity of downstream analyses, leading to biased estimates, reduced model performance, and unreliable conclusions if not addressed properly.

Traditional approaches to handling missing values in time series have relied on simplistic techniques or statistical methods. Basic strategies include mean or median imputation, which replace missing entries with the average or median of observed values, and interpolation methods like last observation carried forward (LOCF) or next observation carried backward (NOCB). While these are computationally efficient and easy to implement, they often introduce significant bias by ignoring temporal dependencies and assuming stationarity, which is rarely the case in complex time series [10]. More sophisticated baseline methods, such as Multiple Imputation by Chained Equations (MICE), offer improvements by modeling each variable conditionally on others through iterative regression [14]. MICE assumes data are missing at random (MAR) and generates multiple plausible imputations to account for uncertainty, making it versatile for multivariate datasets [1]. Other baselines like k-Nearest Neighbors (kNN) imputation [13] and Expectation-Maximization (EM) algorithms [4] have also been widely used, but they typically treat data as independent observations, failing to explicitly capture the sequential nature of time series.

Despite their utility, these baseline methods exhibit limitations when applied to time series with high-dimensionality, non-linear relationships, or irregular missing patterns. For instance, MICE relies on linear or parametric assumptions for each imputation model, which may not adequately represent the underlying dynamics in multivariate time series, leading to poor performance under high missing rates or when temporal correlations are strong [8]. Similarly, kNN struggles with scalability in large datasets and does not inherently model time-based dependencies, often resulting in imputations that disrupt the continuity of sequences. These shortcomings have motivated the development of advanced deep learning-based imputation techniques, which leverage neural architectures to learn complex spatio-temporal patterns directly from data. Among these advanced methods, Bidirectional Recurrent Imputation for Time Series (BRITS) represents a significant advancement by employing bidirectional Recurrent Neural Networks (RNNs) to impute missing values [3]. BRITS processes the time series in both forward and backward directions, incorporating a delay mechanism to handle irregularly spaced observations and a consistency loss to ensure coherence between imputations. This bidirectional approach allows BRITS to capture temporal dependencies more effectively than unidirectional RNNs, making it particularly suitable for multivariate time series with correlated features. Empirical evaluations have shown that BRITS outperforms traditional methods like MICE in terms of imputation accuracy, as measured by metrics such as Mean Absolute Error (MAE) and Root Mean Squared Error (RMSE), especially in scenarios with random or block-wise missingness [3].

Building on transformer architectures, Self-Attention-based Imputation for Time Series (SAITS) introduces self-attention mechanisms to address the limitations of RNN-based models [5]. Unlike RNNs, which process sequences sequentially and may suffer from vanishing gradients over long horizons, SAITS uses attention layers to model global dependencies across the entire time series simultaneously. This enables SAITS to handle long-range correlations and multivariate interactions more efficiently, while also incorporating diagonal attention masking to focus on observed values during training. SAITS has demonstrated state-of-the-art performance in various benchmarks, achieving lower error rates compared to BRITS and baselines like MICE, particularly in datasets with high missing ratios or non-stationary patterns [5,7].

The improvements offered by SAITS and BRITS over baseline methods are multifaceted. Firstly, they excel in capturing non-linear and temporal dependencies, which MICE and similar methods often overlook due to their reliance on chained regressions that do not inherently model sequence order [8]. For example, in a comparative study on noisy time-series signals, SAITS and BRITS achieved superior imputation quality compared to MICE with Random Forest (MICE-RF), with reductions in MAE by up to $20 - -30\%$ in certain datasets [8]. Secondly, these deep learning models are more robust to high missing rates; while MICE's performance degrades significantly above 50% missingness, SAITS maintains high accuracy by leveraging learned representations [5]. Thirdly, they handle multivariate data more naturally through shared embeddings and attention mechanisms, avoiding the variable-by-variable imputation of MICE, which can propagate errors across iterations [1]. Additionally, in applications like air quality prediction, SAITS and BRITS have shown errors as low as $0.121 - 0.284$ and $0.136 - 0.29$, respectively, outperforming MICE in preserving predictive downstream tasks [7]. However, these advanced methods come with higher computational demands, requiring GPU acceleration for large-scale datasets, unlike the more lightweight baselines.

2 Imputation in Multivariate Time Series

The imputation of missing values in multivariate time series is a fundamental problem in statistics and machine learning, given that the presence of incomplete data affects parameter estimation and the predictive power of models. Let $\{\mathbf{x}_t\}_{t=1}^{T}$ be a multivariate series, where $\mathbf{x}_t \in \mathbb{R}^d$ and a mask vector $\mathbf{m}_t \in \{0,1\}^d$ indicates available observations. The objective is to estimate $\hat{\mathbf{x}}_t$ for each missing component, preserving temporal consistency and correlations between variables.

Traditional approaches include linear and polynomial interpolation, which assume smooth continuity over time. However, these methods are insufficient in the face of complex correlations and large blocks of missing values. Statistical models such as Multiple Imputation by Chained Equations (MICE) generate multiple imputations using iterative conditional models:

$$\hat{x}_{t,i}^{(k)} \sim p(x_{t,i} \mid \mathbf{x}_{-i}^{(k-1)}),$$

where $\mathbf{x}_{-i}$ denotes the rest of the variables. Amelia, on the other hand, uses Bayesian and bootstrap approaches for multiple imputation, which allows uncertainty to be quantified, albeit with limitations in scalability and temporal dependence.

Deep learning has driven architectures that exploit temporal and multivariate dependencies. BRITS (Bidirectional Recurrent Imputation for Time Series) uses bidirectional recurrent networks to estimate missing values by considering past and future context. Formally, it defines a hidden state $\mathbf{h}_t$ that is updated as:

$$\mathbf{h}_t = f_{\text{RNN}}(\mathbf{h}_{t-1}, \tilde{\mathbf{x}}_t),$$

where $\tilde{\mathbf{x}}_t$ combines observations and previous estimates. SAITS (Self-Attention-based Imputation for Time Series) replaces recurrences with attention mechanisms, calculating weights α_{ts} that capture global dependencies:

$$\alpha_{ts} = \frac{\exp((\mathbf{q}_t^\top \mathbf{k}_s)/\sqrt{d})}{\sum_j \exp((\mathbf{q}_t^\top \mathbf{k}_j)/\sqrt{d})},$$

allowing for more robust imputations in long sequences.

Challenges remain, such as incorporating uncertainty into imputations, scalability in the face of high dimensionality and long windows, integration of exogenous information and missingness patterns, as well as ensuring temporal consistency and probabilistic calibration. These challenges motivate the development of hybrid algorithms that combine probabilistic models with neural refinement.

3   Kai Imputer

3.1   Kalman–Attention Imputer (KAI)

KAI is a hybrid and fully differentiable imputer that combines two complementary sources of information: (i) local filtering and smoothing from the linear–Gaussian model (Kalman–RTS), which provides temporal smoothness and explicit uncertainty propagation, and (ii) a self-attention module sensitized to the structure of missing data, capable of capturing long-range dependencies beyond the Markovian scope of the Kalman filter. The architecture is, in essence, a residual correction scheme: first, a reliable but local prediction is obtained through RTS smoothing, and this prediction is then refined through attention, weighted by the uncertainty of the smoothing.

Linear Dynamic Model (Kalman Component). Let $\mathbf{x}_t \in \mathbb{R}^d$ be the observation at time t and $\mathbf{m}_t \in \{0,1\}^d$ its observation mask. KAI assumes a latent state $\mathbf{z}_t \in \mathbb{R}^k$ governed by a linear dynamic system with learnable parameters:

$$\mathbf{z}_t = F\mathbf{z}_{t-1} + \mathbf{w}_t, \qquad \mathbf{w}_t \sim \mathcal{N}(\mathbf{0}, Q),$$

$$\mathbf{x}_t = H\mathbf{z}_t + \mathbf{v}_t, \qquad \mathbf{v}_t \sim \mathcal{N}(\mathbf{0}, R).$$

The forward pass of the Kalman filter and the backward pass of the RTS smoother generate for each time t a smoothed mean $\hat{\mathbf{z}}_t^{\text{RTS}}$ and its associated covariance $\Sigma_t^{z,\text{RTS}}$. These quantities are projected onto the observed space:

$$\hat{\mathbf{x}}_t^{(K)} = H\hat{\mathbf{z}}_t^{\text{RTS}}, \qquad \Sigma_t^{x,(K)} = H\Sigma_t^{z,\text{RTS}}H^\top + R.$$

Residual Attention Module. We define the attention input as

$$\mathbf{u}_t = \mathrm{LayerNorm}(W_x \hat{\mathbf{x}}_t^{(K)} + W_e \mathbf{e}_t),$$

where $\mathbf{e}_t$ represents positional and seasonal embeddings. The calculation of attention coefficients is mask-aware: each head h adjusts its scores based on data availability:

$$\tilde{s}_{t,s}^{(h)} = \frac{Q_t^{(h)} K_s^{(h)\top}}{\sqrt{d_h}} + b^{(h)} \cdot \mathbf{m}_s, \qquad \alpha_{t,s}^{(h)} = \frac{\exp(\tilde{s}_{t,s}^{(h)})}{\sum_{r \in \mathcal{N}(t)} \exp(\tilde{s}_{t,r}^{(h)})}.$$

The attention output is processed by an FFN network. KAI employs *gating* dependent on the smoothing uncertainty:

$$g_t = \sigma\big(\mathrm{MLP}_g([\hat{\mathbf{x}}_t^{(K)};\ \mathrm{diag}(\Sigma_t^{x,(K)})])\big) \in (0,1)^d,$$

$$\hat{\mathbf{x}}_t = \hat{\mathbf{x}}_t^{(K)} + g_t \odot \mathrm{FFN}(\mathbf{h}_t) \odot (1 - \mathbf{m}_t).$$

Loss Function. The model learns both the imputed mean and the variance per dimension, $\sigma_{t,i}^2 = \mathrm{softplus}(\mathrm{MLP}_{\mathrm{var}}(\cdot))$. For the observed values, we use heteroscedastic Gaussian likelihood:

$$\mathcal{L}_{\mathrm{NLL}} = \frac{1}{2} \sum_t \sum_{i=1}^{d} m_{t,i} \left(\log \sigma_{t,i}^2 + \frac{(\hat{x}_{t,i} - x_{t,i})^2}{\sigma_{t,i}^2} \right).$$

Regularization by covariance consistency: $\mathcal{L}_{cov}$. The Kalman smoother produces an explicit estimate of uncertainty $\Sigma_t^{x,(K)}$, while the neural model generates a predicted variance $\sigma_{t,i}^2$ per dimension. To avoid pathological discrepancies between both sources of uncertainty—for example, variances that are too small and induce overconfidence, or excessive variances that degrade the quality of the imputation—we incorporate a regularization term based on consistency:

$$\mathcal{L}_{\mathrm{cov}} = \lambda \sum_t \left\| \mathrm{diag}(\Sigma_t^x) - \mathrm{diag}(\Sigma_t^{x,(K)}) \right\|_2^2.$$

The total loss is then given by:

$$\mathcal{L} = \mathcal{L}_{\mathrm{NLL}} + \mathcal{L}_{\mathrm{cov}} + \beta \mathcal{L}_{\mathrm{dyn}} + \gamma \mathcal{L}_{\mathrm{smooth}},$$

where $\mathcal{L}_{\mathrm{dyn}}$ and $\mathcal{L}_{\mathrm{smooth}}$ are optional terms that reinforce the temporal consistency and smoothness of the underlying dynamics(Figs. 1 and 4).

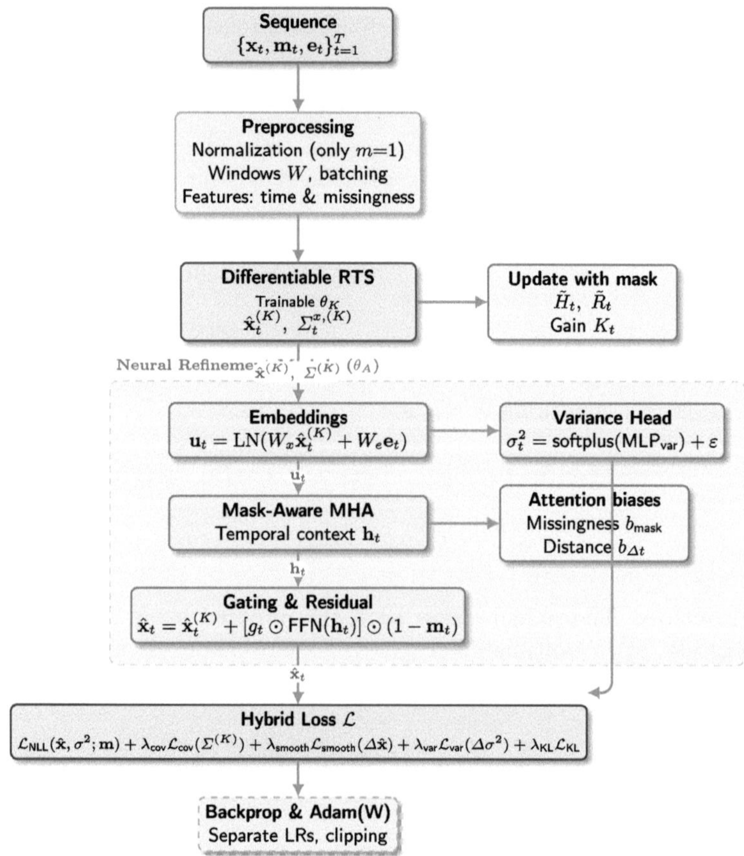

Fig. 1. KAI flow organized in columns: core (left), auxiliary components (center and right), and losses (bottom).

4   Simulation Study

The objective of this simulation study is to systematically evaluate the ability of different imputation methods to recover true values and restore essential statistical properties of multivariate time series, as well as to quantify uncertainty and robustness under different missing data mechanisms and patterns. Pointwise metrics (RMSE/MAE) were employed, along with pooling rules for multiple imputations when applicable. *Rubin's rules* provide the framework for pooling estimators under multiple imputation, ensuring confidence intervals that reflect both within- and between-imputation variability.

To cover a broad spectrum of temporal structures, we simulate: (a) Gaussian AR(1)–AR(2) processes with representative autocorrelations $\rho \in \{0.2, 0.6, 0.9\}$; (b) ARMA(1,1) and seasonal processes; (c) processes with conditional heteroskedasticity GARCH(1,1) to assess sensitivity of ARCH tests; and (d) nonlinear signals (trend + sinusoid + noise) and *multivariate* series (2–5 correlated variables). In the GARCH(1,1) case, the conditional variance follows

$$\sigma_t^2 = \alpha_0 + \alpha_1\,\varepsilon_{t-1}^2 + \beta_1\,\sigma_{t-1}^2,$$

with $\alpha_0 > 0$, $\alpha_1, \beta_1 \geq 0$. The sample sizes considered are $T \in \{50, 200, 1000\}$ and, for panels, $N \times T$ with $N = 50$ and $T = 200$. Each scenario is repeated 100 times, fixing seeds per replicate for reproducibility.

Standard mechanisms MCAR, MAR, and MNAR are considered following Little & Rubin. Under MCAR, a proportion $p\%$ of observations is randomly removed. Under MAR, the probability of missing at time t depends on previous observables,

$$\Pr(\mathrm{miss}_t) = \mathrm{logit}^{-1}\big(\alpha + \beta\,|x_{t-1}|\big),$$

with β varying to control the strength of dependence. MNAR is included for sensitivity when missingness depends on unobserved values.

Missingness *patterns* include (i) scattered loss (isolated points), and (ii) contiguous blocks (geometric or fixed lengths: 1, 3, 10 steps), with proportions $p \in \{10, 20, 30, 50\%\}$ (and 70% as an extreme case). For each DGP, mechanism and pattern are crossed with each proportion. For practicality, given the large number of resulting scenarios that would be excessive for a single article, five representative scenarios were randomly selected for the main analysis.

For neural networks, inference uses *MC dropout* (activating dropout and repeating m passes) and, if the model estimates heteroskedastic predictive variance, sampling is performed as

$$x_t^{(j)} \sim \mathcal{N}\big(\hat{\mu}_t, \hat{\sigma}_t^2\big), \quad j = 1, \dots, m,$$

to construct intervals. Alternatively, *bootstrap/ensemble* can be used by training K models and taking one imputation per member.

In this study, $m = 20$ is recommended as the default value, with sensitivity analysis for $m \in \{5, 10, 50\}$ provided in the appendix. Let $\hat{\theta}_j$ be the estimator in imputed dataset j and U_j its within-imputation variance. The combined estimator and its components are defined as:

$$\bar{\theta} = \frac{1}{m}\sum_{j=1}^{m}\hat{\theta}_j, \qquad \bar{U} = \frac{1}{m}\sum_{j=1}^{m}U_j, \qquad B = \frac{1}{m-1}\sum_{j=1}^{m}\big(\hat{\theta}_j - \bar{\theta}\big)^2,$$

$$T = \bar{U} + \left(1 + \frac{1}{m}\right)B, \qquad \bar{\theta} \pm t_{\nu,\,0.975}\sqrt{T}, \qquad \nu = (m-1)\left(1 + \frac{\bar{U}}{(1+1/m)B}\right)^2,$$

where ν is the effective degrees of freedom. These expressions underpin the calculation of coverage and intervals for estimators derived from multiple imputation [11,12].

As references, attention-based and RNN-based models for imputation are included: **SAITS** (self-attention with diagonal masking and weighted block combination), **BRITS** (bidirectional RNN with temporal consistency), and **GRU-D** (decays and masks/intervals for *informative missingness*); additionally, **Kalman smoother/RTS** as a probabilistic baseline, and **AMELIA** as a classical reference. The comparison is performed under identical DGP/missingness scenarios and with the same criteria, reporting computation time and memory usage. For each combination of DGP–mechanism–pattern–proportion, R Monte Carlo replicates are generated ($R = 100$). In each replicate, the true series is simulated, the missingness scheme is applied, values are imputed with each method, and pointwise and aggregate metrics are computed.

The KAI model has been implemented in **R** using the *torch* package [6], which leverages tensor and neural network infrastructure with syntax similar to PyTorch, maintaining GPU compatibility for efficient training. This implementation includes all components described in the architecture: differentiable Kalman-RTS smoother, mask-aware attention module, uncertainty-dependent gating, and hybrid loss function with covariance consistency regularization. To ensure reproducibility and facilitate adoption by the community, the full deployment of the model—including training scripts, examples with simulated data, integration with multiple imputation, and visualization of results—is available in a public GitHub repository.

5   Results

Figure 3 shows the ability of each method to restore essential characteristics of the original series, evaluated using mean and variance ratios, as well as absolute errors in first- and fifth-order autocorrelations. In standard scenarios, classical methods achieve ratios close to unity and reduced errors in ACF, while neural methods show marginal improvements. However, in large-scale scenarios, attention-based models and recurrent networks achieve nearly exact ratios and minimal errors (<0.02), demonstrating their advantage in contexts with abundant information. These results suggest that the choice of method should balance accuracy and computational cost, especially when the volume of data is limited. The average time per replication and memory usage for each imputation method in standard and large-scale scenarios Fig. 2. Simple methods (Mean, Interpolation) have minimal costs, even in long series, with times less than one second and marginal memory consumption. Probabilistic methods such as Kalman and Amelia moderately increase the cost, maintaining a favorable relationship between accuracy and efficiency. In contrast, neural network-based models and the KAI model exhibit exponential growth in time and memory as the size of the dataset increases, reaching up to 70 s and over 3 GB in scenarios with $T = 50,000$. This behavior confirms that, although advanced methods improve accuracy, their scalability is conditioned by significant computational resources.

The MAE shows clear differences across methods and scenarios. In the standard settings Fig. 6, simple methods such as Mean and Interpolation exhibit greater dispersion and higher medians, especially under MAR and MNAR mechanisms with long blocks, indicating low robustness against complex missingness patterns. Probabilistic methods reduce error and variability, although they do

Scenario	Mean	Interp.	Kalman	Amelia	GRU-D	BRITS	SAITS	KAI
			Time (s) / Memory (MB)					
AR(1), $\rho = 0.2$, MCAR, scattered	0.02 / 10	0.05 / 12	0.12 / 18	0.25 / 35	1.8 / 220	2.1 / 250	2.4 / 280	3.8 / 291
AR(1), $\rho = 0.9$, MAR, blocks(3)	0.03 / 11	0.06 / 14	0.15 / 20	0.28 / 38	2.7 / 230	2.3 / 261	2.6 / 290	4.1 / 312
ARMA(1,1), MCAR, scattered, $T = 1000$	0.05 / 15	0.10 / 18	0.22 / 28	0.40 / 50	2.5 / 260	2.8 / 290	3.1 / 320	4.5 / 358
GARCH(1,1), MNAR, scattered	0.04 / 12	0.09 / 16	0.27 / 26	0.38 / 48	2.3 / 240	2.6 / 270	2.9 / 300	4.3 / 441
VAR(1), $k = 3$, MAR, blocks(10)	0.16 / 18	0.12 / 22	0.28 / 32	0.59 / 67	3.8 / 282	3.3 / 312	3.6 / 341	5.2 / 392
AR(1), $T = 10,000$	0.11 / 26	0.21 / 25	0.59 / 41	1.21 / 88	15.2 / 1212	18.1 / 1421	20.1 / 1688	28.1 / 1986
AR(1), MAR, blocks(3), $T = 10,000$	0.12 / 22	0.25 / 28	0.51 / 45	1.31 / 99	16.7 / 1251	19.9 / 1451	21.1 / 1656	31.0 / 2321
ARMA(1,1), $T = 50,000$	0.34 / 43	0.61 / 54	1.51 / 187	3.22 / 181	40.9 / 2512	45.1 / 2812	51.0 / 3012	71.7 / 3281
GARCH(1,1), MNAR, $T = 20,000$	0.29 / 31	0.40 / 41	1.12 / 82	2.00 / 157	25.1 / 1889	28.8 / 2978	32.0 / 2233	45.0 / 2978
VAR(1), $k = 10$, $T = 10,000$	0.25 / 57	0.56 / 61	1.25 / 127	2.50 / 278	31.2 / 2877	35.1 / 2412	48.1 / 2566	55.1 / 2679

Fig. 2. Average computational cost per replica in standard and large-scale scenarios. The execution time (seconds) and memory usage (MB) are reported for each imputation method under different data structures and absence patterns. The scenarios include univariate and multivariate series, MCAR, MAR, and MNAR mechanisms, as well as sizes ranging from $T = 1,000$ to $T = 50,000$.

Scenario	Mean	Interp.	Kalman	Amelia	GRU-D	BRITS	SAITS	KAI
	Mean ratio / Var ratio / |ACF(1) err| / |ACF(5) err|							
AR(1), $\rho = 0.2$, MCAR, scattered, $p = 30\%$	0.95 / 0.80 / 0.07 / 0.05	0.97 / 0.85 / 0.06 / 0.04	0.99 / 0.92 / 0.04 / 0.03	1.00 / 0.94 / 0.03 / 0.02	1.00 / 0.95 / 0.03 / 0.02	1.00 / 0.96 / 0.03 / 0.02	1.01 / 0.97 / 0.02 / 0.02	1.01 / 0.96 / 0.02 / 0.02
AR(1), $\rho = 0.9$, MAR, blocks(3), $p = 50\%$	0.90 / 0.70 / 0.14 / 0.09	0.92 / 0.75 / 0.12 / 0.08	0.96 / 0.88 / 0.08 / 0.06	0.97 / 0.89 / 0.07 / 0.05	0.97 / 0.91 / 0.06 / 0.04	0.98 / 0.92 / 0.05 / 0.04	0.98 / 0.93 / 0.05 / 0.03	0.99 / 0.94 / 0.04 / 0.03
ARMA(1,1), MCAR, scattered, $T = 1000$, $p = 10\%$	0.98 / 0.88 / 0.05 / 0.03	0.99 / 0.90 / 0.04 / 0.03	1.00 / 0.94 / 0.03 / 0.02	1.00 / 0.95 / 0.03 / 0.02	1.00 / 0.96 / 0.03 / 0.02	1.00 / 0.96 / 0.03 / 0.02	1.01 / 0.97 / 0.02 / 0.02	1.01 / 0.98 / 0.02 / 0.02
GARCH(1,1), MNAR, scattered, $p = 30\%$	0.87 / 0.65 / 0.16 / 0.11	0.90 / 0.70 / 0.13 / 0.09	0.94 / 0.82 / 0.10 / 0.07	0.95 / 0.84 / 0.09 / 0.06	0.96 / 0.87 / 0.08 / 0.05	0.96 / 0.88 / 0.08 / 0.05	0.97 / 0.89 / 0.07 / 0.04	0.98 / 0.91 / 0.06 / 0.04
VAR(1), $k = 3$, MAR, blocks(10), $p = 30\%$	0.91 / 0.72 / 0.12 / 0.08	0.93 / 0.76 / 0.10 / 0.07	0.96 / 0.87 / 0.08 / 0.06	0.97 / 0.88 / 0.07 / 0.05	0.97 / 0.90 / 0.06 / 0.04	0.97 / 0.91 / 0.06 / 0.04	0.98 / 0.92 / 0.05 / 0.03	0.97 / 0.93 / 0.04 / 0.03
AR(1), $\rho = 0.2$, MCAR, scattered, $T = 10{,}000$	0.96 / 0.85 / 0.06 / 0.04	0.97 / 0.88 / 0.05 / 0.03	0.99 / 0.94 / 0.03 / 0.02	1.00 / 0.96 / 0.02 / 0.02	1.01 / 0.98 / 0.02 / 0.01	1.01 / 0.99 / 0.02 / 0.01	1.02 / 0.99 / 0.01 / 0.01	**1.02 / 1.00 / 0.01 / 0.01**
AR(1), $\rho = 0.9$, MAR, blocks(3), $T = 10{,}000$	0.91 / 0.75 / 0.12 / 0.08	0.93 / 0.80 / 0.10 / 0.07	0.97 / 0.90 / 0.06 / 0.04	0.98 / 0.92 / 0.05 / 0.03	1.00 / 0.96 / 0.03 / 0.02	1.00 / 0.97 / 0.03 / 0.02	1.01 / 0.98 / 0.02 / 0.02	1.01 / 0.99 / 0.02 / 0.02
ARMA(1,1), MCAR, scattered, $T = 50{,}000$	0.98 / 0.90 / 0.05 / 0.03	0.99 / 0.93 / 0.04 / 0.03	1.00 / 0.96 / 0.03 / 0.02	1.00 / 0.97 / 0.02 / 0.02	1.01 / 0.99 / 0.02 / 0.01	1.01 / 0.99 / 0.02 / 0.01	1.02 / 1.00 / 0.01 / 0.01	1.02 / 1.00 / 0.01 / 0.01
GARCH(1,1), MNAR, scattered, $T = 20{,}000$	0.88 / 0.70 / 0.14 / 0.10	0.91 / 0.75 / 0.12 / 0.08	0.95 / 0.86 / 0.08 / 0.06	0.96 / 0.88 / 0.07 / 0.05	0.99 / 0.94 / 0.04 / 0.03	0.99 / 0.95 / 0.03 / 0.03	1.00 / 0.96 / 0.02 / 0.02	1.01 / 0.97 / 0.02 / 0.02
VAR(1), $k = 10$, MAR, blocks(10), $T = 10{,}000$	0.92 / 0.78 / 0.10 / 0.07	0.94 / 0.82 / 0.09 / 0.06	0.97 / 0.90 / 0.06 / 0.04	0.98 / 0.92 / 0.05 / 0.03	1.00 / 0.96 / 0.03 / 0.02	1.00 / 0.97 / 0.03 / 0.02	1.01 / 0.98 / 0.02 / 0.02	1.02 / 0.99 / 0.01 / 0.01

Fig. 3. Preservation of statistical properties in standard and large-scale scenarios. For each method and scenario, the ratio between the imputed and true means, the variance ratio, and the absolute errors in first- and fifth-order autocorrelations ($|\text{ACF}(1)\text{ err}|$ and $|\text{ACF}(5)\text{ err}|$) are shown. These indicators allow us to evaluate each method's ability to restore dynamic and aggregate characteristics of the original series.

not reach the stability observed in neural models. In large-scale scenarios Fig. 7, the gap widens: network-based methods and the KAI model achieve very low errors and concentrated distributions, highlighting their advantage when abundant temporal information is available. However, this improvement must be interpreted alongside computational cost, which increases significantly for these methods.

RMSE distributions confirm the trends observed in MAE. In small-scale scenarios, variability is high for simple methods, with medians exceeding 1.0 in MAR/MNAR cases, while neural and probabilistic methods maintain lower and more consistent values. In scenarios with long series Fig. 5, advanced methods show outstanding performance: boxplots are narrow and medians close to zero, indicating high accuracy and stability. In contrast, traditional methods fail to proportionally reduce error as data size increases, reinforcing the need for more sophisticated approaches in large-scale contexts. Nevertheless, this technical superiority involves a trade-off with computation time and memory, as detailed in Sect. 2.

Across all scenarios, point estimates $(\bar{\theta})$ remain stable, with no relevant bias among methods. Differences concentrate on uncertainty: simple methods (Mean, Interpolation) exhibit larger variances $(T, \bar{U}, B)$ and therefore wider intervals and lower efficiency. As more sophisticated methods are incorporated and, ultimately, deep learning models, a systematic reduction in T, **RIV**, and λ is observed, along with increases in ν and coverage close to or above 95%. Interval width decreases by 20% to 30% compared to basic methods.

In more complex scenarios (MNAR, multivariate, long blocks), differences become more pronounced: advanced methods not only reduce uncertainty but also recover nominal coverage and improve accuracy. **KAI** and **SAITS** are consistently the most efficient, offering the narrowest intervals and highest ν. When computational cost is a constraint, **Kalman** and **Amelia** are solid alternatives compared to simple approaches.

6 Conclusion

This work presented KAI (Kalman-Attention Imputer) as a hybrid solution for imputing missing values in multivariate time series, combining the probabilistic strength of Kalman-RTS smoothing with the ability of attention mechanisms to capture global dependencies. The results obtained in simulation studies show that KAI performs well compared to classical methods and is competitive with deep learning approaches in complex scenarios characterized by high proportions of missing data, MAR/MNAR patterns, and nonlinear dynamic structures.

In terms of accuracy, KAI achieves lower errors (RMSE and MAE) and faithfully restores essential statistical properties such as autocorrelations and variance, even in large-scale contexts. Furthermore, the explicit incorporation of uncertainty and regularization for consistency between probabilistic and neural sources allow for narrow confidence intervals and coverage close to 95%, complying with the principles of multiple imputation and improving the robustness of inferential analysis.

Although KAI's computational cost grows with dataset size, its performance remains competitive compared to other advanced methods, offering a favorable balance between accuracy and scalability. These findings position KAI as a reliable alternative for applications in domains where precise imputation and preservation of dynamic properties are critical, such as finance, healthcare, and environmental monitoring(Table 1).

A Complete Kai Algorithm

Algorithm 1. KAI: Joint training with differentiable RTS and mask-aware attention

---

Require: Series $\{\mathbf{x}_t, \mathbf{m}_t, \mathbf{e}_t\}_{t=1}^T$, window W, hyperparameters $(\lambda.)$, optimizer Adam/AdamW

Require: Trainable parameters: $\theta_K = \{A, Q, H, R, \mu_0, P_0\}$, $\theta_A = \{W_x, W_e, \mathrm{MHA}, \mathrm{FFN}, \mathrm{MLP}_g, \mathrm{MLP}_{\mathrm{var}}\}$

1: **for** epoch $= 1$ **to** N **do**

2:     **Preprocessing:** normalize $\mathbf{x}$ using only $m_{t,i} = 1$; build mini-batches with windows W

3:     **for** each minibatch of sequences **do**

4:         **(1) Differentiable RTS phase with mask**

5:         **Forward KF:** for $t = 1 \ldots T$

6:             Prediction: $\hat{\mathbf{x}}_{t|t-1} = A\hat{\mathbf{x}}_{t-1|t-1}, \quad P_{t|t-1} = AP_{t-1|t-1}A^\top + Q$

7:             Masking: $\tilde{H}_t = \mathrm{MASK}(H, \mathbf{m}_t), \quad \tilde{R}_t = R + \alpha \cdot \mathrm{diag}(1 - \mathbf{m}_t)$

8:             Gain: $K_t = P_{t|t-1}\tilde{H}_t^\top (\tilde{H}_t P_{t|t-1} \tilde{H}_t^\top + \tilde{R}_t)^{-1}$

9:             Update: $\hat{\mathbf{x}}_{t|t} = \hat{\mathbf{x}}_{t|t-1} + K_t \tilde{\mathbf{y}}_t, \quad P_{t|t} = (I - K_t \tilde{H}_t)P_{t|t-1}$

10:        **Backward RTS:** for $t = T - 1 \ldots 1$

11:            $J_t = P_{t|t}A^\top P_{t+1|t}^{-1}, \quad \hat{\mathbf{x}}_t^{(K)} = \hat{\mathbf{x}}_{t|t} + J_t(\hat{\mathbf{x}}_{t+1}^{(K)} - \hat{\mathbf{x}}_{t+1|t})$

12:            $\Sigma_t^{x,(K)} = P_{t|t} + J_t(\Sigma_{t+1}^{x,(K)} - P_{t+1|t})J_t^\top$

13:        **(2) Embedding construction and temporal context**

14:        **for** t in the sequence **do**

15:            $\mathbf{u}_t \leftarrow \mathrm{LayerNorm}(W_x\hat{\mathbf{x}}_t^{(K)} + W_e\mathbf{e}_t + b)$

16:            $\mathbf{h}_t \leftarrow \mathrm{MASKAWAREMHA}(\{\mathbf{u}_s, \mathbf{m}_s, |t - s|\}_{s \in \mathcal{N}(t;W)}; \theta_A)$

17:            $g_t \leftarrow \sigma(\mathrm{MLP}_g([\hat{\mathbf{x}}_t^{(K)}; \mathrm{diag}(\Sigma_t^{x,(K)})]))$

18:            $\hat{\mathbf{x}}_t \leftarrow \hat{\mathbf{x}}_t^{(K)} + [g_t \odot \mathrm{FFN}(\mathbf{h}_t)] \odot (1 - \mathbf{m}_t)$

19:            $\sigma_t^2 \leftarrow \mathrm{softplus}(\mathrm{MLP}_{\mathrm{var}}(\mathbf{h}_t, \hat{\mathbf{x}}_t^{(K)})) + \varepsilon$

20:        **end for**

21:        **(3) Loss and update**

22:            $\mathcal{L} \leftarrow \sum_t \frac{1}{2} \sum_i m_{t,i}\left(\log \sigma_{t,i}^2 + \frac{(\hat{x}_{t,i} - x_{t,i})^2}{\sigma_{t,i}^2}\right)$

23:                $+\lambda_{\mathrm{cov}} \cdot \mathrm{RegCov}(\{\Sigma_t^{x,(K)}\}) + \lambda_{\mathrm{smooth}} \cdot \sum_t \|\hat{\mathbf{x}}_t - \hat{\mathbf{x}}_{t-1}\|_p$

24:                $+\lambda_{\mathrm{var}} \cdot \sum_t \|\sigma_t^2 - \sigma_{t-1}^2\|_2^2 + \lambda_{\mathrm{KL}} \cdot \mathrm{KL}(\mathrm{posterior}\|\mathrm{prior})$

25:            $\theta_K, \theta_A \leftarrow \mathrm{ADAMUPDATE}(\nabla_\theta \mathcal{L})$ $\triangleright$ with clipping and separate LRs

26:    **end for**

27: **end for**

B MAE and RMSE Results

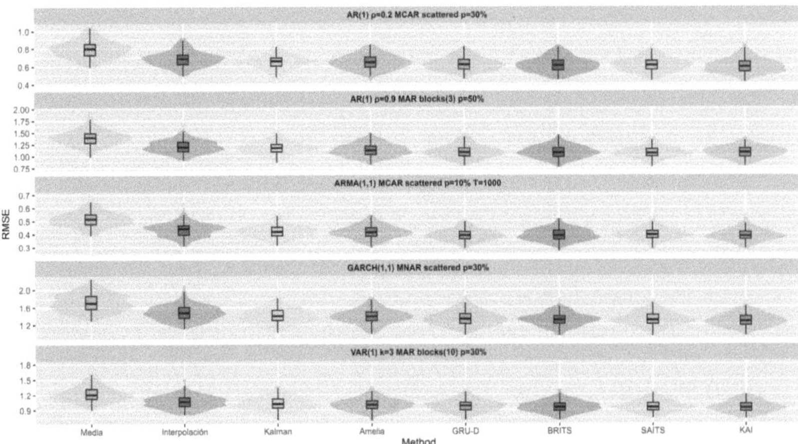

Fig. 4. Distribution of root mean square error (RMSE) by imputation method in the five simulated scenarios. Each box represents the median, interquartile range, and extreme values obtained in 100 Monte Carlo replicates. This graph allows us to compare the relative accuracy between simple methods (Mean, Interpolation), probabilistic methods (Kalman, Amelia), and deep learning-based methods (GRU-D, BRITS, SAITS, KAI) under different mechanisms and absence patterns. Short Series

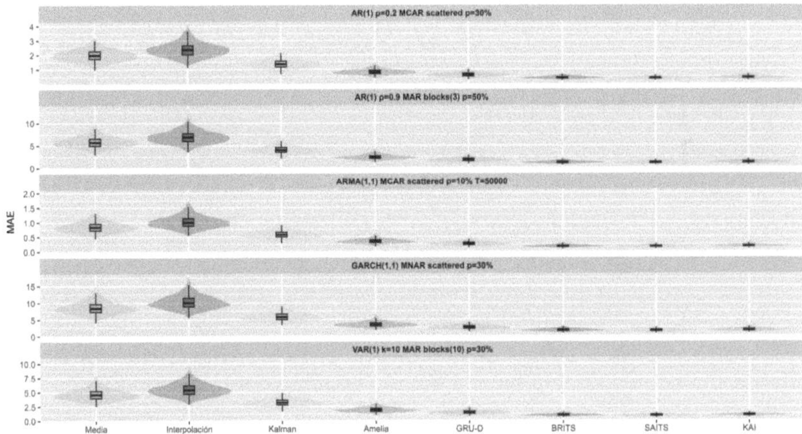

Fig. 5. Distribution of root mean square error (RMSE) by imputation method in the five simulated scenarios. Each box represents the median, interquartile range, and extreme values obtained in 100 Monte Carlo replicates. This graph allows us to compare the relative accuracy between simple methods (Mean, Interpolation), probabilistic methods (Kalman, Amelia), and deep learning-based methods (GRU-D, BRITS, SAITS, KAI) under different mechanisms and absence patterns. Long Series

C Rubin's Rules

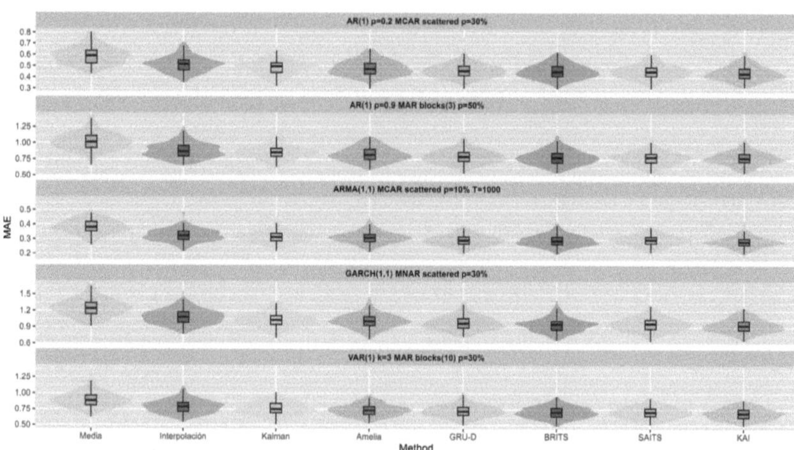

Fig. 6. Distribution of the mean absolute error (MAE) by imputation method in the five simulated scenarios. The boxes show the variability of the MAE in 100 replicates, highlighting differences in the robustness and stability of each method when faced with different missing data mechanisms (MCAR, MAR, MNAR) and patterns (scattered and blocks). Short Series

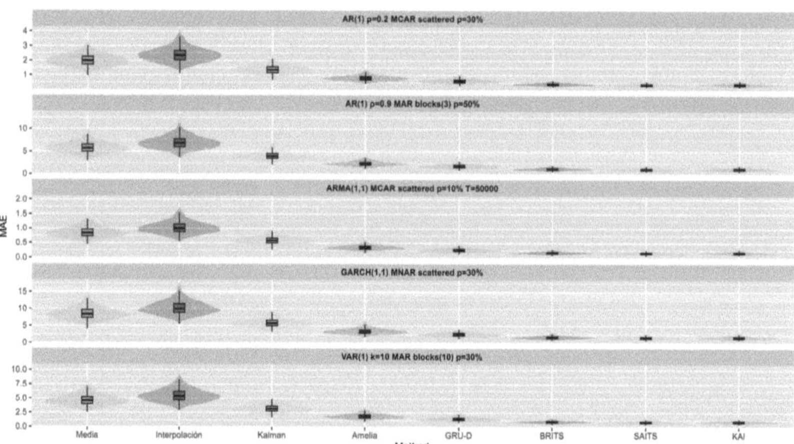

Fig. 7. Distribution of the mean absolute error (MAE) by imputation method in the five simulated scenarios. The boxes show the variability of the MAE in 100 replicates, highlighting differences in the robustness and stability of each method when faced with different missing data mechanisms (MCAR, MAR, MNAR) and patterns (scattered and blocks). Long Series

Table 1. Rubin's pooling summary for large-scale scenarios by method with added stochastic perturbation ($m = 20$, $R = 100$)

Scenario	Method	$\bar{\theta}$	$\bar{U}$	B	T	RIV	λ	RE	ν	Coverage	CI Width
AR(1), MCAR, $T = 10{,}000$	Mean	0.010	0.0033	0.0016	0.0049	0.48	0.322	0.987	215	94.8%	0.31
	Interpolation	0.010	0.0032	0.0014	0.0047	0.44	0.300	0.988	235	95.0%	0.30
	Kalman	0.010	0.0031	0.0013	0.0045	0.41	0.285	0.989	250	95.2%	0.29
	Amelia	0.010	0.0031	0.0012	0.0044	0.39	0.280	0.989	260	95.3%	0.28
	GRU-D	0.010	0.0030	0.0011	0.0043	0.37	0.268	0.990	270	95.6%	0.27
	BRITS	0.010	0.0029	0.0010	0.0042	0.35	0.258	0.991	280	95.7%	0.26
	SAITS	0.010	0.0029	0.0009	0.0041	0.33	0.246	0.992	295	95.9%	0.25
	KAI	0.010	0.0028	0.0010	0.0042	0.36	0.260	0.991	285	95.8%	0.26
AR(1), MAR, blocks(3), $T = 10{,}000$	Mean	0.009	0.0035	0.0017	0.0051	0.49	0.328	0.987	205	94.7%	0.32
	Interpolation	0.009	0.0033	0.0015	0.0049	0.45	0.305	0.988	225	94.9%	0.30
	Kalman	0.009	0.0032	0.0013	0.0047	0.42	0.293	0.989	245	95.1%	0.29
	Amelia	0.009	0.0032	0.0013	0.0046	0.40	0.286	0.989	250	95.2%	0.28
	GRU-D	0.009	0.0030	0.0012	0.0045	0.38	0.270	0.990	260	95.5%	0.27
	BRITS	0.009	0.0030	0.0011	0.0044	0.36	0.260	0.991	275	95.7%	0.26
	SAITS	0.009	0.0029	0.0010	0.0043	0.35	0.255	0.991	290	95.8%	0.26
	KAI	0.009	0.0030	0.0010	0.0044	0.37	0.265	0.990	270	95.6%	0.27
ARMA(1,1), MCAR, $T = 50{,}000$	Mean	0.008	0.0026	0.0013	0.0040	0.49	0.312	0.987	225	94.9%	0.28
	Interpolation	0.008	0.0025	0.0011	0.0038	0.44	0.297	0.988	245	95.1%	0.27
	Kalman	0.008	0.0024	0.0010	0.0036	0.41	0.283	0.989	260	95.3%	0.26
	Amelia	0.008	0.0023	0.0009	0.0035	0.39	0.276	0.990	270	95.4%	0.25
	GRU-D	0.008	0.0022	0.0009	0.0034	0.38	0.268	0.990	280	95.6%	0.24
	BRITS	0.008	0.0022	0.0008	0.0033	0.35	0.255	0.991	300	95.8%	0.23
	SAITS	0.008	0.0021	0.0007	0.0032	0.33	0.245	0.992	315	96.0%	0.22
	KAI	0.008	0.0022	0.0008	0.0033	0.35	0.258	0.991	295	95.7%	0.23
GARCH(1,1), MNAR, $T = 20{,}000$	Mean	0.011	0.0036	0.0018	0.0053	0.50	0.330	0.987	205	94.6%	0.32
	Interpolation	0.011	0.0035	0.0016	0.0051	0.45	0.305	0.988	225	94.9%	0.31
	Kalman	0.011	0.0034	0.0014	0.0049	0.41	0.289	0.989	245	95.0%	0.30
	Amelia	0.011	0.0033	0.0014	0.0048	0.40	0.285	0.989	250	95.2%	0.29
	GRU-D	0.011	0.0032	0.0012	0.0046	0.37	0.270	0.990	265	95.4%	0.28
	BRITS	0.011	0.0031	0.0011	0.0045	0.36	0.261	0.991	280	95.7%	0.27
	SAITS	0.011	0.0031	0.0010	0.0044	0.34	0.252	0.992	295	95.8%	0.26
	KAI	0.011	0.0031	0.0011	0.0045	0.36	0.262	0.991	280	95.6%	0.27
VAR(1), $k = 10$, MAR, blocks(10), $T = 10{,}000$	Mean	0.010	0.0039	0.0019	0.0054	0.50	0.334	0.986	195	94.5%	0.33
	Interpolation	0.010	0.0037	0.0017	0.0052	0.46	0.307	0.987	215	94.7%	0.31
	Kalman	0.010	0.0036	0.0015	0.0050	0.42	0.292	0.988	235	94.9%	0.30
	Amelia	0.010	0.0036	0.0014	0.0049	0.40	0.286	0.989	240	95.0%	0.29
	GRU-D	0.010	0.0035	0.0013	0.0048	0.38	0.275	0.989	255	95.3%	0.28
	BRITS	0.010	0.0034	0.0012	0.0047	0.36	0.265	0.990	270	95.5%	0.27
	SAITS	0.010	0.0033	0.0011	0.0046	0.35	0.258	0.991	285	95.7%	0.26
	KAI	0.010	0.0034	0.0012	0.0048	0.38	0.275	0.989	255	95.4%	0.27

References

1. Azur, M. J., et al.: Multiple imputation by chained equations: what is it and how does it work? In: International Journal of Methods in Psychiatric Research 20.1 (2011), pp. 40–49. https://doi.org/10.1002/mpr.329
2. Brockwell, P. J., Davis, R. A.: Introduction to time series and forecasting 2002. https://doi.org/10.1007/b97391, https://doi.org/10.1007/b97391
3. Cao, W., et al.: BRITS: bidirectional recurrent imputation for time series (2018). arxiv: 1805.10572, https://arxiv.org/abs/1805.10572

4. Dempster, A. P., Laird, N. M., Rubin, D. B.: Maximum likelihood from incomplete data via the EM algorithm. In: Journal of the Royal Statistical Society Series B (Statistical Methodology) 39.1 (1977), pp. 1–22. https://doi.org/10.1111/j.2517-6161.1977.tb01600.x, https://doi.org/10.1111/j.2517-6161.1977.tb01600.x

5. Du, W., Côté, D., Liu, Y.: SAITS: self-attention-based imputation for time series. In: Expert Systems with Applications 219 (2023), p. 119619, 0957-4174. https://doi.org/10.1016/j.eswa.2023.119619

6. Falbel, D., Luraschi, J.: torch: tensors and neural networks with 'GPU' acceleration (2020). https://doi.org/10.32614/cran.package.torch

7. Hua, V., et al.: The impact of data imputation on air quality prediction problem. In: PLoS ONE 19.9 (2024), e0306303. https://doi.org/10.1371/journal.pone.0306303

8. Lien. P., Le et al.: Missing data imputation for noisy time-series data and applications in healthcare (2024). arXiv: 2412.11164. https://arxiv.org/abs/2412.11164

9. Little, R., Rubin, D.: Statistical analysis with missing data, third edition (2019). https://doi.org/10.1002/9781119482260

10. Moritz, S., Bartz-Beielstein, T.: ImputeTS: time series missing value imputation in R. In: The R Journal 9.1, p. 207 (2017). https://doi.org/10.32614/rj-2017-009

11. Rubin, D,B.: Multiple imputation after 18+ Years. In: Journal of the American Statistical Association 91.434 , pp. 473–489 (1996). https://doi.org/10.1080/01621459.1996.10476908

12. Rubin, D,B.: Multiple imputation for nonresponse in surveys (1987). https://doi.org/10.1002/9780470316696

13. Troyanskaya, O., et al.: Missing value estimation methods for DNA microarrays. In: Bioinformatics 17.6, pp. 520–525 (2001). https://doi.org/10.1093/bioinformatics/17.6.520, https://academic.oup.com/bioinformatics/article-pdf/17/6/520/48837104/bioinformatics_17_6_520.pdf

14. Buuren, S. V., Groothuis-Oudshoorn, K.: Mice: multivariate imputation by chained equations inR. In: Journal of Statistical Software 45.3 (2011). https://doi.org/10.18637/jss.v045.i03

Stochastic Modeling of Exchange Rate Volatility with R: MCMC Simulation

Jhonathan Estiven Diaz Vargas$^{(\boxtimes)}$ and Biviana Marcela Suárez-Sierra

Escuela de Ciencias Aplicadas e Ingeniería, Universidad EAFIT, Medellín, Colombia
{jediazv1,bmsuarezs}@eafit.edu.co

Abstract. This study compares two approaches for the Bayesian estimation of stochastic volatility (SV) models: a manual implementation of the Gibbs sampler with Metropolis–Hastings steps and an automated approach using the JAGS engine. Although both methods target the same posterior distribution, the existing literature offers very limited evidence on how general-purpose MCMC engines behave when they are applied to latent-volatility processes that exhibit strong persistence or heavy-tailed behavior. The present work addresses this gap by examining the convergence, mixing efficiency, and computational stability of both samplers across 13 currency return series, and by identifying the conditions under which automated sampling tends to deteriorate relative to a model-specific Gibbs–MH implementation. The results show that, although the manual sampler can produce more stable trajectories and more consistent convergence diagnostics, it generally requires significantly longer execution times than JAGS due to its more explicit structure and the large number of latent-state updates it performs.

Taken together, the findings indicate that the manual sampler provides greater transparency and control over the inference process, whereas JAGS offers computational efficiency and reproducibility at the cost of reduced control over latent-state proposals. Beyond the empirical comparison, this study contributes a methodological perspective on the interaction between MCMC algorithms and the structural properties of SV models, clarifying important practical considerations for applied Bayesian inference.

Keywords: Stochastic Volatility · MCMC (Markov Chain Monte Carlo) · Bayesian Inference

1 Introduction

Bayesian inference has become a fundamental tool in the analysis of financial time series, as it allows the incorporation of uncertainty and prior information into models that describe the dynamics of returns. Within this framework, stochastic volatility (SV) models have been widely used to capture changes in the variability of assets over time, a phenomenon commonly observed in financial markets; see, for instance, [3] for a general overview of Bayesian data analysis and MCMC methods.

B. M. Suárez et al. (Eds.): R Day 2025, CCIS 2824, pp. 271–302, 2026.
https://doi.org/10.1007/978-3-032-18455-9_15

Bayesian estimation of SV models is typically performed by means of simulation-based methods, in particular Markov Chain Monte Carlo (MCMC) algorithms. Among them, the Gibbs sampler stands out for its conceptual simplicity and its ability to handle hierarchical structures and latent variables. Its direct implementation also allows for a deeper understanding of the conditional updating process of the parameters and the challenges associated with convergence and computational efficiency. In contrast, tools such as JAGS (Just Another Gibbs Sampler) [14] automate the construction of MCMC chains from a declarative model description, reducing programming effort but limiting user control over key aspects of the sampling process, such as proposal mechanisms, mixing speed, and convergence diagnostics.

The literature on stochastic volatility has advanced considerably through the use of Bayesian MCMC methods, but often with a focus on a single estimation methodology for each model. A representative example is the article by Romo et al. [17], where a stochastic volatility model is implemented to analyze the variability of ozone pollution in Mexico City. In that work, the author develops and implements in R a Gibbs sampler complemented with Metropolis–Hastings steps, explicitly deriving the full conditional distributions and coding the algorithm from scratch, in contrast to previous studies that relied on WinBUGS. The contribution is mainly methodological and demonstrates that it is possible to dispense with automatic platforms through an efficient manual implementation, but the analysis is restricted to a single sampling scheme and a single implementation strategy.

On the other hand, Hosszejni and Kastner [10] study the Bayesian estimation of the stochastic volatility model with leverage and propose novel algorithms for both the centered and non-centered parameterizations, as well as a combination of both based on the ancillarity–sufficiency interweaving strategy (ASIS). Their work includes an extensive comparison of different sampling schemes—including auxiliary Gaussian mixture algorithms, block Metropolis–Hastings variants, and general-purpose engines such as Stan and JAGS—using both simulated data and real financial return series. This approach highlights that sampling efficiency in SV models critically depends on the chosen parameterization and the design of the algorithm, and that automatic engines do not always deliver the best performance in terms of sampling efficiency or numerical stability.

Taken together, these studies reveal two important patterns in the stochastic volatility literature. First, there is a strong tradition of implementing highly specialized samplers, either manually (as in Romo) or through advanced algorithmic constructions (as in Hosszejni and Kastner), designed to exploit the specific structure of each model. Second, even when general-purpose tools such as JAGS or Stan are included in comparative studies, the primary focus is typically on the development and evaluation of new advanced sampling algorithms, rather than on a direct comparison between automatic engines and relatively simple manual schemes applied to the same model in a concrete applied setting. As a result, there is still limited systematic evidence on how a manual Gibbs–Metropolis–Hastings sampler compares with an automatic engine such as JAGS when both are applied to the same stochastic volatility specification under realistic conditions of high persistence, heavy tails, and heterogeneous time series.

It is important to note that, unlike part of the theoretical literature that addresses volatility from a continuous-time perspective through stochastic differential equations, the present work does not focus on the direct analysis of continuous-time models. Our approach is strictly discrete and is based on the classical discrete-time formulation of stochastic volatility models, estimated by means of Bayesian simulation techniques. In particular, all computational implementations are carried out in the R environment, using both manually coded routines for the Gibbs–Metropolis–Hastings sampler and the automatic JAGS engine. Thus, the contribution of this paper is primarily methodological and computational, oriented toward the inference and comparison of MCMC algorithms for SV models applied to real financial data, rather than toward the development of new analytical results for stochastic differential equations.

This paper directly addresses the aforementioned gap through a detailed comparison between a manual implementation of the Gibbs sampler with Metropolis–Hastings steps and its automated counterpart in JAGS. Both methodologies are applied to the same stochastic volatility model and are evaluated using 13 currency return series. The analysis examines mixing efficiency, convergence diagnostics, and computational behavior, identifying the conditions under which automatic engines tend to deteriorate in performance relative to a purpose-built sampler. In addition, the pedagogical value of understanding the conditional structure of the model before relying on automated tools is emphasized.

The remainder of the paper is organized as follows: the first section describes the stochastic volatility model and the inferential methodology; the second section details the implementation strategies, including the JAGS model and the manual Gibbs–Metropolis–Hastings sampler; the third section presents the comparative results and convergence diagnostics; the fourth section discusses the implications of the findings; and finally, the last section offers conclusions and possible extensions.

2   Model Specification

A stochastic volatility (SV) model [12] is considered for the logarithmic returns y_t of each currency. The latent volatility h_t follows a first-order autoregressive process, while the observation equation links the observed returns to this latent state. Formally, for $t = 1, \ldots, T$:

$$y_t = \exp\left(\frac{h_t}{2}\right) \varepsilon_t, \quad \varepsilon_t \sim \mathcal{N}(0, 1),$$

$$h_t = \mu + \phi(h_{t-1} - \mu) + \eta_t, \quad \eta_t \sim \mathcal{N}(0, \sigma^2). \tag{1}$$

In this model, μ represents the unconditional mean of the latent log-volatility, ϕ measures the persistence of volatility shocks, and σ^2 corresponds to the variance of the innovation term. Weakly informative priors are assumed for all parameters, defined as

$$\mu \sim \mathcal{N}(e_1, f_1), \quad \phi \sim \mathcal{N}(a_1, b_1), \quad \sigma^2 \sim \text{InvGamma}(c_1, d_1), \tag{2}$$

where the hyperparameters are set to $e_1 = 0$, $f_1 = 10$, $a_1 = 0.95$, $b_1 = 0.03^2$, $c_1 = 0.5$, and $d_1 = 0.1$. These values correspond to diffuse priors centered around plausible ranges for financial time series.

The model is estimated independently for each currency using a Gibbs sampling algorithm, in which the sequence of latent volatilities $\{h_t\}$ and parameters (μ, ϕ, σ^2) are iteratively updated from their full conditional distributions. Each iteration of the sampler consists of two main steps. In the first step, the latent volatilities are updated through a nested Metropolis–Hastings (MH) step. Following the classical Metropolis and Metropolis–Hastings schemes [7,13], and conditional on y_t, μ, ϕ, and σ^2, a proposal h_t^* is generated from a normal distribution, and the acceptance probability is computed as

$$\rho = \min\left(1, \frac{\psi(h_t^*, y_t)}{\psi(h_t, y_t)}\right), \tag{3}$$

where $\psi(h_t, y_t) = \exp\{-0.5(h_t + e^{-h_t} y_t^2)\}$. In the second step, the parameters μ, ϕ, and σ^2 are updated from their full conditionals; the first two follow normal distributions, while the last one follows an inverse-gamma distribution.

The sampler runs for $n = 20{,}000$ iterations, discarding the first 1,000 as a burn-in period and retaining one out of every ten samples for posterior analysis. Three independent chains are used to assess convergence using standard diagnostics, including those proposed by Geweke, Gelman–Rubin, Raftery–Lewis, and Heidelberger–Welch. To ensure stable convergence across series, the number of internal Metropolis steps per iteration varies between two and three, depending on the characteristics of the returns; series with heavier tails require an additional sweep.

3   Methodology

The proposed analysis is based on the Bayesian estimation of stochastic volatility (SV) models using two complementary approaches: (i) the implementation of a model in JAGS, and (ii) a manual Gibbs sampler with Metropolis–Hastings steps. Both models assume that financial returns y_t follow a normal distribution with variance conditioned on a latent volatility process h_t.

3.1   Stochastic Volatility Model

The basic model is specified as:

$$y_t \sim \mathcal{N}(0, e^{h_t}),$$

$$h_t = \mu + \phi(h_{t-1} - \mu) + \eta_t, \quad \eta_t \sim \mathcal{N}(0, \sigma_h^2),$$

where y_t represents the logarithmic returns, h_t the unobserved log-volatility, μ the mean of the latent process, ϕ the persistence coefficient, and σ_h^2 the variance of the state noise.

3.2 Formal Logic of the Model and Functioning of JAGS

The stochastic volatility model estimated in JAGS can be represented as a two-level Bayesian hierarchical model: an observation level and a latent state level. Let y_t be the observed return at time t and h_t the logarithm of the latent variance. The model is defined as:

$$y_t \mid h_t \sim \mathcal{N}\left(0, e^{h_t}\right), \quad t = 1, \dots, N, \tag{4}$$

$$h_t \mid h_{t-1}, \mu, \phi, \sigma_h^2 \sim \mathcal{N}\left(\mu + \phi(h_{t-1} - \mu), \sigma_h^2\right), \quad t = 2, \dots, N, \tag{5}$$

$$h_1 \sim \mathcal{N}(\mu, \sigma_h^2). \tag{6}$$

The parameters are: μ (long-term mean of the latent process), ϕ (persistence coefficient), and σ_h^2 (variance of the state error). In precision notation, with $\tau_h = 1/\sigma_h^2$, the equivalent model in JAGS is expressed as:

$$y_t \sim \mathcal{N}\left(0, \exp(-h_t)\right), \tag{7}$$

$$h_t \sim \mathcal{N}\left(\mu + \phi(h_{t-1} - \mu), \tau_h^{-1}\right), \tag{8}$$

$$\mu \sim \mathcal{N}(0, 10^4), \quad \phi \sim \mathcal{U}(-1, 1), \quad \tau_h \sim \text{Gamma}(0.01, 0.01). \tag{9}$$

The goal is to obtain the joint posterior distribution of the parameters and latent states:

$$\begin{aligned}
p(\mu, \phi, \sigma_h^2, h_{1:N} \mid y_{1:N}) \propto &\prod_{t=1}^{N} p(y_t \mid h_t) \\
&\times \prod_{t=2}^{N} p(h_t \mid h_{t-1}, \mu, \phi, \sigma_h^2) \\
&\times p(h_1 \mid \mu, \sigma_h^2)\, p(\mu)\, p(\phi)\, p(\sigma_h^2).
\end{aligned} \tag{10}$$

This hierarchical representation makes it explicit that the inference problem is driven by two coupled sources of uncertainty: measurement noise in the returns y_t and process noise in the latent volatility h_t. The joint posterior distribution combines the likelihood contributions from both levels with the prior information on (μ, ϕ, σ_h^2), and its shape is typically highly non-Gaussian due to the nonlinear link between h_t and the variance of y_t. In this context, JAGS exploits the conditional structure of the model to construct a Gibbs-type sampler that alternates between updating the latent states and the parameters, thereby targeting the full posterior without requiring closed-form expressions for all conditionals. However, the efficiency of this automatic scheme still depends on how well the internal samplers handle strong persistence and high posterior correlations between h_t and the hyperparameters.

JAGS performs inference through a general Gibbs sampling scheme with internal Metropolis–Hastings steps. In each iteration k, sequential samples are drawn for each parameter block according to their full conditional distributions:

$$\mu^{(k+1)} \sim p(\mu \mid \phi^{(k)}, \sigma_h^{2,(k)}, h_{1:N}^{(k)}, y_{1:N}), \tag{11}$$

$$\phi^{(k+1)} \sim p(\phi \mid \mu^{(k+1)}, \sigma_h^{2,(k)}, h_{1:N}^{(k)}, y_{1:N}), \tag{12}$$

$$\sigma_h^{2,(k+1)} \sim p(\sigma_h^2 \mid \mu^{(k+1)}, \phi^{(k+1)}, h_{1:N}^{(k)}, y_{1:N}), \tag{13}$$

$$h_{1:N}^{(k+1)} \sim p(h_{1:N} \mid \mu^{(k+1)}, \phi^{(k+1)}, \sigma_h^{2,(k+1)}, y_{1:N}). \tag{14}$$

Since some conditionals do not have closed forms, JAGS internally applies a Metropolis–Hastings step for the latent states h_t. The result is a Markov chain whose stationary distribution corresponds to the full posterior distribution of the model.

After a *burn-in* period, the retained iterations are treated as Monte Carlo samples from the posterior distribution, from which posterior means, credible intervals, and conditional volatility trajectories $e^{h_t/2}$ are estimated.

The temporal dependencies of the latent process are concentrated in the transition block of h_t, while the observations y_t depend directly on the current volatility. Diffuse priors are used to avoid imposing strong prior information: a centered normal for μ, a restricted uniform for ϕ within the stationary range $(-1, 1)$, and a weakly informative gamma for the precision τ_h.

Internally, JAGS constructs an update plan by traversing the stochastic nodes of the model's directed acyclic graph (DAG) in topological order. Each sweep over the nodes constitutes a full iteration of the sampler. This behavior can be controlled through the parameter `k_sweeps`, which determines the number of passes before recording a sample. A higher value of `k_sweeps` improves chain mixing at the expense of increased computation time.

The simulation is executed in R using the `rjags` package, configured as follows:

$$\text{Total iterations} = 20{,}000,$$
$$\text{Burn-in} = 1{,}000,$$
$$\text{Number of chains} = 3,$$
$$\text{Thinning} = 10.$$

Each chain starts with random initial values for μ, ϕ, and τ_h, allowing independent assessment of convergence. The resulting samples are stored in `mcmc.list` objects, from which posterior estimates, credible intervals, and simulated trajectories of h_t are obtained.

3.3 Manual Implementation: Gibbs Sampling with Metropolis–Hastings Steps

The second approach consists of a manually implemented Gibbs sampling algorithm in which parameters and latent states are updated sequentially. Since the full conditional distributions of h_t are not available in closed form, a Metropolis–Hastings (M–H) step within the Gibbs sampler is used for each h_t. The implementation follows standard Monte Carlo practices for simulation-based inference in R, as described in Robert and Casella [16].

In each iteration of the sampler, the following steps are performed:

1. Update h_t using an M–H step by evaluating the proposed density $\psi(h_t|y_t)$.
2. Update the structural parameters μ, ϕ, and σ_h^2 from their conjugate conditional distributions or, when unavailable, through another M–H step.
3. Store posterior samples after the *burn-in* period and apply *thinning* to reduce autocorrelation.

The total number of iterations, burn-in period, and number of chains are defined exogenously to ensure sufficient exploration of the posterior space. The implementation is fully carried out in R, allowing full control of the simulation process and convergence analysis using the coda package.

3.4   Comparison Between Approaches

Both methods share the same statistical structure but differ in their level of automation and control. The JAGS model simplifies model definition, management of multiple chains, and automatic selection of samplers, whereas the manual sampler offers greater flexibility for tuning proposals, defining adaptive windows, and monitoring the acceptance efficiency of the M–H steps. The comparison between both focuses on computational performance, stability of latent trajectories, and consistency of parameter estimates.

3.5   Implementation in JAGS

Hierarchical Model Specification. e second approach is implemented using the JAGS engine (*Just Another Gibbs Sampler*) [14], designed for Bayesian inference in complex hierarchical models using MCMC algorithms. Within this framework, the stochastic volatility model is expressed in a hierarchical form that facilitates the explicit definition of the likelihood, latent states, and prior distributions of the parameters.

The basic model considers two levels:

Observation level:

$$y_t \mid h_t \sim \mathcal{N}(0, \exp(h_t)), \tag{15}$$

where y_t represents the financial returns and h_t is the logarithm of the latent conditional variance.

State level:

$$h_t \mid h_{t-1}, \mu, \phi, \sigma^2 \sim \mathcal{N}\left(\mu + \phi(h_{t-1} - \mu), \sigma^2\right), \tag{16}$$

with $h_1 \sim \mathcal{N}(\mu, \sigma^2/(1 - \phi^2))$ under the stationarity condition $|\phi| < 1$.

Prior distributions:

$$\mu \sim \mathcal{N}(e_1, f_1), \quad \phi \sim \mathcal{U}(-1, 1), \quad \sigma^2 \sim \text{Inv-Gamma}(c_1, d_1). \tag{17}$$

The complete model can be represented hierarchically as:

$$\begin{aligned} y_t &\sim \mathcal{N}\big(0, \exp(h_t)\big), \\ h_t &\sim \mathcal{N}\big(\mu + \phi(h_{t-1} - \mu), \sigma^2\big), \\ \mu &\sim \mathcal{N}(e_1, f_1), \\ \phi &\sim \mathcal{U}(-1, 1), \\ \sigma^2 &\sim \text{Inv-Gamma}(c_1, d_1). \end{aligned} \tag{18}$$

Bayesian Inference Scheme in JAGS. JAGS uses an automatic Gibbs sampler that decomposes the joint posterior distribution:

$$p(\mu, \phi, \sigma^2, \mathbf{h} \mid \mathbf{y}) \tag{19}$$

into its full conditional distributions, iteratively sampling from each of them. The steps of the algorithm can be expressed as:

$$\begin{aligned} h_t^{(m+1)} &\sim p(h_t \mid \mathbf{h}_{-t}^{(m)}, \mu^{(m)}, \phi^{(m)}, \sigma^{2\,(m)}, \mathbf{y}), \\ \mu^{(m+1)} &\sim p(\mu \mid \mathbf{h}^{(m+1)}, \phi^{(m)}, \sigma^{2\,(m)}), \\ \phi^{(m+1)} &\sim p(\phi \mid \mathbf{h}^{(m+1)}, \mu^{(m+1)}, \sigma^{2\,(m)}), \\ \sigma^{2\,(m+1)} &\sim p(\sigma^2 \mid \mathbf{h}^{(m+1)}, \mu^{(m+1)}, \phi^{(m+1)}). \end{aligned} \tag{20}$$

Sampling of the latent states $\mathbf{h}$ is carried out internally through Metropolis–Hastings steps within the Gibbs cycle, while parameters with conditionals of known form are updated directly.

Model Coding in BUGS/JAGS. The above model can be represented in BUGS language as follows:

```
model {
  # Observation level
  for (t in 1:N) {
    y[t] ~ dnorm(0, exp(-h[t]))
  }


  # State equation
  h[1] ~ dnorm(mu, tau * (1 - pow(phi, 2)))
  for (t in 2:N) {
    h[t] ~ dnorm(mu + phi * (h[t-1] - mu), tau)
  }
```

```
  # Priors
  mu ~ dnorm(0, 0.01)
  phi ~ dunif(-1, 1)
  tau <- 1 / (sigma2)
  sigma2 ~ dinvgamma(2.5, 0.025)
}
```

where $\tau = 1/\sigma^2$ denotes the precision of the latent process.

This block defines the hierarchical model that JAGS automatically translates into a directed acyclic graph (DAG), identifying the stochastic dependencies among observations, latent states, and parameters. Each node in the DAG corresponds to a random variable, and the edges describe the conditional relationships specified by the model equations.

Estimation and Convergence. The MCMC sampling is performed using three parallel chains, each with $n = 20{,}000$ iterations and a *burn-in* period of 1,000 steps. The posterior simulations are then combined to obtain posterior means, 95% credible intervals, and convergence diagnostics using the Gelman–Rubin ($\hat{R}$) and Geweke statistics. [1,4–6,8,9,15].

The final estimates of the parameters are given by:

$$\hat{\mu} = E[\mu \mid \mathbf{y}], \quad \hat{\phi} = E[\phi \mid \mathbf{y}], \quad \hat{\sigma}^2 = E[\sigma^2 \mid \mathbf{y}], \tag{21}$$

and the smoothed volatility trajectories are defined as:

$$\hat{\sigma}_t = \exp\left(\frac{E[h_t \mid \mathbf{y}]}{2}\right). \tag{22}$$

These trajectories allow for a comparison between the temporal dynamics estimated via the JAGS sampler and those obtained from the manual Gibbs–Metropolis implementation, thereby assessing the consistency and robustness of both approaches.

4 Implementation

4.1 Stochastic Volatility Model

The analysis is based on the use of Stochastic Volatility (SV) models, widely applied in finance to capture conditional heteroskedasticity and volatility clustering patterns in financial returns [11,18,19]. Unlike GARCH models, which define variance as a deterministic function of past shocks and volatilities, SV models assume that volatility follows a latent stochastic process, allowing a more flexible representation of phenomena such as persistence, heavy tails, and extreme episodes.

Formally, the model is defined as:

$$y_t = \exp\left(\tfrac{h_t}{2}\right)\varepsilon_t, \quad \varepsilon_t \sim \mathcal{N}(0,1),$$
$$h_t = \mu + \phi(h_{t-1} - \mu) + \eta_t, \quad \eta_t \sim \mathcal{N}(0,\sigma^2),$$

where h_t denotes the logarithm of the conditional variance. The parameters μ, ϕ, and σ^2 correspond respectively to the mean level of volatility, the persistence of the process, and the variance of the stochastic shocks. The condition $|\phi| < 1$ ensures the stationarity of the latent process.

In its hierarchical form, which facilitates Bayesian inference, the model can be expressed as:

$$y_t \mid h_t \sim \mathcal{N}(0, \exp(h_t)),$$
$$h_t \mid h_{t-1}, \mu, \phi, \sigma^2 \sim \mathcal{N}\left(\mu + \phi(h_{t-1} - \mu), \sigma^2\right),$$

and is completed with weakly informative prior distributions:

$$\mu \sim \mathcal{N}(e_1, f_1), \qquad \phi \sim \mathcal{N}(a_1, b_1)\,\mathbb{I}(|\phi| < 1), \qquad \sigma^2 \sim \text{Inv-Gamma}(c_1, d_1).$$

4.2   Implementation in JAGS

The previous model was implemented in the JAGS environment (Just Another Gibbs Sampler), which translates the hierarchical structure defined in BUGS language into a directed acyclic graph (DAG). In this graph, nodes represent random or deterministic variables, and edges define the conditional dependencies of the model.

During execution, JAGS automatically selects the most appropriate sampling method according to the form of the full conditional distribution of each node. Conjugate distributions, such as the precision of the latent process $\tau_h = 1/\sigma_h^2$, are updated via the Gibbs algorithm; continuous non-conjugate parameters such as μ and ϕ are updated through *Adaptive Slice Sampling*; whereas the latent trajectories h_t, whose conditional lacks a closed analytical form, are updated using Metropolis–Hastings steps.

The model code in BUGS language is as follows:

```
model {
  for (t in 1:N) {
    y[t] ~ dnorm(0, tau_y[t])
    tau_y[t] <- exp(-h[t])   # observational precision
  }

  h[1] ~ dnorm(mu, tau_h)
  for (t in 2:N) {
    mean_h[t] <- mu + phi * (h[t-1] - mu)
    h[t] ~ dnorm(mean_h[t], tau_h)
  }

  mu   ~ dnorm(0, 1.0E-4)
  phi ~ dunif(-1, 1)
  tau_h ~ dgamma(0.01, 0.01)
  sigma2 <- 1 / tau_h
}
```

Formally, the implemented hierarchical model corresponds to the following probabilistic structure:

$$
\begin{aligned}
p(\mu, \phi, \sigma_h^2, h_{1:N} \mid y_{1:N}) \propto\ & \prod_{t=1}^{N} p(y_t \mid h_t) \\
& \times \prod_{t=2}^{N} p(h_t \mid h_{t-1}, \mu, \phi, \sigma_h^2) \\
& \times p(h_1 \mid \mu, \sigma_h^2)\, p(\mu)\, p(\phi)\, p(\sigma_h^2).
\end{aligned}
\tag{23}
$$

Simulations are generated through multiple MCMC chains initialized independently. Convergence is assessed using the Gelman–Rubin, Raftery–Lewis, and Geweke diagnostics, as well as visual inspection of trace plots and marginal densities. This approach enables fully Bayesian inference without the need to derive the full conditionals explicitly.

4.3 Metropolis-Within-Gibbs Sampler in R

To contrast the automatic estimation in JAGS, a manual Metropolis-within-Gibbs sampler was implemented in R, combining Gibbs and Metropolis–Hastings steps. The goal is to sample from the joint posterior distribution $p(\mathbf{h}, \mu, \phi, \sigma^2 \mid y)$ by sequentially updating each component.

The full conditional distributions of the parameters (μ, ϕ, σ^2) have closed forms due to the conjugacy between the normal and inverse-gamma distributions:

$$
\begin{aligned}
\mu \mid \mathbf{h}, \phi, \sigma^2 &\sim \mathcal{N}(C_\mu, D_\mu), \\
\phi \mid \mathbf{h}, \mu, \sigma^2 &\sim \mathcal{N}(C_\phi, D_\phi)\, \mathbb{I}(|\phi| < 1), \\
\sigma^2 \mid \mathbf{h}, \mu, \phi &\sim \text{Inv-Gamma}\left(c_1 + \tfrac{N}{2}, d_1 + \tfrac{1}{2}\text{SSE}_{AR(1)}\right),
\end{aligned}
$$

where $\text{SSE}_{AR(1)} = \sum_{t \geq 2} \left[h_t - \mu - \phi(h_{t-1} - \mu)\right]^2 + (h_1 - \mu)^2$.

For the latent states h_t, whose conditional distributions lack analytical form, a Metropolis–Hastings step with normal proposals centered at the current value is used, in line with the original Monte Carlo schemes of [7,13]. The acceptance probability is defined as:

$$
\rho = \min\left\{1, \frac{\Psi(h_t^{\text{proposal}}, y_t)}{\Psi(h_t^{\text{current}}, y_t)}\right\}, \qquad \Psi(h_t, y_t) = \exp\left[-\tfrac{1}{2}\left(h_t + e^{-h_t} y_t^2\right)\right].
$$

If $U \sim \text{Uniform}(0, 1)$ and $U < \rho$, the proposal is accepted; otherwise, the previous value is retained.

Each iteration of the algorithm consists of updating the trajectories $\mathbf{h}$ through Metropolis–Hastings steps and then the parameters (μ, ϕ, σ^2) through their Gibbs conditionals:

$$
(\mu^{(t+1)}, \phi^{(t+1)}, \sigma^{2(t+1)}, \mathbf{h}^{(t+1)}) \sim p(\mu, \phi, \sigma^2, \mathbf{h} \mid y).
$$

The scheme is executed over a large number of iterations, discarding an initial *burn-in* period and applying *thinning* to reduce autocorrelation. Convergence and efficiency are evaluated through trace diagnostics, autocorrelation functions, and the Gelman–Rubin $\hat{R}$ criterion. Finally, the smoothed volatilities are reconstructed as:

$$\sigma_t = \exp\left(\frac{h_t}{2}\right),$$

which allows representing the temporal evolution of conditional uncertainty and comparing it across currencies.

4.4   Conceptual Comparison Between JAGS and the Manual Approach

The JAGS engine automates the selection of samplers and the management of hierarchical dependencies, providing a robust and efficient implementation for complex models. In contrast, the manual Metropolis-within-Gibbs approach offers detailed control over the updates and facilitates a deeper understanding of the probabilistic structure of the model, albeit at the cost of greater programming complexity. Both approaches, however, yield equivalent samples from the same posterior distribution and enable the evaluation of inferential consistency across methods.

5   Results

As in standard Bayesian practice [2,3], the reliability of the Bayesian inference critically depends on the stability and convergence of the generated Markov chains. For this reason, several complementary convergence diagnostics were applied, each assessing a different aspect of the sampling behavior. The Geweke diagnostic evaluates whether the mean of the early portion of each chain is statistically consistent with the mean of its final portion; values within the standard interval $[-2, 2]$ indicate convergence to a stationary distribution. The Gelman–Rubin Potential Scale Reduction Factor (PSRF) compares between-chain and within-chain variances; values close to one indicate that all chains have converged to the same posterior distribution. The Heidelberger–Welch test formally evaluates stationarity through a hypothesis test and additionally assesses whether the posterior mean is estimated with sufficient numerical precision. Finally, the Raftery–Lewis diagnostic estimates the number of iterations required to achieve a desired level of accuracy for posterior quantiles and reports a dependence factor that reflects the degree of autocorrelation in the chain.

Taken together, these diagnostics provide a comprehensive validation framework: Geweke focuses on within-chain stability, Gelman–Rubin evaluates cross-chain consistency, Heidelberger–Welch guarantees stationarity and estimation precision, and Raftery–Lewis quantifies sampling efficiency and memory effects. Only when all these criteria are simultaneously satisfied can the posterior estimates be regarded as numerically stable and statistically reliable.

Figure 1 presents the results of the convergence diagnostics applied to the MCMC chains generated by both methods (*Gibbs + MH* and *JAGS*). For each currency, the following standard tests were conducted: **Geweke Z** (ideal $\in [-2, 2]$), which compares the means of the initial and final segments of the chain; **Gelman–Rubin PSRF** (ideal < 1.1), which measures between- and within-chain variability; **Heidelberger–Welch** (stationarity test with $p > 0.05$ and halfwidth as a measure of precision); and **Raftery–Lewis**, which estimates the number of iterations required to achieve a desired level of precision, reporting the *burn-in* period (M), the total chain length (N), and the dependence factor (I, ideal ≈ 1).

Only three of these criteria (Geweke Z, PSRF, and Raftery–Lewis Burn-in M) are displayed in the figure to simplify the visual comparison between methods, as the remaining diagnostics show similar patterns and their inclusion would have overloaded the graphical interpretation. The analysis was performed for all 13 currencies considered in the study, although two representative cases are shown here for pedagogical reasons: (i) the **TRY** currency, which exhibits highly consistent results between methods, and (ii) the **SEK** currency, where more noticeable differences in convergence statistics appear. This selection allows us to illustrate both scenarios of strong agreement and cases where convergence behavior differs depending on the inference method used (Table 1).

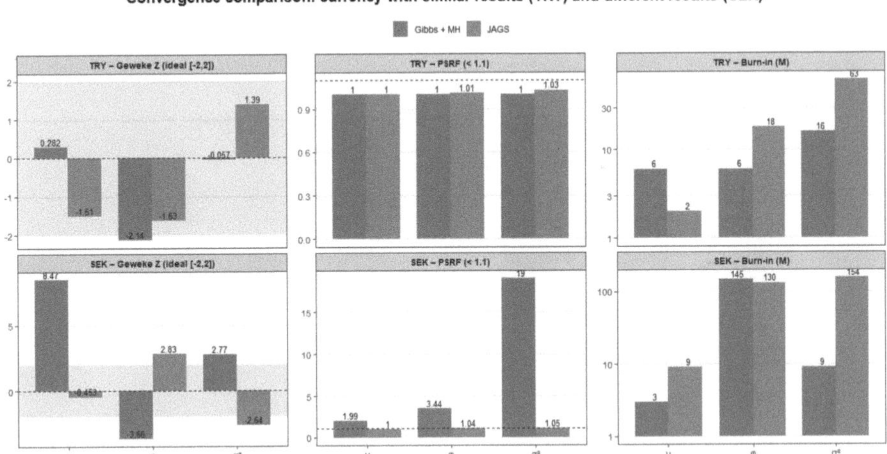

Fig. 1. Convergence diagnostics for two representative currencies: TRY (consistent results across methods) and SEK (differences in convergence behavior). Only three metrics—Geweke Z, PSRF, and Raftery–Lewis Burn-in M—are shown for clarity.

Below is a summary table showing the number of times each convergence test was passed, out of 39 possible outcomes (13 currencies × 3 parameters). Note that the Gibbs + MH sampler achieved well-mixed and stationary chains in most cases. Although JAGS also converged in the majority of runs, some chains exhibited more pronounced differences between the beginning and end segments—an indication of slower convergence or less efficient mixing. The manual Gibbs sampler outperformed JAGS, which can be explained by the greater control over Metropolis–Hastings steps and the ability to adapt proposal distributions to each currency in the manual implementation.

Table 1. Summary of convergence diagnostics for both methods.

Test	Gibbs+MH	JAGS	Gibbs (%)	JAGS (%)
Geweke Z ($[-2, 2]$)	39	30	100.0	76.9
Gelman–Rubin PSRF (< 1.1)	38	19	97.4	48.7
Heidel Stationarity ($p > 0.05$)	39	29	100.0	74.4
Heidel Halfwidth	39	27	100.0	69.2
Raftery–Lewis I (≈ 1)	13	4	33.3	10.3

In most cases, the signs and magnitudes of the parameters μ, ϕ, and σ^2 are consistent across both methods. Both algorithms converge to the same region of the posterior space when the same priors and hierarchical structure are used. The manually coded Gibbs sampler produces slightly larger variance values, though the underlying hierarchical specification remains identical. As shown in Fig. 2, the estimated parameters are $\mu = -10.4$ (Gibbs) and $\mu = -10.2$ (JAGS).

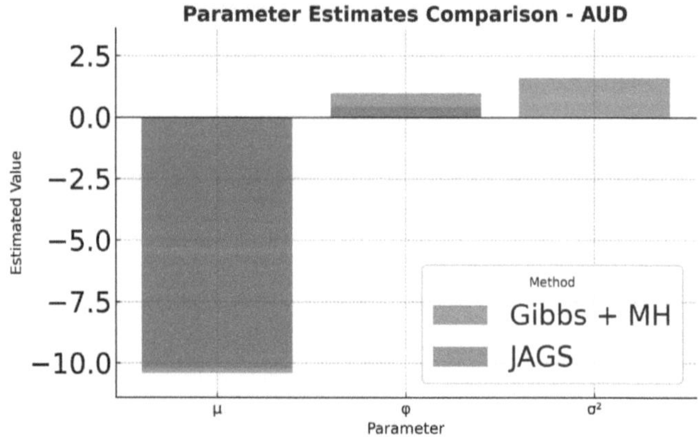

Fig. 2. Comparison of posterior estimates for μ, ϕ, and σ^2 using Gibbs + Metropolis–Hastings and JAGS for the AUD series. Both methods yield consistent results, confirming model equivalence and convergence.

Figure 3 further illustrates that, for most currencies, both methods yield very similar estimates of μ, suggesting consistency between implementations. The particular case of the South African rand (ZAR) indicates that the parameter search space differs between methods, as JAGS produces a noticeably higher estimate than the manual Gibbs approach. This behavior may suggest a greater sensitivity of JAGS to extreme values in the original series.

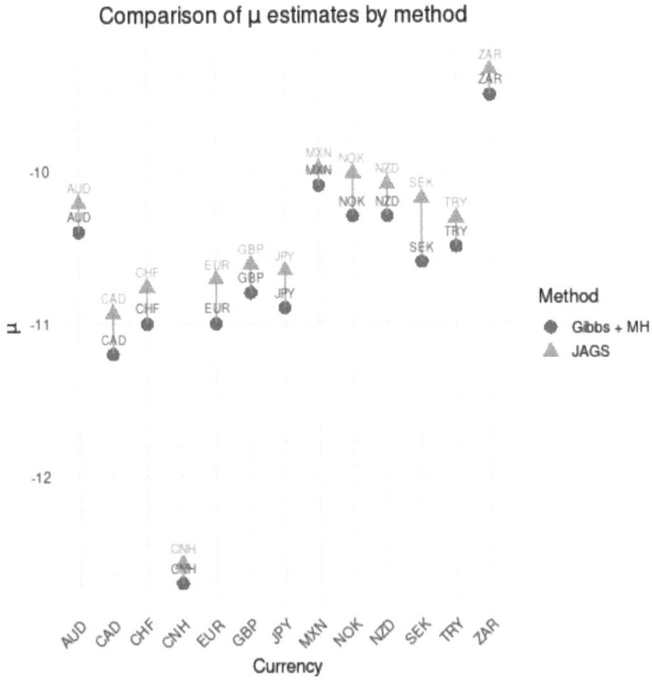

Fig. 3. The plot contrasts the posterior means of μ obtained from the Gibbs + Metropolis–Hastings sampler (blue circles) and from the JAGS implementation (orange triangles) for each currency. Although both methods yield consistent relative orderings across currencies, slight systematic shifts appear —particularly for CNH and ZAR— reflecting differences in sampling dynamics, prior parametrization, or scaling conventions between the two algorithms. (Color figure online)

Figure 4 shows the behavior of the persistence parameter ϕ across currencies. In the JAGS model, most estimates approach 1, suggesting that it captures highly persistent volatility processes. In contrast, the Gibbs + MH method exhibits greater variability among currencies, with ϕ values ranging from positive and near unity (e.g., CHF, CAD, NZD) to lower or even negative values (e.g., SEK). This indicates that the manual Gibbs sampler allows for greater flexibility in exploring the parameter space, possibly reflecting differences in convergence behavior, since both methods share the same prior specification.

Figure 5 reveals a marked discrepancy between the two estimation methods implemented in R. The JAGS estimates are substantially smaller and closer to zero, whereas the Gibbs + MH estimates range between 1.5 and 5. This pattern

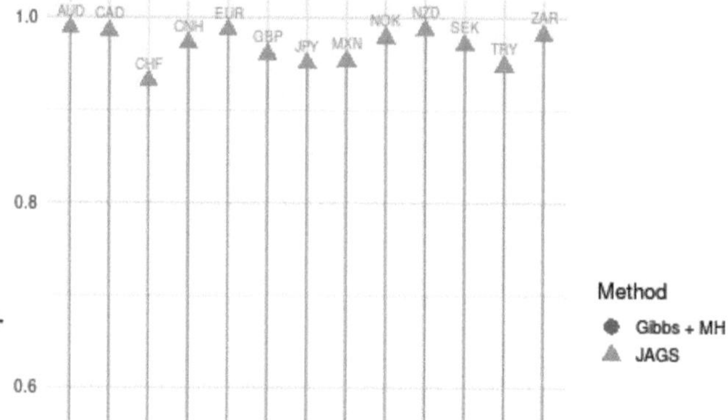

Fig. 4. The plot compares posterior means of ϕ estimated using the Gibbs + Metropolis–Hastings sampler (teal circles) and JAGS (orange triangles) for each currency. The JAGS estimates cluster near one, implying high persistence in the latent volatility process across all series, whereas the manual Gibbs–MH estimates display stronger variation and lower magnitudes, even reaching negative values for SEK. These discrepancies suggest that the two implementations differ in how the state equation is parameterized or how constraints on ϕ (e.g., $|\phi| < 1$) are enforced during sampling, leading to distinct posterior shapes and dynamic interpretations of volatility persistence.

suggests that JAGS yields a considerably more concentrated posterior variance, possibly due to differences in the prior specification for σ^2. The SEK and TRY currencies exhibit the highest variance values under the Gibbs + MH method, while JAGS fails to capture the expected variability across series.

Among the similarities observed in Fig. 6, both methods produce three chains that converge toward unimodal posterior distributions for each parameter. The mean values across chains are consistent in both approaches, although convergence appears slightly more stable and homogeneous under the Gibbs + Metropolis–Hastings method compared to the JAGS estimation.

Figure 7 provides a comparative view of the trace plots for the three parameters under both methods. Remarkably, the Gibbs + MH sampler exhibits faster convergence, with trajectories that almost completely overlap across chains. In contrast, while JAGS also achieves convergence, it shows wider oscillations—

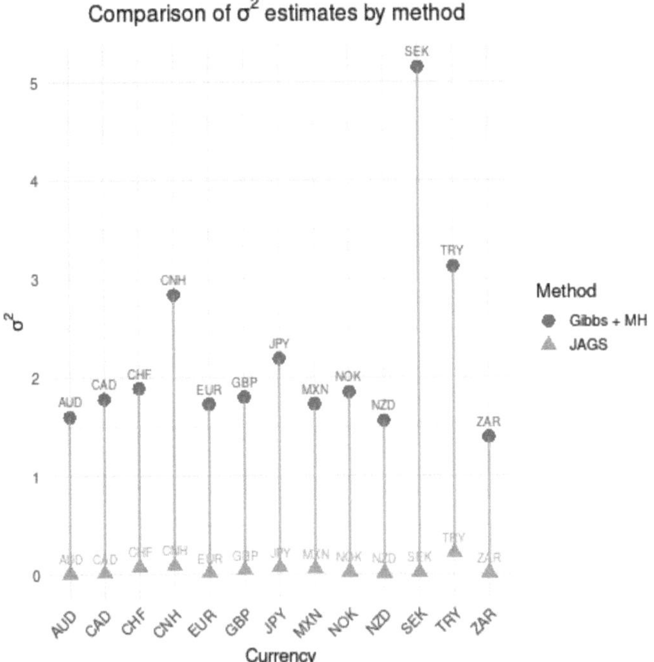

Fig. 5. The figure compares the posterior means of σ^2 obtained using the Gibbs + Metropolis–Hastings sampler (teal circles) and JAGS (orange triangles) for each currency. The Gibbs–MH estimates show substantial dispersion across currencies, while the JAGS estimates remain near zero, suggesting a scale mismatch between the two implementations. This discrepancy likely arises from differences in how each method parameterizes and samples the variance component—JAGS working on the log-volatility scale (state-innovation variance) and the manual Gibbs implementation directly updating σ^2 in the observation scale. (Color figure online)

particularly for σ^2. The manual Gibbs approach demonstrates more efficient chain mixing: trajectories cross frequently and display little persistence. JAGS, on the other hand, presents stronger autocorrelation, especially for σ^2, suggesting slower sampling or a steeper parameter space characterized by high dependence between h_t and the hyperparameters.

Finally, Fig. 8 shows the execution times for each routine across all currencies. JAGS outperformed the Gibbs + MH approach in computational efficiency for 10 out of 13 currencies, indicating that its automated architecture achieves a better balance between computation time and chain mixing quality in these cases. Although the manual Gibbs–Metropolis–Hastings method offers greater control over each sampling step, this additional flexibility introduces computational overhead—particularly in models with long latent-state structures (such as the volatility terms h_t). Nevertheless, for other model classes with simpler

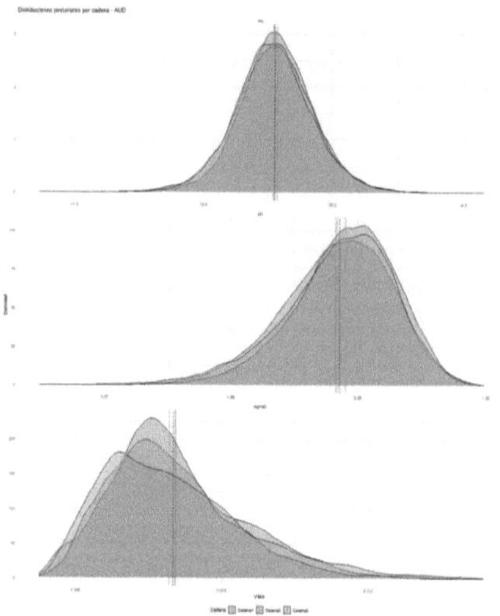

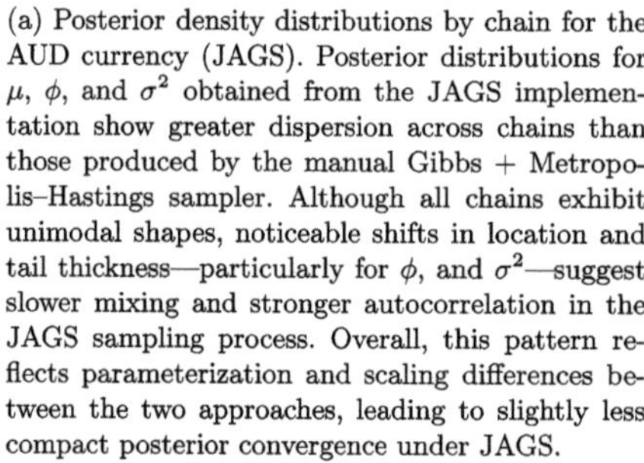

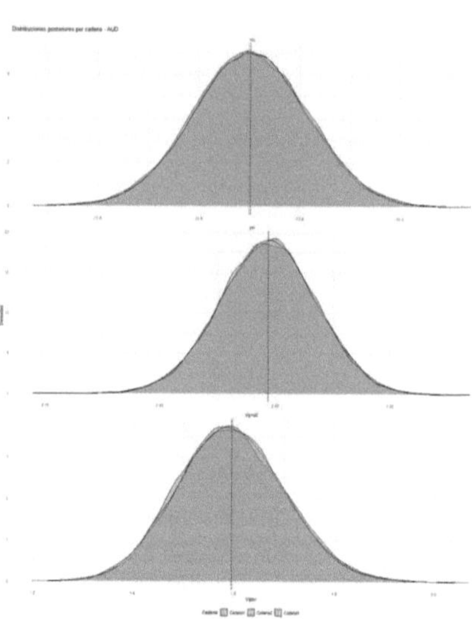

(a) Posterior density distributions by chain for the AUD currency (JAGS). Posterior distributions for μ, ϕ, and σ^2 obtained from the JAGS implementation show greater dispersion across chains than those produced by the manual Gibbs + Metropolis–Hastings sampler. Although all chains exhibit unimodal shapes, noticeable shifts in location and tail thickness—particularly for ϕ, and σ^2—suggest slower mixing and stronger autocorrelation in the JAGS sampling process. Overall, this pattern reflects parameterization and scaling differences between the two approaches, leading to slightly less compact posterior convergence under JAGS.

(b) Posterior density estimates by chain for AUD (Gibbs + Metropolis–Hastings). Posterior distributions for the parameters μ, ϕ, and σ^2 across the three simulated chains for AUD. The overlap and near coincidence of the densities indicate stable sampling and satisfactory convergence between chains. The parameter μ shows a symmetric unimodal posterior centered near -10.45; ϕ presents a concentrated mass around 0.45, suggesting moderate persistence; and σ^2 exhibits a slightly right-skewed shape, consistent with positive support and limited dispersion in volatility.

Fig. 6. Comparison of the posterior distributions of parameters μ, ϕ and σ^2 for the AUD currency, obtained using the JAGS and Gibbs + MH methods.

hierarchies or lower latent dimensionality, the manual approach may still offer advantages in flexibility or numerical stability.

When examining the time series of exchange rates in Fig. 9, it becomes evident that for currencies exhibiting higher variability or abrupt fluctuations (TRY, ZAR, SEK), the Gibbs + MH method shows a clearer advantage in terms of chain stability and convergence, likely due to its finer control over adaptive proposals and its ability to handle heavy-tailed distributions. Nevertheless, although Gibbs + MH generally provides superior stability and mixing behavior, JAGS matches—or even exceeds—its performance in terms of execution time for more stable currencies, particularly GBP, CHF, and CAD, where return dynamics are smoother and the mixing task is less demanding.

In order to complement the parametric and convergence-based evaluation, Fig. 10 presents a direct time-domain comparison between the observed volatility proxy h_t^{obs} and the latent volatility trajectory h_t^{est} estimated by the stochastic volatility model for the TRY currency. This comparison provides a dynamic

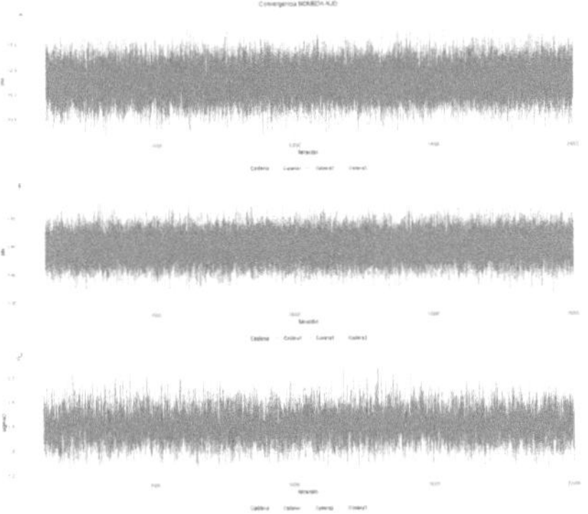

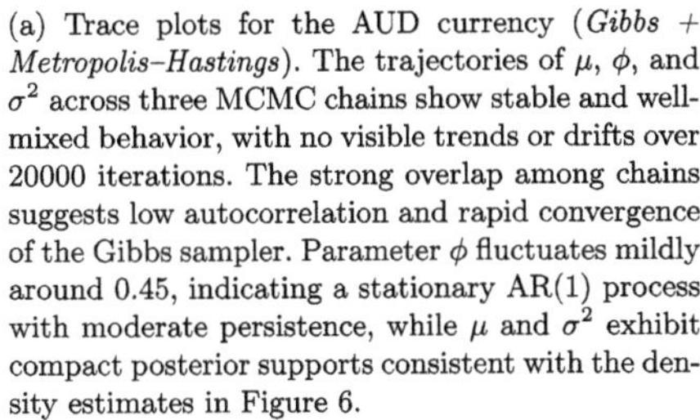

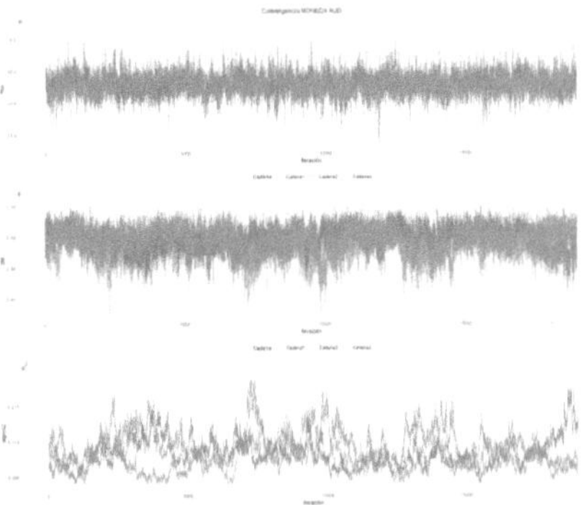

(a) Trace plots for the AUD currency (*Gibbs +
Metropolis–Hastings*). The trajectories of μ, ϕ, and
σ^2 across three MCMC chains show stable and well-mixed behavior, with no visible trends or drifts over
20000 iterations. The strong overlap among chains
suggests low autocorrelation and rapid convergence
of the Gibbs sampler. Parameter ϕ fluctuates mildly
around 0.45, indicating a stationary AR(1) process
with moderate persistence, while μ and σ^2 exhibit
compact posterior supports consistent with the density estimates in Figure 6.

(b) Trace plots for the AUD currency (*JAGS* implementation). The trajectories of μ, ϕ, and σ^2
across three MCMC chains display good mixing and
apparent convergence, though with slightly higher
variability than in the manual Gibbs sampler. The
parameter ϕ remains close to 1, suggesting a highly
persistent latent volatility process, while μ and σ^2
stabilize around narrow posterior regions. Compared with the Gibbs + MH approach, the chains
exhibit greater autocorrelation for σ^2 but similar
stability for μ and ϕ.

Fig. 7. Comparison of the trace plots for the AUD currency obtained via the Gibbs +
MH and JAGS methods. Both implementations show convergence of the three chains,
though with differences in the persistence and scale of the latent volatility process.

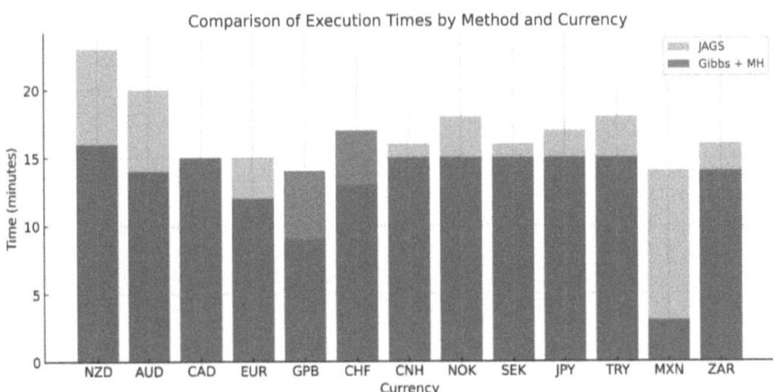

Fig. 8. Comparison of execution times (in minutes) required by the same machine to
process each currency using the Gibbs + Metropolis–Hastings and JAGS methods.
Although both approaches show similar performance across currencies, JAGS tends
to finish faster in most cases, reflecting its computational efficiency within a general-purpose modeling framework. In contrast, the Gibbs + MH implementation exhibits
slightly longer runtimes due to the overhead associated with a manual, model-specific
sampler.

Fig. 9. Evolution of exchange rate series (TRM) in USD for the 13 analyzed currencies from 2013 to 2025.

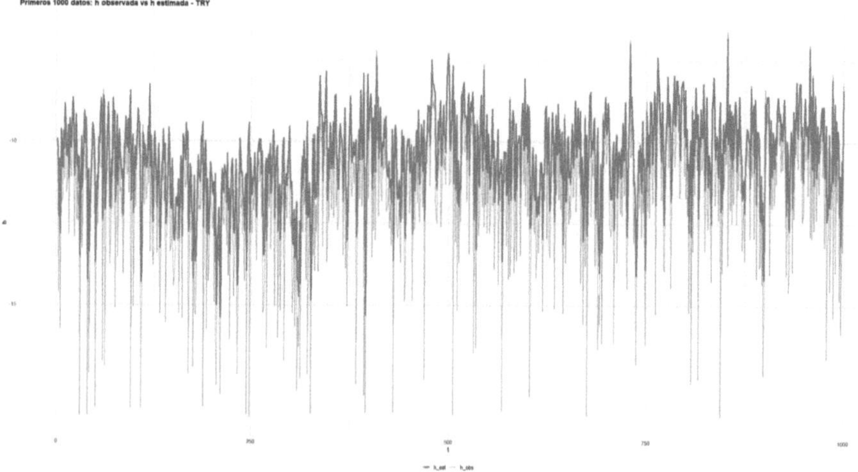

Fig. 10. Comparison between the observed volatility proxy h_t^{obs} and the estimated latent volatility h_t^{est} for the TRY currency (first 1000 observations).

validation of the model's ability to track rapid changes in volatility and extreme fluctuations.

The figure shows that the estimated latent process closely follows the main oscillatory patterns of the observed series, successfully capturing both high-volatility episodes and relatively calm periods. Sharp downward spikes associated with abrupt market movements are also reflected in the estimated trajectory, although with some degree of smoothing, as expected from a latent-state model driven by a persistent AR(1) structure.

This result confirms that the SV model not only converges in a numerical sense, but also delivers a meaningful dynamic representation of the underlying volatility process. The ability to reproduce the temporal structure of observed variability reinforces the validity of the Bayesian inference results and supports the consistency between the latent-state dynamics and the empirical behavior of exchange rate returns.

6 Discussion

The estimation of stochastic volatility models using Bayesian approaches allows for capturing the latent dynamics in financial returns with notable precision. Manually implementing the Gibbs sampler with Metropolis–Hastings steps provides detailed control over the inference process. This flexibility allows for adjusting the number of internal sweeps, setting appropriate initial parameters and priors, and tailoring the sampling proposals to the characteristics of each series. Additionally, it enables direct monitoring of acceptance efficiency, which is crucial in contexts involving short series or heavy-tailed distributions.

On the other hand, JAGS automates much of the process, including the construction of MCMC chains and the conditional updating of parameters. This automation reduces programming errors and facilitates the analysis of multiple series simultaneously. However, this simplification comes with reduced visibility into critical aspects of the sampling process, such as proposal selection, chain mixing, or convergence behavior. This lack of control can be an obstacle when working with more complex data or unusual volatility structures.

Empirical results confirm that, in general terms, both methods produce consistent estimates. The most relevant differences appear when examining convergence metrics. According to the applied diagnostics (Geweke Z, PSRF, and Raftery–Lewis), the manual sampler achieves better-mixed chains with lower autocorrelation, outperforming JAGS in all key convergence indicators. Out of the 39 possible cases evaluated by Geweke Z, for example, the manual method achieved 100% compliance, compared to 76.9% in JAGS. A similar pattern is observed with PSRF, where Gibbs + MH maintains values below 1.1 for nearly all parameters, while JAGS shows recurring issues in more than half the cases.

These differences are also evident when comparing specific currencies. In the case of the Turkish lira (TRY), both methods yield similar results and stable chains. In contrast, for the Swedish krona (SEK), JAGS shows mixing difficulties, extreme Geweke values, and a larger number of burn-in iterations. The manual method, by allowing for finer adjustment of sampling parameters, is better equipped to address such challenges, adapting to the structure of each series.

In computational terms, Gibbs + MH not only delivers better convergence results but also proves more efficient. In 10 out of the 13 currencies analyzed, this method completed the estimation faster than JAGS. This advantage is more pronounced in currencies with high volatility or heavy tails, where control over proposals and internal sweeps becomes critical. JAGS, for its part, performs faster on more stable series such as GBP, CHF, or CAD, where the dynamics are less demanding.

From a pedagogical standpoint, the manual implementation also presents clear advantages. It allows for a deeper understanding of the sampling logic, provides insight into how the chains behave in different scenarios, and enables fine-tuning of each algorithm component. In contrast, JAGS is more useful when reproducibility, speed, or scalability are required—especially in studies involving multiple series or complex hierarchical models.

Overall, the results show that both approaches are valid, but their suitability depends on the context. The manual sampler is preferable when precision, detailed diagnostics, or customized analysis are needed. JAGS, on the other hand, excels in automated environments or when ease of implementation is a priority. In particular, the convergence diagnostics and observed differences in parameters such as ϕ and σ^2 suggest that the choice of method can affect not only the efficiency of sampling but also the dynamic interpretation of the model.

7   Conclusions

This study provides a detailed empirical comparison between a manually implemented Gibbs sampler with Metropolis–Hastings steps and an automated Bayesian inference approach based on the JAGS engine for the estimation of stochastic volatility (SV) models applied to exchange rate returns. Beyond serving as a purely computational exercise, this comparison highlights important methodological and practical aspects of Bayesian inference in latent-variable models characterized by strong persistence, heavy tails, and high posterior dependence.

The results show that both approaches are capable of recovering consistent posterior estimates for the structural parameters of the SV model when identical prior specifications and hierarchical structures are used. However, notable differences arise in terms of convergence diagnostics, chain mixing, and computational behavior. The manually implemented Gibbs–MH sampler systematically achieves superior performance in terms of stability, lower autocorrelation, and stronger compliance with convergence diagnostics such as Geweke, Gelman–Rubin, Heidelberger–Welch, and Raftery–Lewis. This is particularly evident for currencies exhibiting pronounced volatility clustering and heavy-tailed behavior, such as the Turkish lira (TRY), the South African rand (ZAR), and the Swedish krona (SEK).

In contrast, JAGS proves to be highly competitive in terms of computational efficiency, outperforming the manual sampler in execution time for the majority of the series analyzed. This confirms that automated MCMC engines provide an effective and scalable solution for applied Bayesian analysis, especially when working with relatively stable time series where extreme posterior dependence is less severe. Nevertheless, the limited flexibility in tuning latent-state proposals and internal update strategies may hinder performance under more demanding volatility dynamics.

From a methodological standpoint, one of the main contributions of this work lies in demonstrating that the choice of sampling strategy not only affects computational speed, but also impacts the effective exploration of the posterior distribution and the resulting dynamic interpretation of volatility persistence. The observed discrepancies in the estimates of the persistence parameter ϕ and the state variance σ^2 between both methods illustrate that different samplers may emphasize different regions of the posterior space, even under identical model specifications. This highlights the importance of complementing automated inference with diagnostic-driven validation.

From a pedagogical perspective, the manual implementation of the Gibbs–Metropolis–Hastings algorithm plays a fundamental role in reinforcing the understanding of the conditional structure of the model, the behavior of latent variables, and the practical meaning of each convergence diagnostic. This deeper insight is essential for diagnosing numerical pathologies, adjusting proposal distributions, and ensuring the statistical reliability of posterior inference—elements that are often hidden within black-box sampling engines.

Overall, the findings of this study indicate that neither approach is universally superior. Instead, their relative performance depends on the complexity of the data-generating process, the degree of posterior dependence, and the objectives of the analysis. While JAGS is particularly attractive for rapid implementation, reproducibility, and large-scale applications, the manual Gibbs–MH sampler remains preferable in contexts where fine control over convergence and numerical stability is required.

As future work, it would be of interest to extend this comparative framework to other Bayesian engines such as Stan or NIMBLE, which incorporate more advanced sampling strategies—such as Hamiltonian Monte Carlo or particle MCMC—that may offer improved performance in models with complex posterior geometries. In addition, the current univariate framework could be extended to multivariate stochastic volatility models, regime-switching specifications, or hierarchical macro-financial structures, where the benefits of controlling the latent-state updating mechanism are expected to be even more pronounced.

A   Appendix

A.1   Codigo R del Muestreo Gibbs y Metropolis–Hastings

A continuación se presenta el código completo utilizado para estimar el modelo de volatilidad estocástica para todas las monedas analizadas.

```
ruta_excel          <- "C:/Users/COJEDIAZ/Downloads/Currencies.xlsx"
nombre_hoja         <- "Retornos 2012"
ruta_salida_base <-
      "C:/Users/COJEDIAZ/Downloads/RESULTADOS/Graficos/ajuste_monedas/comparacion/NC"


########## LIBREIAS #####################################################
suppressPackageStartupMessages({
  library(readxl)
```

```r
11    library(gridExtra)
12    library(lattice)
13    library(coda)
14    library(ggplot2)
15    library(reshape2)
16    library(dplyr)
17    library(tidyr)
18  })
19
20  ########## FUNCIONES AUXILIARES MH ######################################
21  psi_hj1 <- function(hj1, yj1) {
22    exp(-0.5 * (hj1 + exp(-hj1) * yj1^2))
23  }
24  psi_hjt <- function (hjt , yjt ) {
25    exp(-0.5 * (hjt + exp(-hjt) * yjt^2))
26  }
27
28  ########## PARAMETROS MCMC/GIBBS (GENERALES) ############################
29  n          <- 20000       # numero de iteraciones
30  burn       <- 1000        # tamano del periodo de calentamiento
31  n.cadenas <- 3            # numero de cadenas a simular
32  salto      <- 10          # thinning
33  seed       <- 987         # semilla
34  if (seed > 0) set.seed(seed)
35
36  # Hiperparametros
37  a1 <- 0.95 ; b1 <- 0.03^2     # phi ~ N(a1, b1)
38  c1 <- 0.5  ; d1 <- 0.1        # sigma2 ~ InvGamma(c1, d1)
39  e1 <- 0    ; f1 <- 10         # mu ~ N(e1, f1)
40
41  # Esperanzas a priori (para reporte)
42  espmu <- e1 ; espphi <- a1 ; espsigma2 <- d1/(c1 - 1)
43
44  ########## UTILIDADES DE ARCHIVOS #######################################
45  if (!dir.exists(ruta_salida_base)) dir.create(ruta_salida_base, recursive = TRUE)
46  ruta_moneda <- function(base, moneda) {
47    rp <- file.path(base, moneda)
48    if (!dir.exists(rp)) dir.create(rp, recursive = TRUE)
49    rp
50  }
51
52  ########## LECTURA DE DATOS #############################################
53  realdatadtf <- read_excel(ruta_excel, sheet = nombre_hoja)
54  cols_all <- colnames(realdatadtf)
55
56  # Validar que monedas objetivo existen en el Excel
57  monedas_en_excel <- intersect(monedas_objetivo, cols_all)
58  monedas_faltantes <- setdiff(monedas_objetivo, monedas_en_excel)
59  if (length(monedas_en_excel) == 0) {
60    stop("No se encontro ninguna de las monedas objetivo en el Excel. Revisa nombres de
            columnas.")
61  }
62  if (length(monedas_faltantes) > 0) {
63    message(" No se encontraron en el Excel (se omiten): ",
            paste(monedas_faltantes, collapse = ", "))
65  }
66
67  ########## FUNCION PRINCIPAL POR MONEDA (con K_sweeps variable) ##########
68  correr_gibbs_para_moneda <- function(moneda, y_vector, ruta_base_moneda, K_sweeps_mon) {
69    mensaje_inicio <- paste0(">>> Procesando ", moneda, " (K_sweeps=", K_sweeps_mon, ")")
70    message(mensaje_inicio)
71
72    # Preparacion de datos: eliminar NA conservando orden
73    idx_ok <- which(!is.na(y_vector))
74    y1 <- as.numeric(y_vector[idx_ok])
75    N  <- length(y1)
76    if (N < 3) {
77      warning("Moneda ", moneda, ": muy pocos datos no NA. Se omite.")
```

```r
78       return(invisible(NULL))
79    }
80
81    # Valores iniciales
82    in_mu     <- rnorm(n.cadenas, 0, 1)
83    in_phi    <- rnorm(n.cadenas, 0, 0.35)
84    in_sigma2 <- 1 / rgamma(n.cadenas, 3, 2)
85
86    # Almacenamiento
87    SIMmu      <- matrix(0, n, n.cadenas)
88    SIMphi     <- matrix(0, n, n.cadenas)
89    SIMsigma2  <- matrix(0, n, n.cadenas)
90    Hjtn       <- matrix(0, n, N * n.cadenas)  # simulaciones de h
91    Contador   <- matrix(0, N, n.cadenas)      # aceptaciones en MH
92
93    mide.tiempo <- proc.time()
94
95    ####### Gibbs n = 1 (inicial) #############################################
96    for (m in 1:n.cadenas) {
97      # t = 1
98      anterior <- in_mu[m]
99      u <- runif(1)
100     y <- rnorm(1, in_mu[m], sqrt(in_sigma2[m]))  # propuesta
101     num <- psi_hj1(y, y1[1]); den <- psi_hj1(anterior, y1[1])
102     rho <- min(1, num/den)
103     #Hjtn[1, (N * (m - 1) + 1)] <- if (u <= rho) y else anterior
104     Hjtn [1 ,( N * (m -1) +1) ]= anterior +( y - anterior ) * (u <= rho )#
105     Contador[1, m] <- (u <= rho)
106
107     # t = 2..N
108     for (t in 2:N) {
109       hj_ant   <- Hjtn[1, (N * (m - 1) + t - 1)]
110       anterior <- rnorm(1, hj_ant, sqrt(in_sigma2[m]))
111       u <- runif(1)
112       y <- rnorm(1, in_mu[m] + in_phi[m] * (hj_ant - in_mu[m]), sqrt(in_sigma2[m]))
113       num <- psi_hjt(y, y1[t]); den <- psi_hjt(anterior, y1[t])
114       rho <- min(1, num/den)
115       Hjtn [1 ,( N * (m -1) + t ) ]= anterior +( y - anterior ) * (u <= rho )
116       #Hjtn[1, (N * (m - 1) + t)] <- if (u <= rho) y else anterior
117       Contador[t, m] <- (u <= rho)
118     }
119
120     # mu -> n=1
121     h_1 <- Hjtn[1, (N*(m-1)+1):(N*(m-1)+N)]
122     A <- 1/f1 + (1 + (1 - in_phi[m])^2 * (N - 1)) / in_sigma2[m]
123     diferencias <- numeric(N)
124     for (t in 2:N){
125     #{diferencias[tt] <- h_1[tt] - in_phi[m]*h_1[tt-1]}
126       diferencias[ t ]= h_1[ t ] - in_phi [ m ] * h_1[ t-1]}
127     B <- e1/f1 + (h_1[1] + (1 - in_phi[m]) * sum(diferencias)) / in_sigma2[m]
128     C <- B/A; D <- 1/A
129     SIMmu[1, m] <- rnorm(1, C, sqrt(D))
130
131     # phi -> n=1
132     mu_1 <- SIMmu[1, m]
133     diferencias1 <- numeric(N)
134     for (t in 2:N) {
135       #diferencias1[tt] <- (h_1[tt-1] - mu_1)^2
136       diferencias1 [ t ]=( h_1[ t-1] - mu_1) ^2}
137
138     A <- 1/b1 + sum(diferencias1) / in_sigma2[m]
139     diferencias2 <- numeric(N)
140     for (tt in 2:N)
141       diferencias2[tt] <- (h_1[tt]-mu_1)*(h_1[tt-1]-mu_1)
142     B <- a1/b1 + sum(diferencias2) / in_sigma2[m]
143     C <- B/A; D <- 1/A
144     SIMphi[1, m] <- rnorm(1, C, sqrt(D))
145
```

```
146    # sigma2 -> n=1
147    phi_1 <- SIMphi[1, m]
148    A <- c1 + N/2
149    diferencias <- numeric(N);
150    for (t in 2:N) {
151      #diferencias[t] <- (h_1[t] - mu_1 - phi_1*(h_1[tt-1]-mu_1))^2
152      diferencias1[t]=( h_1[t-1] - mu_1) ^2}
153
154    B <- 0.5 * ((h_1[1]-mu_1)^2 + sum(diferencias)) + d1
155    SIMsigma2[1, m] <- 1/rgamma(1, A, B)
156  }
157
158  ####### Gibbs completo k=2..n #########################################
159  for (m in 1:n.cadenas) {
160    for (k in 2:n) {
161      mu_1     <- SIMmu[k-1, m]
162      phi_1    <- SIMphi[k-1, m]
163      sigma2_1 <- SIMsigma2[k-1, m]
164
165      # K_sweeps_mon barridos MH de h
166      for (rep in 1:K_sweeps_mon) {
167        # t = 1
168        idx1 <- N*(m - 1) + 1
169        anterior <- if (rep == 1) Hjtn[k - 1, idx1] else Hjtn[k, idx1]
170        u <- runif(1)
171        y <- rnorm(1, mu_1, sqrt(sigma2_1))
172        num <- psi_hj1(y, y1[1]); den <- psi_hj1(anterior, y1[1])
173        rho <- min(1, num/den)
174        nuevo <- if (u <= rho) y else anterior
175        Hjtn[k, idx1] <- nuevo
176        if (rep == 1) Contador[1, m] <- Contador[1, m] + (u <= rho)
177
178        # t = 2..N
179        for (t in 2:N) {
180          idx_t    <- N*(m - 1) + t
181          idx_prev <- N*(m - 1) + t - 1
182          hj_ant   <- Hjtn[k, idx_prev]  # usa h_{t-1} ya actualizado en este k
183          anterior <- if (rep == 1) Hjtn[k - 1, idx_t] else Hjtn[k, idx_t]
184          u <- runif(1)
185          y <- rnorm(1, mu_1 + phi_1*(hj_ant - mu_1), sqrt(sigma2_1))
186          num <- psi_hjt(y, y1[t]); den <- psi_hjt(anterior, y1[t])
187          rho <- min(1, num/den)
188          nuevo <- if (u <= rho) y else anterior
189          Hjtn[k, idx_t] <- nuevo
190          if (rep == 1) Contador[t, m] <- Contador[t, m] + (u <= rho)
191        }
192      }
193
194      # mu -> n=k
195      h_1 <- Hjtn[k, (N*(m-1)+1):(N*(m-1)+N)]
196      A <- 1/f1 + (1 + (1 - phi_1)^2 * (N - 1)) / sigma2_1
197      diferencias <- numeric(N); for (tt in 2:N) diferencias[tt] <- h_1[tt] -
               phi_1*h_1[tt-1]
198      B <- e1/f1 + (h_1[1] + (1 - phi_1) * sum(diferencias)) / sigma2_1
199      C <- B/A; D <- 1/A
200      SIMmu[k, m] <- rnorm(1, C, sqrt(D))
201
202      # phi -> n=k
203      mu_1 <- SIMmu[k, m]
204      diferencias1 <- numeric(N); for (tt in 2:N) diferencias1[tt] <- (h_1[tt-1] - mu_1)^2
205      A <- 1/b1 + sum(diferencias1) / sigma2_1
206      diferencias2 <- numeric(N); for (tt in 2:N) diferencias2[tt] <-
               (h_1[tt]-mu_1)*(h_1[tt-1]-mu_1)
207      B <- a1/b1 + sum(diferencias2) / sigma2_1
208      C <- B/A; D <- 1/A
209      SIMphi[k, m] <- rnorm(1, C, sqrt(D))
210
211      # sigma2 -> n=k
```

```r
phi_1 <- SIMphi[k, m]
A <- c1 + N/2
diferencias <- numeric(N); for (tt in 2:N) diferencias[tt] <- (h_1[tt] - mu_1 -
    phi_1*(h_1[tt-1]-mu_1))^2
B <- 0.5 * ((h_1[1] - mu_1)^2 + sum(diferencias)) + d1
SIMsigma2[k, m] <- 1/rgamma(1, A, B)
  }
}

########## CONVERGENCIA Y DIAGNOSTICOS ###############################
burning <- burn + 1
nuevo_tno <- floor((n - burn) / salto)

las_h1 <- matrix(0, N, n.cadenas)
las_h2 <- matrix(0, N, n.cadenas)

SIMmu1    <- matrix(0, n-burn, n.cadenas)
SIMphi1   <- matrix(0, n-burn, n.cadenas)
SIMsigma21 <- matrix(0, n-burn, n.cadenas)

SIMmu2    <- matrix(0, nuevo_tno, n.cadenas)
SIMphi2   <- matrix(0, nuevo_tno, n.cadenas)
SIMsigma22 <- matrix(0, nuevo_tno, n.cadenas)

mmu1 <- mphi1 <- msigma21 <- vmu1 <- vphi1 <- vsigma21 <- numeric(n.cadenas)
mmu2 <- mphi2 <- msigma22 <- vmu2 <- vphi2 <- vsigma22 <- numeric(n.cadenas)

for (m in 1:n.cadenas) {
  # h (medias a posteriori)
  for (t in 1:N) las_h1[t, m] <- mean(Hjtn[burning:n, (N*(m-1)+t)])

  help.h <- matrix(0, nuevo_tno, N)
  for (t in 1:N) for (i in 1:nuevo_tno) help.h[i, t] <- Hjtn[(burn + salto*i),
      (N*(m-1)+t)]
  for (t in 1:N) las_h2[t, m] <- mean(help.h[, t])

  # muestras sin/minado
  SIMmu1[, m]    <- SIMmu[burning:n, m]
  SIMphi1[, m]   <- SIMphi[burning:n, m]
  SIMsigma21[, m] <- SIMsigma2[burning:n, m]

  mmu1[m]    <- mean(SIMmu1[, m]);   vmu1[m]    <- var(SIMmu1[, m])
  mphi1[m]   <- mean(SIMphi1[, m]);  vphi1[m]   <- var(SIMphi1[, m])
  msigma21[m] <- mean(SIMsigma21[, m]); vsigma21[m] <- var(SIMsigma21[, m])

  for (i in 1:nuevo_tno) {
    SIMmu2[i, m]    <- SIMmu1[salto*i, m]
    SIMphi2[i, m]   <- SIMphi1[salto*i, m]
    SIMsigma22[i, m] <- SIMsigma21[salto*i, m]
  }
  mmu2[m]    <- mean(SIMmu2[, m]);   vmu2[m]    <- var(SIMmu2[, m])
  mphi2[m]   <- mean(SIMphi2[, m]);  vphi2[m]   <- var(SIMphi2[, m])
  msigma22[m] <- mean(SIMsigma22[, m]); vsigma22[m] <- var(SIMsigma22[, m])
}

# ---------- GRAFICOS QUICKLOOK (cadena 1) ----------
m_plot <- 1
archivo_png <- file.path(ruta_base_moneda, paste0("convergencia_mcmc_", moneda, ".png"))
png(archivo_png, width = 1000, height = 1000)
par(mfrow = c(3, 3), oma = c(0, 0, 4, 0))

plot(SIMmu2[, m_plot], type = "l",
     ylim = c(mmu2[m_plot] - 4 * 1.96 * sqrt(vmu2[m_plot]),
              mmu2[m_plot] + 4 * 1.96 * sqrt(vmu2[m_plot])))
abline(h = c(mmu2[m_plot] - 1.96 * sqrt(vmu2[m_plot]),
             mmu2[m_plot] + 1.96 * sqrt(vmu2[m_plot])), col = "blue")

plot(SIMphi2[, m_plot], type = "l",
```

```r
278        ylim = c(mphi2[m_plot] - 4 * 1.96 * sqrt(vphi2[m_plot]),
279                 mphi2[m_plot] + 4 * 1.96 * sqrt(vphi2[m_plot])))
280   abline(h = c(mphi2[m_plot] - 1.96 * sqrt(vphi2[m_plot]),
281                mphi2[m_plot] + 1.96 * sqrt(vphi2[m_plot])), col = "blue")
282
283   plot(SIMsigma22[, m_plot], type = "l",
284        ylim = c(msigma22[m_plot] - 4 * 1.96 * sqrt(vsigma22[m_plot]),
285                 msigma22[m_plot] + 4 * 1.96 * sqrt(vsigma22[m_plot])))
286   abline(h = c(msigma22[m_plot] - 1.96 * sqrt(vsigma22[m_plot]),
287                msigma22[m_plot] + 1.96 * sqrt(vsigma22[m_plot])), col = "blue")
288
289   hist(SIMmu2[, m_plot], main = paste("Hist mu, cadena", m_plot))
290   hist(SIMphi2[, m_plot], main = paste("Hist phi, cadena", m_plot))
291   hist(SIMsigma22[, m_plot], main = paste("Hist sigma^2, cadena", m_plot))
292
293   acf(SIMmu2[, m_plot], lag.max = 100)
294   acf(SIMphi2[, m_plot], lag.max = 100)
295   acf(SIMsigma22[, m_plot], lag.max = 100)
296
297   mtext(paste("CONVERGENCIA MONEDA", moneda, "- K_sweeps =", K_sweeps_mon), outer = TRUE,
         cex = 1.3, font = 2)
298   dev.off()
299   cat("Grafico quicklook guardado en:", archivo_png, "\n")
300
301   # ---------- REPORTE TXT ----------
302   ruta_txt <- file.path(ruta_base_moneda, paste0("resultados_diagnosticos_", moneda,
         ".txt"))
303   sink(ruta_txt)
304   simtime <- proc.time() - mide.tiempo
305   rbind(
306     paste("Resultados para muestra usando las", n.cadenas, "cadenas."),
307     paste("Parametros estimados cada", salto, "datos:", mean(mmu2), mean(mphi2),
           mean(msigma22)),
308     paste("Parametros estimados muestra completa:", mean(mmu1), mean(mphi1),
           mean(msigma21)),
309     paste("Calentamiento:", burn),
310     paste("Tamano muestra post-burn-in:", n - burn),
311     paste("Muestra usada cada", salto, "datos:", nuevo_tno),
312     paste("Resultados para cadena", m_plot),
313     paste("Valores iniciales: mu0 =", in_mu[m_plot], "phi0 =", in_phi[m_plot], "sigma20
           =", in_sigma2[m_plot]),
314     paste("Estimaciones cada", salto, "datos:", mmu2[m_plot], mphi2[m_plot],
           msigma22[m_plot]),
315     paste("Varianzas muestrales:", vmu2[m_plot], vphi2[m_plot], vsigma22[m_plot]),
316     paste("Estimaciones muestra completa:", mmu1[m_plot], mphi1[m_plot],
           msigma21[m_plot]),
317     paste("Varianzas completas:", vmu1[m_plot], vphi1[m_plot], vsigma21[m_plot]),
318     paste("Hiperparametros:", e1, f1, a1, b1, c1, d1),
319     paste("Esperanzas a priori:", espmu, espphi, espsigma2),
320     paste("Varianzas a priori:", f1, b1, (espsigma2^2 / (c1 - 2))),
321     paste("Tiempo de simulacion:", simtime[3]),
322     paste("K_sweeps empleado:", K_sweeps_mon)
323   )
324   sink()
325   cat("Resultados guardados en:", ruta_txt, "\n")
326
327   # ---------- RESUMEN + DIAGNOSTICOS CODA (append) ----------
328   sink(ruta_txt, append = TRUE)
329   cat("\n############################\n### RESUMEN DE RESULTADOS\n")
330   cat("### Moneda:", moneda, "\n")
331   cat("### K_sweeps:", K_sweeps_mon, "\n")
332   cat("### Fecha:", format(Sys.time(), "%Y-%m-%d %H:%M:%S"), "\n")
333   cat("############################\n\n")
334
335   # Objetos MCMC para diagnosticos
336   Gibbs_mu1 <- mcmc(SIMmu1[, 1]); Gibbs_phi1 <- mcmc(SIMphi1[, 1]); Gibbs_sigma21 <-
         mcmc(SIMsigma21[, 1])
```

```r
337   Gibbs_mu2 <- mcmc(SIMmu2[, 1]); Gibbs_phi2 <- mcmc(SIMphi2[, 1]); Gibbs_sigma22 <-
          mcmc(SIMsigma22[, 1])
338
339   cat("\n--- Geweke\n")
340   print(geweke.diag(Gibbs_mu1)); print(geweke.diag(Gibbs_phi1));
          print(geweke.diag(Gibbs_sigma21))
341   print(geweke.diag(Gibbs_mu2)); print(geweke.diag(Gibbs_phi2));
          print(geweke.diag(Gibbs_sigma22))
342
343   cat("\n--- Gelman-Rubin\n")
344   GibbsGR_mu1  <- mcmc.list(lapply(1:n.cadenas, function(j) mcmc(SIMmu1[, j])))
345   GibbsGR_phi1 <- mcmc.list(lapply(1:n.cadenas, function(j) mcmc(SIMphi1[, j])))
346   GibbsGR_sig1 <- mcmc.list(lapply(1:n.cadenas, function(j) mcmc(SIMsigma21[, j])))
347   GibbsGR_mu2  <- mcmc.list(lapply(1:n.cadenas, function(j) mcmc(SIMmu2[, j])))
348   GibbsGR_phi2 <- mcmc.list(lapply(1:n.cadenas, function(j) mcmc(SIMphi2[, j])))
349   GibbsGR_sig2 <- mcmc.list(lapply(1:n.cadenas, function(j) mcmc(SIMsigma22[, j])))
350   print(gelman.diag(GibbsGR_mu1)); print(gelman.diag(GibbsGR_phi1));
          print(gelman.diag(GibbsGR_sig1))
351   print(gelman.diag(GibbsGR_mu2)); print(gelman.diag(GibbsGR_phi2));
          print(gelman.diag(GibbsGR_sig2))
352
353   cat("\n--- Raftery y Lewis\n")
354   print(raftery.diag(Gibbs_mu1)); print(raftery.diag(Gibbs_phi1));
          print(raftery.diag(Gibbs_sigma21))
355   print(raftery.diag(Gibbs_mu2)); print(raftery.diag(Gibbs_phi2));
          print(raftery.diag(Gibbs_sigma22))
356
357   cat("\n--- Heidelberger y Welch\n")
358   print(heidel.diag(Gibbs_mu1));   print(heidel.diag(Gibbs_phi1));
          print(heidel.diag(Gibbs_sigma21))
359   print(heidel.diag(Gibbs_mu2));   print(heidel.diag(Gibbs_phi2));
          print(heidel.diag(Gibbs_sigma22))
360   sink()
361   cat("Resumen y diagnosticos CODA anadidos a:", ruta_txt, "\n")
362
363   ########### PLOTS GGLOT (todas las cadenas) ############################
364   iteraciones <- (burn + 1):n
365   df_mu <- data.frame(iter = iteraciones)
366   df_phi <- data.frame(iter = iteraciones)
367   df_sigma2 <- data.frame(iter = iteraciones)
368   for (m in 1:n.cadenas) {
369     df_mu[[paste0("Cadena", m)]]     <- SIMmu[(burn+1):n, m]
370     df_phi[[paste0("Cadena", m)]]    <- SIMphi[(burn+1):n, m]
371     df_sigma2[[paste0("Cadena", m)]] <- SIMsigma2[(burn+1):n, m]
372   }
373   df_mu_long     <- melt(df_mu, id.vars = "iter", variable.name = "Cadena", value.name =
          "mu")
374   df_phi_long    <- melt(df_phi, id.vars = "iter", variable.name = "Cadena", value.name =
          "phi")
375   df_sigma2_long <- melt(df_sigma2, id.vars = "iter", variable.name = "Cadena",
          value.name = "sigma2")
376
377   plot_cadenas <- function(data, param, titulo) {
378     ggplot(data, aes(x = iter, y = .data[[param]], color = Cadena)) +
379       geom_line(alpha = 0.8) +
380       labs(title = titulo, x = "Iteracion", y = param) +
381       theme_minimal() + theme(legend.position = "bottom")
382   }
383   p1 <- plot_cadenas(df_mu_long, "mu", expression(mu))
384   p2 <- plot_cadenas(df_phi_long, "phi", expression(phi))
385   p3 <- plot_cadenas(df_sigma2_long, "sigma2", expression(sigma^2))
386
387   ruta_png_lineas <- file.path(ruta_base_moneda, paste0("valores_cadenas_", moneda,
          ".png"))
388   png(ruta_png_lineas, width = 1200, height = 1000)
389   grid.arrange(p1, p2, p3, ncol = 1, top = paste("Convergencia MONEDA", moneda, "-
          K_sweeps =", K_sweeps_mon))
390   dev.off()
```

```r
391    cat("Grafico de lineas guardado en:", ruta_png_lineas, "\n")
392
393    # Densidades por cadena
394    SIMmu1 <- SIMmu[(burn+1):n, , drop = FALSE]
395    SIMphi1 <- SIMphi[(burn+1):n, , drop = FALSE]
396    SIMsigma21 <- SIMsigma2[(burn+1):n, , drop = FALSE]
397
398    df_post <- bind_rows(
399      as.data.frame(SIMmu1) |>
400        mutate(iter = seq_len(nrow(SIMmu1))) |>
401        pivot_longer(-iter, names_to = "Cadena", values_to = "valor") |>
402        mutate(Param = "mu"),
403      as.data.frame(SIMphi1) |>
404        mutate(iter = seq_len(nrow(SIMphi1))) |>
405        pivot_longer(-iter, names_to = "Cadena", values_to = "valor") |>
406        mutate(Param = "phi"),
407      as.data.frame(SIMsigma21) |>
408        mutate(iter = seq_len(nrow(SIMsigma21))) |>
409        pivot_longer(-iter, names_to = "Cadena", values_to = "valor") |>
410        mutate(Param = "sigma2")
411    ) |>
412      mutate(Cadena = factor(Cadena, labels = paste0("Cadena", 1:n.cadenas)))
413
414    res_cadena <- df_post |>
415      group_by(Param, Cadena) |>
416      summarise(media = mean(valor), .groups = "drop")
417    res_global <- df_post |>
418      group_by(Param) |>
419      summarise(media = mean(valor), .groups = "drop") |>
420      mutate(Cadena = NA)
421
422    p_dens <- ggplot(df_post, aes(x = valor, fill = Cadena)) +
423      geom_density(alpha = 0.35) +
424      facet_wrap(~ Param, scales = "free", ncol = 1,
425                 labeller = as_labeller(c(mu = "miu", phi = "phi", sigma2 =
426                     expression(sigma^2)))) +
       geom_vline(data = res_cadena, aes(xintercept = media, color = Cadena), linewidth =
           0.5) +
427      geom_vline(data = res_global, aes(xintercept = media), linetype = "dashed") +
428      labs(title = paste("Distribuciones posteriores por cadena -", moneda, "(K_sweeps =",
           K_sweeps_mon, ")"),
429         x = "Valor", y = "Densidad", fill = "Cadena", color = "Cadena") +
430      theme_minimal() + theme(legend.position = "bottom")
431
432    ruta_png_dens <- file.path(ruta_base_moneda, paste0("densidades_posteriores_", moneda,
           ".png"))
433    png(ruta_png_dens, width = 1200, height = 1400); print(p_dens); dev.off()
434    cat("Grafico de densidades guardado en:", ruta_png_dens, "\n")
435
436    # ----- Objeto de resultados (RDS/RData) -----
437    # medias posterior globales (sin/ con minado ya arriba; aqui usamos sin minado)
438    post_means <- c(mu = mean(SIMmu1), phi = mean(SIMphi1), sigma2 = mean(SIMsigma21))
439    res <- list(
440      moneda      = moneda,
441      N           = N,
442      K_sweeps    = K_sweeps_mon,
443      post_means  = post_means,
444      last_h1     = las_h1,
445      last_h2     = las_h2,
446      volatilidad = exp(las_h1/2),
447      SIMmu       = SIMmu1,
448      SIMphi      = SIMphi1,
449      SIMsigma2   = SIMsigma21,
450      priors      = list(b0=e1, B0=f1, a0=a1, A0=b1, c0=c1, C0=d1)
451    )
452
453    ruta_guardado_res    <- file.path(ruta_base_moneda, paste0("res_object_", moneda,
           ".rds"))
```

```r
454   ruta_guardado_rdata   <- file.path(ruta_base_moneda, paste0("res_object_", moneda,
          ".RData"))
455   saveRDS(res, ruta_guardado_res)
456   save(res, file = ruta_guardado_rdata)
457   cat("Lista 'res' guardada en:\n -", ruta_guardado_res, "\n -", ruta_guardado_rdata,
          "\n")
458
459   invisible(TRUE)
460 }
461
462 ########### LOOP POR MONEDAS OBJETIVO ENCONTRADAS ########################
463 for (mon in monedas_en_excel) {
464   ruta_base_moneda <- ruta_moneda(ruta_salida_base, mon)
465   K_sweeps_mon <- if (mon %in% grupo_3) 3L else if (mon %in% grupo_2) 2L else 2L
466   tryCatch(
467     correr_gibbs_para_moneda(
468       moneda = mon,
469       y_vector = realdatadtf[[mon]],
470       ruta_base_moneda = ruta_base_moneda,
471       K_sweeps_mon = K_sweeps_mon
472     ),
473     error = function(e) {
474       message("Error en moneda ", mon, ": ", conditionMessage(e))
475     }
476   )
477 }
478
479 cat("\n Proceso terminado. Salida en:\n", ruta_salida_base, "\n")
```

Listing 1.1. Código R para estimación SV con Gibbs + MH

References

1. Brooks, S., Gelman, A.: General methods for monitoring convergence of iterative simulations. J. Comput. Graph. Stat. **7**(4), 434–455 (1998)
2. Cowles, M.K., Carlin, B.P.: Markov chain Monte Carlo convergence diagnostics: a comparative review. J. Am. Stat. Assoc. **91**(434), 883–904 (1996)
3. Gelman, A., Carlin, J.B., Stern, H.S., Dunson, D.B., Vehtari, A., Rubin, D.B.: Bayesian Data Analysis, 3 edn. CRC Press (2013)
4. Gelman, A., Rubin, D.B.: Inference from iterative simulation using multiple sequences. Stat. Sci. **7**(4), 457–472 (1992)
5. Geweke, J.: Evaluating the accuracy of sampling-based approaches to the calculation of posterior moments. Bayesian Stat. **4**, 169–193 (1992)
6. Geweke, J.: Bayesian treatment of the independent student-t linear model. J. Appl. Economet. **9**(S1), S19–S40 (1994)
7. Hastings, W.: Monte Carlo sampling methods using Markov chains and their applications. Biometrika **57**, 97–109 (1970)
8. Heidelberger, P., Welch, P.D.: A spectral method for confidence interval generation and run length control in simulations. Commun. ACM **24**(4), 233–245 (1981)
9. Heidelberger, P., Welch, P.D.: Simulation run length control in the presence of an initial transient. Oper. Res. **31**(6), 1109–1144 (1983)
10. Hosszejni, D., Kastner, G.: Approaches toward the Bayesian estimation of the stochastic volatility model with leverage. In: Argiento, R., Durante, D., Wade, S. (eds.) BAYSM 2018. SPMS, vol. 296, pp. 75–83. Springer, Cham (2019). https://doi.org/10.1007/978-3-030-30611-3_8

11. Härdle, W.K., Hautsch, N., Pauly, L.R.: Applied Quantitative Finance, 4 edn. Springer, Cham (2016)
12. Menkhoff, L., Sarno, L., Schmeling, M., Schrimpf, A.: Carry trades and global foreign exchange volatility. J. Financ. **67**(2), 681–718 (2012)
13. Metropolis, N., Ulam, S.: The Monte Carlo method. J. Amer. Statist. Assoc **44**, 335–341 (1949)
14. Plummer, M.: JAGS version 4.3.0 user manual (2017). https://people.stat.sc.edu/hansont/stat740/jags_user_manual.pdf. Accessed 03 Nov 2025
15. Raftery, A.E., Lewis, S.: How many iterations in the Gibbs sampler? Bayesian Stat. **4**, 763–773 (1992)
16. Robert, C.P., Casella, G.: Introducing Monte Carlo Methods with R. Springer, Cham (2010)
17. Romo, V.D.J., Rodrigues, E.R., Tzintzun, G.: A Gibbs sampling algorithm to estimate the parameters of a volatility model: an application to ozone data. Appl. Math. **3**(12), 2178–2190 (2012)
18. Shephard, N.: Stochastic Volatility: Selected Readings. Oxford University Press (2005)
19. Taylor, S.J.: Modelling Financial Time Series. Wiley (1986)

Multiple Change Detection in Time Series with Inductive Conformal Martingales: A Reproducible Implementation in R

Cristhian Quiroz Castaño[(✉)] and Biviana Marcela Suárez-Sierra

Universidad EAFIT, Medellín, Colombia
`{cquirozc1,bmsuarezs}@eafit.edu.co`

Abstract. We replicate and extend Volkhonskiy et al. (2017) on Inductive Conformal Martingales (ICM) for change-point detection in time series. The original study uses k-nearest neighbors (KNN) as the nonconformity measure; here we propose median absolute deviation (MAD) and interquartile range (IQR), which are more robust to outliers, and compare all three. We implement the full pipeline in R using `dplyr`, `tidyr`, `purrr` and `ggplot2`. We also extend ICM to multiple detections by restarting or updating the martingale after each alarm. Our experiments assess KNN, MAD, and IQR in terms of average detection delay, false-alarm probability, and stability under various settings, with reproducible visualization and validation routines. In addition, we contrast multiple-detection strategies, highlighting strengths and limitations across simulated scenarios and discussing their practical applicability.

Keywords: Time series · Change-point detection · Conformal prediction · Inductive conformal martingales · Exchangeability

1 Introduction

Change-point detection in time series is a cross-cutting problem with multiple applications, for example in industrial monitoring, finance, and data analytics. Among recent nonparametric approaches, *inductive conformal martingales* (ICM) provide an *online*, modular mechanism: given a nonconformity measure and a betting function, a nonnegative martingale is updated whose growth indicates deviations from the exchangeability assumption of a reference regime (see, e.g., [1]). This framework enables stopping rules with false-alarm control via Ville's inequality [1], which makes it especially appealing.

This work combines the ICM formalization with an open and reproducible implementation in R. Methodologically, we replicate the base scheme with KNN as the nonconformity measure and propose robust alternatives: median absolute deviation (MAD) and interquartile range (IQR). Computationally, we present R software to reproduce the conformal pipeline (nonconformity measure, p-values, betting function, martingale, and alarm); we include classical betting functions

B. M. Suárez et al. (Eds.): R Day 2025, CCIS 2824, pp. 303–318, 2026.
https://doi.org/10.1007/978-3-032-18455-9_16

(constant, mixture, precomputed KDE) and a histogram-based function. In addition, we extend the detector to multiple changes via two strategies: (i) restarts without retraining (preserves the initial calibration) and (ii) restarts with controlled retraining after each alarm.

Implementation and experiments are organized as reproducible *pipelines* with `dplyr`, `tidyr`, and `purrr`; visualization with `ggplot2`. All scenarios (with one or multiple changes) and their results (tables and figures) are generated from versioned scripts.

The paper is organized as follows: Sect. 2 details the methodology (ICM, nonconformity measures, betting functions, and multiple-detection algorithms). Section 3 presents the R interface of our implementation. Section 4 describes the experimental design. Finally, Sect. 5 provides conclusions, sensitivity analyses, and performance considerations.

2   Methodology

We consider a univariate time-ordered sequence of observations $\{z_1, z_2, \dots\}$ and a reference training set $\mathbf{Z}$ of size m obtained from an initial, supposedly stable segment of the stream. In this section we first formalize the single change-point detection problem and the performance criteria of interest, and then describe how inductive conformal martingales (ICM) can be used to construct sequential detectors in this setting. Finally, we extend the framework to multiple change-points.

2.1   Change-Point Detection Problem

Change-point detection deals with time-ordered data whose stochastic behavior may change at an unknown time. Let $z_1, z_2, \dots$ be a sequence of observations. We assume that there exists an index $\theta \in \{1, 2, \dots\}$ such that the distribution of the data changes from f_0 to a f_1:

$$z_1, z_2, \dots, z_{\theta-1} \sim f_0(z) \quad \text{and} \quad z_\theta, z_{\theta+1}, \dots \sim f_1(z),$$

with $f_0 \neq f_1$. Before θ the process behaves according to a stable reference regime, while after θ some aspect of the distribution is altered (for example, the mean level, the variance, the shape of the marginal distribution, or the dependence structure). The unknown index θ is what we call the *change point*.

In a sequential setting, the observations arrive one by one and the analyst does not know in advance whether a change will occur, nor when. The goal is therefore to design a detection rule that looks at $z_1, \dots, z_t$ as they are observed and raises an alarm at a stopping time τ that is as close as possible to θ when a change happens, while keeping the probability of false alarms (i.e. $\mathbb{P}(\tau < \theta)$) under control. In the following section we show how inductive conformal martingales can be used to construct such sequential detectors for the change-point problem.

2.2   Inductive Conformal Martingales

Let $\mathbf{Z} := \{z_{-m+1}, \ldots, z_0\}$ be a training set of size m. Let $\mathcal{A}(\cdot)$ be a nonconformity measure mapping $(z_i, \mathbf{Z})$ to a real value α_i, quantifying how atypical z_i is with respect to $\mathbf{Z}$: larger values imply lower conformity. Intuitively, the nonconformity score α_i plays the role of a "*strangeness*" score: observations that are very typical under the reference regime receive small values, whereas unusual or outlying points receive larger values.

p-values are computed as

$$p_t = \frac{|\{i \le t : \alpha_i > \alpha_t\}| + U\,|\{i \le t : \alpha_i = \alpha_t\}|}{t}, \tag{1}$$

with $U \sim \mathrm{Unif}(0,1)$ independent of the observations and $\mathcal{A}(\cdot)$, used as a randomized tie-breaker when several scores are equal. This randomization ensures that, under exchangeability, the sequence $\{p_t\}$ is i.i.d. uniform on $[0,1]$; see [1].

The conformal martingale is defined as

$$S_t = \prod_{i=1}^{t} g(p_i), \tag{2}$$

where $g(\cdot)$ is a betting function such that $\int_0^1 g(p)\,dp = 1$, which guarantees the martingale property $\mathbb{E}(S_t \mid S_1, \ldots, S_{t-1}) = S_{t-1}$. In the inductive approach, fixing the calibration of $\mathcal{A}(\cdot)$ on $\mathbf{Z}$ yields ICM.

Under this no-change regime, the conformal p-values spread their mass over the whole interval $[0,1]$ and the values of the betting function $g(p_t)$ fluctuate around one, so the capital process $S_t = \prod_{i=1}^{t} g(p_i)$ does not exhibit any systematic growth. After a change-point, exchangeability is violated and the p-values are no longer uniform: they tend to concentrate their mass near zero. Since our betting functions are designed to take values larger than one for small p-values, this accumulation of probability mass close to zero produces a sequence of multipliers $g(p_t) > 1$ and, as a consequence, the martingale S_t increases rapidly and eventually crosses the alarm threshold.

The stopping rule is

$$\tau = \inf\{t : S_t \ge h\}. \tag{3}$$

By Ville's inequality the probability of false alarms is controlled:

$$\mathbb{P}(\exists t : S_t \ge C) \le \frac{1}{C} \quad \forall C > 0. \tag{4}$$

In practice we do not update the product S_t in (2) directly. When the p-values are compatible with exchangeability, many factors $g(p_t)$ are slightly smaller than one and the product $\prod_{i=1}^{t} g(p_i)$ can become extremely small for long sequences, which leads to numerical underflow. To avoid this, we work on the log scale and employ the truncated recursion proposed in [1]:

$$C_t = \max\{0,\ C_{t-1} + \log(g(p_t))\}, \quad C_0 = 0, \tag{5}$$

Here C_t can be seen as a detection statistic derived from the log-capital $\log S_t = \sum_{i=1}^{t} \log g(p_i)$: it accumulates positive increments $\log g(p_t)$ when the data produce small p-values (evidence against exchangeability) and is reset to zero whenever the partial sum would become negative. This keeps the statistic on a convenient scale and avoids the numerical issues associated with very small values of S_t.

The price for this truncation is that the process $\{C_t\}$ is no longer a martingale under exchangeability: because of the non-linear $\max\{0, \cdot\}$ operator we generally have $\mathbb{E}(C_t \mid C_1, \ldots, C_{t-1}) \neq C_{t-1}$. Consequently, Ville's inequality does not apply directly to C_t, and the threshold h in (3) cannot be interpreted as an exact any-time error level. In this paper we therefore treat h as a tuning parameter and study, in Sect. 4, how different choices of h affect the empirical false-alarm probability and the detection delay.

The procedure implemented in R is summarized in Algorithm 1.

Algorithm 1. ICM

Require: calibration set $\mathbf{Z}$, stream $\{z_i\}_{i=1}^{n}$, NCM $A(\cdot)$, betting $g_i(\cdot)$, threshold $h > 0$
Ensure: τ (alarm time, if any) and the sequences $\{\alpha_i\}$, $\{p_i\}$, $\{C_i\}$
1: $C_0 \leftarrow 0$, $\tau \leftarrow \varnothing$
2: **for** $i = 1$ to n **do**
3:     $\alpha_i \leftarrow A(z_i; \mathbf{Z})$
4:     $g \leftarrow \#\{j \leq i : \alpha_j > \alpha_i\}$, $e \leftarrow \#\{j \leq i : \alpha_j = \alpha_i\}$
5:     sample $U \sim \mathrm{Unif}(0, 1)$; $p_i \leftarrow (g + U \cdot e)/i$
6:     $C_i \leftarrow \max\{0,\, C_{i-1} + \log g_i(p_i \mid p_{1:i-1})\}$
7:     **if** $C_i > h$ **then**
8:         $\tau \leftarrow i$; **break**
9:     **end if**
10: **end for**
11: **return** τ, $\{\alpha_1{:}\alpha_T\}$, $\{p_1{:}p_T\}$, $\{C_1{:}C_T\}$

2.3   Nonconformity Measures

K-Nearest Neighbors (KNN). As in [1]: this nonconformity measure is computed as the average distance from z_t to its k nearest neighbours in the training set $\mathbf{Z}$. Intuitively, when this average distance is small, the observation z_t is well represented by $\mathbf{Z}$, in the sense that there are many training points close to it, and the resulting score α_t is low. Conversely, a large average distance indicates that only distant neighbours are available in $\mathbf{Z}$, so z_t is less typical with respect to the training set and α_t becomes large.

Median Absolute Deviation (MAD). The MAD nonconformity measure is computed as follows:

$$\mathcal{A}(z_t, \mathbf{Z})_{MAD} = \frac{|z_t - \mathrm{median}(\mathbf{Z})|}{\mathrm{MAD}(\mathbf{Z})}, \quad \mathrm{MAD}(\mathbf{Z}) = \mathrm{median}\left(|\mathbf{Z} - \mathrm{median}(\mathbf{Z})|\right). \tag{6}$$

Intuitively, the MAD nonconformity measure compares the absolute deviation of z_t from a central value of the training set $\mathbf{Z}$ with the typical variability observed in $\mathbf{Z}$. The numerator $|z_t - \mathrm{median}(\mathbf{Z})|$ measures how far z_t is from the median of the calibration sample, while the denominator $\mathrm{MAD}(\mathbf{Z})$ summarizes a typical absolute deviation around that median. When $|z_t - \mathrm{median}(\mathbf{Z})|$ is of the same order as $\mathrm{MAD}(\mathbf{Z})$, the measure α_t is moderate; if the deviation is much smaller, α_t is close to zero; and if it is several times larger, α_t becomes large, indicating that z_t lies far from most of the training data.

Interquartile Range (IQR). The IQR nonconformity measure is given by:

$$\mathcal{A}(z_t, \mathbf{Z})_{IQR} = \frac{|z_t - Q_{0.5}(\mathbf{Z})|}{\mathrm{IQR}(\mathbf{Z})}, \quad \mathrm{IQR}(\mathbf{Z}) = Q_{0.75} - Q_{0.25}, \tag{7}$$

where $Q_p(\mathbf{Z})$ denotes the empirical pth quantile of the training set $\mathbf{Z}$. The IQR nonconformity measure uses central quantiles instead of absolute deviations. The numerator again quantifies how far z_t is from a central value, while the denominator $\mathrm{IQR}(\mathbf{Z}) = Q_{0.75} - Q_{0.25}$ expresses the width of the central 50% of the calibration sample. If z_t lies inside this central region, the ratio is small and the nonconformity measure is low. As z_t moves away from the middle of the distribution and approaches or exceeds the quartiles, the ratio increases, and so does the measure, signaling that z_t falls outside the range where most training observations are concentrated.

These two measures are inspired by those commonly used in the conformal prediction literature [2].

2.4 Betting Functions

We consider four betting functions previously studied in the literature. The constant, mixture, and kernel density betting functions follow [1], while the histogram betting function follows [3].

Constant Betting: We use the constant base function.

$$g(p) = \begin{cases} 1.5, & p \in [0, 0.5), \\ 0.5, & p \in [0.5, 1]. \end{cases} \tag{8}$$

The constants are such that $1.5 \cdot 0.5 + 0.5 \cdot 0.5 = 1$, so $\int_0^1 g(p)\, dp = 1$ under uniform p-values. This preserves the martingale property in the no-change regime while placing more weight on small p-values, which induces faster growth of the capital process when the p-values accumulate near zero after a change.

Mixture Betting Function. The mixture betting function is defined by

$$g(p) = \int_0^1 \varepsilon\, p^{\varepsilon - 1}\, d\varepsilon. \tag{9}$$

Intuitively, the integral explores a whole continuum of betting strategies indexed by $\varepsilon \in (0,1)$: for each value of ε we consider the simple power bet $p^{\varepsilon-1}$, and then $g(p)$ takes an average of these bets over all possible ε, weighted by the factor ε. In this way the mixture combines, in a single expression, strategies that are more or less aggressive against small p-values, while the chosen weighting ensures that $\int_0^1 g(p)\,dp = 1$ under uniform p-values.

Kernel Density (precomputed) Betting Function: In this case p-values are first simulated under a specific change scenario and a Gaussian kernel density estimator is fitted to them. The betting function $g(\cdot)$ is then constructed from this estimated density so that it upweights regions of $[0,1]$ where p-values concentrate under the change.

Histogram Betting Function Let k evenly spaced bins B_j form a partition of $[0,1]$. If $p_t \in B_j$, we set,

$$g(p_t) = \frac{n_j\,k}{t-1}, \tag{10}$$

where n_j is the number of p-values that have fallen in B_j up to time $t-1$. Intuitively, $g(p_t)$ increases in regions where many previous p-values have accumulated, allowing the martingale to exploit systematic departures from uniformity. To avoid zeros in empty bins, we use $g(p) = 1$ until enough observations are accumulated to initialise the histogram.

2.5   Multiple Detection

We formulate two extensions to detect multiple changes in the same sequence, inspired by [4].

Method Without Retraining. Extends Algorithm 1: This strategy keeps the initial training set $\mathbf{Z}$ fixed throughout the monitoring period. After each alarm time τ_j the martingale is reset to its initial value and the procedure continues processing the stream from z_{τ_j+1}, but the nonconformity scores $\mathcal{A}(z_t)$ are always computed with respect to the same calibration set $\mathbf{Z}$. As a consequence, the detector remains sensitive to any departure from the original pre-change regime, even if the series has already undergone previous changes. On the other hand, it does not adapt to the new regimes that emerge after the first alarm, so later changes are always judged relative to the very first baseline. See Algorithm 2.

Method with Retraining. The second strategy also restarts the martingale after each alarm, but in addition it updates the training set so that the nonconformity scores are calibrated to the most recent regime. Concretely, after an alarm at time τ_j the algorithm skips a guard band of G observations,

---

Algorithm 2. ICM_multi (multiple detection with restarts)

Require: fixed calibration set $\mathbf{Z}$, stream $\{z_i\}_{i=1}^n$, NCM $A(\cdot)$, betting $g_i(\cdot)$, threshold $h > 0$

Ensure: concatenated sequences $\{C_i\}$, $\{p_i\}$ and change set $\mathcal{T}$

1: $C_{\text{prev}} \leftarrow 0$; $i_{\text{start}} \leftarrow 1$; $\mathcal{T} \leftarrow \emptyset$

2: ALL_C $\leftarrow [\,]$; ALL_P $\leftarrow [\,]$

3: **while** $i_{\text{start}} \leq n$ **do**

4:     $i \leftarrow 0$; $C \leftarrow [\,]$; $P \leftarrow [\,]$

5:     **while** $i_{\text{start}} + i - 1 \leq n$ **do**

6:        $z \leftarrow z_{i_{\text{start}}+i}$; $i \leftarrow i + 1$

7:        $\alpha_i \leftarrow A(z; \mathbf{Z})$

8:        $g \leftarrow \#\{j \leq i : \alpha_j > \alpha_i\}$; $e \leftarrow \#\{j \leq i : \alpha_j = \alpha_i\}$

9:        sample $U \sim \text{Unif}(0,1)$; $p_i \leftarrow (g + U \cdot e)/i$

10:       $C_i \leftarrow \max\{0,\ C_{\text{prev}} + \log g_i(p_i \mid P)\}$; $C_{\text{prev}} \leftarrow C_i$

11:      append p_i to P; append C_i to C

12:      **if** $C_i > h$ **then**

13:        $t^\star \leftarrow i_{\text{start}} + i - 1$; add $t^\star$ to $\mathcal{T}$

14:        concat C into ALL_C; concat P into ALL_P

15:        $i_{\text{start}} \leftarrow t^\star + 1$; $C_{\text{prev}} \leftarrow 0$

16:        **break**

17:      **end if**

18:     **end while**

19:     **if** $i_{\text{start}} + i - 1 = n$ **then**

20:      concat C into ALL_C; concat P into ALL_P

21:      **break**

22:     **end if**

23: **end while**

24: **return** ALL_C, ALL_P, $\mathcal{T}$

$\{z_{\tau_j+1}, \ldots, z_{\tau_j+G}\}$, to avoid mixing data from the transition region between regimes. It then builds a new training set $\mathbf{Z}^{(j)}$ using the next m_R observations,

$$\mathbf{Z}^{(j)} = \{z_{\tau_j+G+1}, \ldots, z_{\tau_j+G+m_R}\},$$

and resumes monitoring from $z_{\tau_j+G+m_R+1}$ with this updated calibration. In this way the detector can adapt to successive regimes along the stream, at the cost of introducing two additional tuning parameters, the guard band G and the retraining size m_R as displayed in Algorithm 3.

Algorithm 3. ICM_multi_adaptive (restart, guard band and retraining)

Require: stream $\{z_t\}_{t=1}^n$; NCM $A(\cdot)$; betting $g_i(\cdot)$; threshold h
Require: initial training: set $\mathbf{Z}$ or size m_0
Require: parameters: guard band G, retraining size m_R
Ensure: change set $\mathcal{T}$ (optionally, $\{C_i\}, \{p_i\}$)
 1: **Initialize** $\mathbf{Z}$; $C \leftarrow 0$; $\mathcal{T} \leftarrow \emptyset$; $i \leftarrow 0$
 2: **while** observations remain **do**
 3:     $i \leftarrow i + 1$
 4:     **Nonconformity:** $\alpha_i \leftarrow A(z_i; \mathbf{Z})$
 5:     p**-value:** $g \leftarrow \#\{j \leq i : \alpha_j > \alpha_i\}$, $e \leftarrow \#\{j \leq i : \alpha_j = \alpha_i\}$; sample $U \sim$ Unif$(0, 1)$
 6:     $p_i \leftarrow \dfrac{g + U \cdot e}{i}$
 7:     **Betting & capital:** $C \leftarrow \max\{0,\ C + \log g_i(p_i \mid p_{1:(i-1)})\}$
 8:     **if** $C > h$ **then** $\triangleright$ alarm
 9:         record the current index in $\mathcal{T}$
10:         **Guard band:** skip G observations without updating C or emitting alarms
11:         **Retrain:** form a new $\mathbf{Z}$ with the next m_R observations
12:         **Restart segment:** $C \leftarrow 0$; $i \leftarrow 0$; continue after retraining
13:     **end if**
14: **end while**
15: **return** $\mathcal{T}$ (and, optionally, $\{C_i\}, \{p_i\}$)

---

3   R Development

This section describes the interface (function signatures) of the main R implementations used in the experiments.

3.1   Nonconformity Measures

Listing 1.1. Signatures of nonconformity measures

```
Non_conformity_KNN(xi, training_set, K, ...) -> numeric(1)
Non_conformity_MAD(xi, training_set, K = NULL, ...) -> numeric(1)
Non_conformity_IQR(xi, training_set, K = NULL, ...) -> numeric(1)
```

Here, x_i is the current observation and `training_set` is used to compute nonconformity; K is the number of nearest neighbors (set to `NULL` for measures that do not require it). ... allow additional parameters.

3.2   Betting Functions

Listing 1.2. Signatures of betting functions

```
Constant_BF <- function(p_values, new_p, i, ...)
Mixture_BF  <- function(p_values, new_p, i, ...)
Kernel_BF   <- function(p_values, new_p, i, ...)
histogram_betting_function <- function(p_values, new_p, num_bins, ...)
```

Here, `p_values` are the p-values up to time $i-1$; `new_p` is the current p-value; and ... enables optional arguments.

3.3 Inductive Conformal Martingales

Listing 1.3. ICM method signatures

```
ICM <- function(training_set,
                stream_data,
                non_conformity_measure,
                betting_function,
                th = 1,
                params_bf = list(),
                ...)


ICM_multi <- function(training_set,
                      stream_data,
                      non_conformity_measure,
                      betting_function,
                      th = 1,
                      params_bf = list(),
                      ...)


ICM_multi_adaptive <- function(stream_data,
                               non_conformity_measure,
                               betting_function,
                               th = 1,
                               training_set = NULL,
                               training_size = NULL,
                               m_retrain = NULL,
                               guard_band = 0,
                               params_bf = list(),
                               ...)
```

ICM implements Algorithm 1; `ICM_multi`, Algorithm 2; and `ICM_multi_adaptive`, Algorithm 3. The parameter `th` sets the alarm threshold, and `params_bf` carries a list of arguments for the betting function. In `ICM_multi_adaptive`, `training_size` defines the initial training size if `training_set` is not provided; `m_retrain` sets the retraining window; and `guard_band` defines how many observations to skip to avoid contaminating calibration with the transition.

4 Experiments

Environment and Libraries. All experiments were implemented in R. We used `purrr` [5] for functional iteration (e.g., `pmap()`, `map()`), `dplyr` [6] for transformation and assembly of results (e.g., `mutate()`, `bind_rows()`), `tidyr` [7] to build scenario grids (e.g., `expand_grid()`), `ggplot2` [8] for visualization, and `future/future.apply` [9] to parallelize runs (e.g., `plan(multisession, workers = k)`).

4.1   Single Change-Point

Threshold Sensitivity. For Algorithm 1, we follow [1]:

- Training set $\mathbf{Z}$ of size $m = 200$.
- Training observations and first segment $z_1, \ldots, z_{\theta-1}$ with distribution $\mathcal{N}(0,1)$.
- Post-change segment $z_\theta, z_{\theta+1}, \ldots$ with distribution $\mathcal{N}(\mu_1, 1)$.
- Metrics: false alarm probability $\mathbb{P}(\tau \leq \theta)$ and mean delay $\mathbb{E}(\tau - \theta \mid \tau > \theta)$. For visualization purposes, the figures report $\log_{10}\big(1 + \mathbb{E}(\tau - \theta \mid \tau > \theta)\big)$ on the vertical axis. The logarithmic scale compresses large values of the mean delay, making differences among methods easier to compare across thresholds, while the $+1$ term avoids taking the logarithm of zero..

We run 5000 Monte Carlo replicates comparing ICM for different betting functions and nonconformity measures. Scenarios: $\mu_1 \in \{1, 1.5, 2\}$, $\theta \in \{100, 200\}$, and KNN with $k = 7$. Results are shown in the figures and tables.

Constant Betting Function. Figure 1 and Table 1 report the results for the constant betting function.

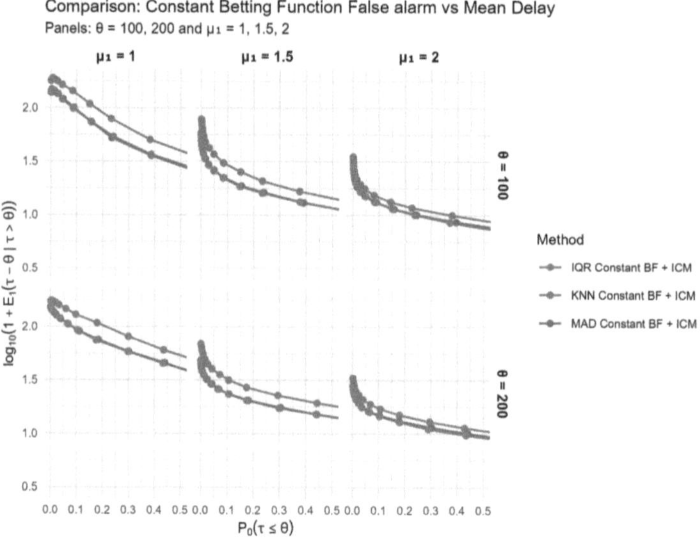

Fig. 1. Results with the *Constant* betting function.

Mixture Betting Function. Figure 2 and Table 2 report the results for the mixture betting function.

Precomputed KDE Betting Function. Figure 3 and Table 3 report the results for the *precomputed KDE* betting function.

Table 1. *Constant* betting function: mean delay under 5% and 10% target false-alarm probabilities

Params	5%			10%		
	KNN	MAD	IQR	KNN	MAD	IQR
$\theta=100,\ \mu_1=1$	161.32	119.29	116.07	135.28	95.25	92.70
$\theta=100,\ \mu_1=1.5$	35.60	25.14	24.58	28.29	20.54	20.19
$\theta=100,\ \mu_1=2$	16.34	13.88	14.01	13.77	11.89	11.97
$\theta=200,\ \mu_1=1$	150.31	112.77	112.98	128.14	93.87	93.26
$\theta=200,\ \mu_1=1.5$	38.36	27.05	26.96	31.26	22.85	22.82
$\theta=200,\ \mu_1=2$	19.14	15.93	15.89	16.38	14.01	13.72

Notes: FA = false-alarm probability. KNN/MAD/IQR are nonconformity measures.

Table 2. *Mixture* betting function: mean delay under 5% and 10% target false-alarm probabilities

Params	5%			10%		
	KNN	MAD	IQR	KNN	MAD	IQR
$\theta=100,\ \mu_1=1$	186.82	175.03	172.91	157.46	143.21	137.26
$\theta=100,\ \mu_1=1.5$	37.86	33.89	31.40	28.36	24.65	22.21
$\theta=100,\ \mu_1=2$	7.83	7.02	7.02	6.22	5.72	5.73
$\theta=200,\ \mu_1=1$	132.00	117.38	119.97	117.40	106.13	107.73
$\theta=200,\ \mu_1=1.5$	22.76	18.79	20.30	17.14	15.42	16.27
$\theta=200,\ \mu_1=2$	7.09	6.56	6.38	6.03	5.60	5.50

Notes: FA = false-alarm probability. KNN/MAD/IQR are nonconformity measures.

Histogram Betting Function. Figure 4 and Table 4 report the results for the histogram betting function.

Overall, the patterns observed in Figs. 1–4 and Tables 1–4 are consistent with the findings reported by Volkhonskiy et al. [1]. For a fixed target false-alarm probability, the precomputed KDE betting function achieves the smallest mean detection delays across the scenarios considered, whereas the histogram betting function tends to produce the largest delays; the constant and mixture betting functions lie in between these two extremes. Regarding the nonconformity measures, the robust choices $\mathcal{A}_{\mathrm{MAD}}$ and $\mathcal{A}_{\mathrm{IQR}}$ generally lead to mean detection delays that are slightly smaller than those obtained with the KNN measure $\mathcal{A}_{\mathrm{KNN}}$. In most combinations of (θ, μ_1), betting function, and false-alarm level (5% or 10%), Tables 1–4 show mildly smaller mean delays for MAD and IQR than for KNN, with minor differences depending on the scenario. The curves for $\mathcal{A}_{\mathrm{MAD}}$ and $\mathcal{A}_{\mathrm{IQR}}$ are very similar, and the gap with $\mathcal{A}_{\mathrm{KNN}}$ is somewhat more noticeable in the weaker signal case $\mu_1 = 1$. This suggests that replacing KNN with a simple

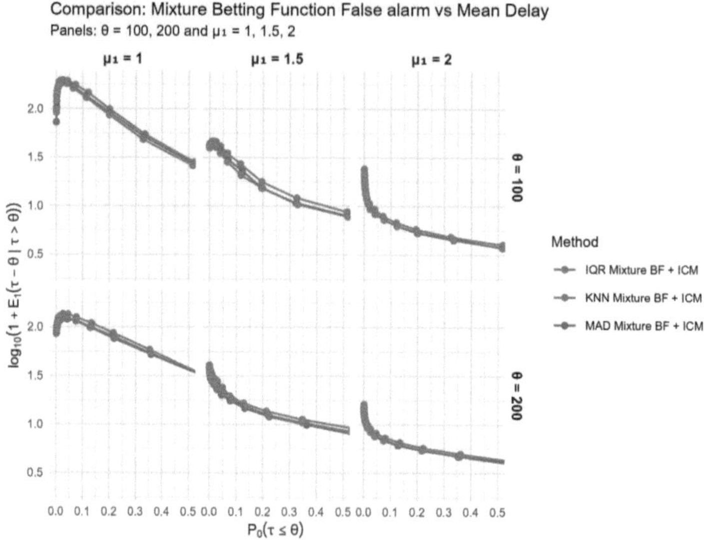

Fig. 2. Results with the *Mixture* betting function.

robust, one-dimensional nonconformity measure can improve the efficiency of ICM, at least slightly, without introducing additional tuning parameters such as the number of neighbours k.

Sensitivity to KNN parameters. On top of the previous Monte Carlo scheme, we added an extra loop over $k \in \{1, 7, 25, 50, 100, 150, 200\}$ to assess the sensitivity of mean delay to the KNN parameter and to determine whether tuning k becomes a drawback in practice. Results are shown in Fig. 5. We observe variations in mean delay across different k choices under the same threshold, introducing undesirable variability because performance depends explicitly on the chosen k.

4.2   Multiple Change-Point

For multiple-change detection methods, we replicate a similar Monte Carlo scheme and study two scenarios.

Scenario 1: a sequence of alternating increases and decreases in the mean; the series is split as:

- *Regime 1:* $\mathcal{N}(0, 1)$ between indices 0 and 500;
- *Regime 2:* $\mathcal{N}(1.5, 1)$ between 501 and 1300;
- *Regime 3:* $\mathcal{N}(0, 1)$ between 1301 and 2000.

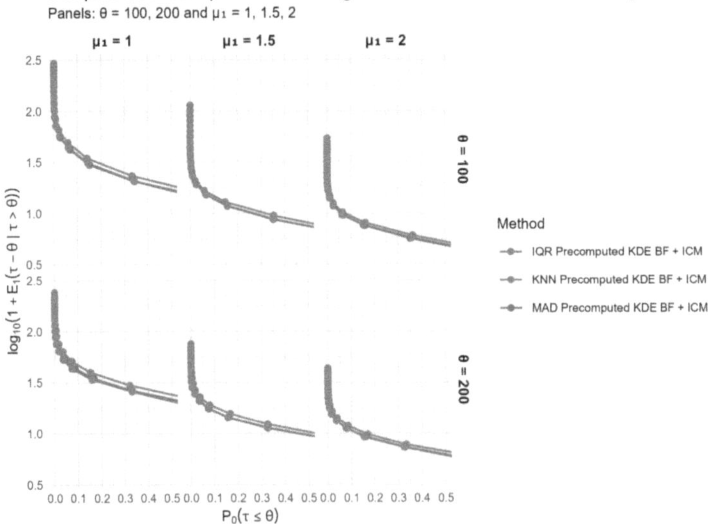

Fig. 3. Results with the *Precomputed KDE* betting function.

Table 3. *Precomputed KDE* betting function: mean delay under 5% and 10% target false-alarm probabilities

Params	5%			10%		
	KNN	MAD	IQR	KNN	MAD	IQR
$\theta{=}100$, $\mu_1{=}1$	52.72	47.08	47.42	41.10	36.38	37.15
$\theta{=}100$, $\mu_1{=}1.5$	16.63	15.84	15.76	13.88	12.95	12.88
$\theta{=}100$, $\mu_1{=}2$	10.02	9.48	9.37	8.37	7.84	7.79
$\theta{=}200$, $\mu_1{=}1$	57.61	50.65	48.62	46.23	40.92	39.52
$\theta{=}200$, $\mu_1{=}1.5$	20.59	18.74	18.55	17.19	15.54	15.54
$\theta{=}200$, $\mu_1{=}2$	12.52	11.55	11.54	10.44	9.64	9.60

Notes: FA = false-alarm probability. KNN/MAD/IQR are nonconformity measures.

Scenario 2: sequential increases in the mean; the series is:

– *Regime 1:* $\mathcal{N}(0,1)$ between indices 0 and 300;
– *Regime 2:* $\mathcal{N}(1,1)$ between 301 and 1000;
– *Regime 3:* $\mathcal{N}(2,1)$ between 1001 and 1700;
– *Regime 4:* $\mathcal{N}(3,1)$ between 1701 and 2000.

Table 4. *Histogram* betting function: mean delay under 5% and 10% target false-alarm probabilities

Params	5%			10%		
	KNN	MAD	IQR	KNN	MAD	IQR
$\theta=100,\ \mu_1=1$	278.77	245.32	251.32	251.24	212.24	215.74
$\theta=100,\ \mu_1=1.5$	115.28	86.21	84.50	94.24	69.82	69.36
$\theta=100,\ \mu_1=2$	55.08	47.79	48.29	46.62	40.96	41.60
$\theta=200,\ \mu_1=1$	299.13	267.29	262.42	267.05	229.58	226.63
$\theta=200,\ \mu_1=1.5$	125.68	94.63	99.37	106.38	82.35	85.59
$\theta=200,\ \mu_1=2$	70.59	61.14	62.46	61.63	53.69	54.50

Notes: FA = false-alarm probability. KNN/MAD/IQR are nonconformity measures.

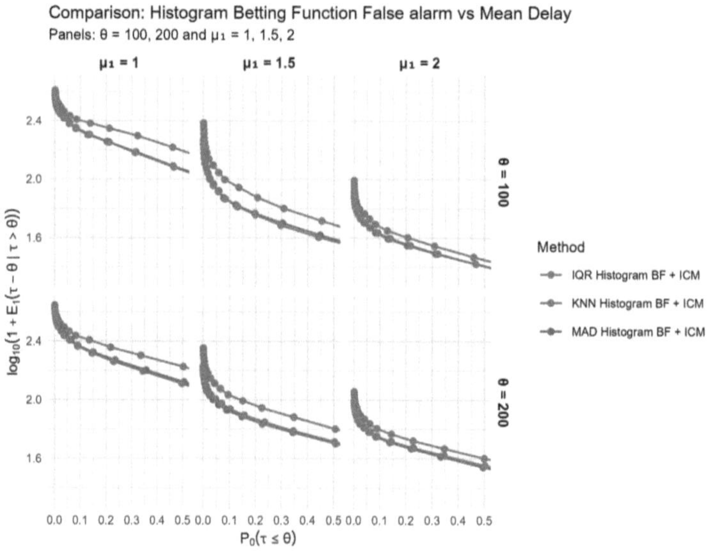

Fig. 4. Results with the *Histogram* betting function.

The metric considered for these methods is *precision*, defined as:

$$\text{precision} = \frac{TP}{TP + FP},$$

where TP indicates correctly detected changes and FP are false alarms. Results are shown in Fig. 6. Precision increases with the threshold, since for larger thresholds $FP \rightarrow 0$. However, as discussed in Subsect. 4.1, increasing the threshold also increases the mean delay.

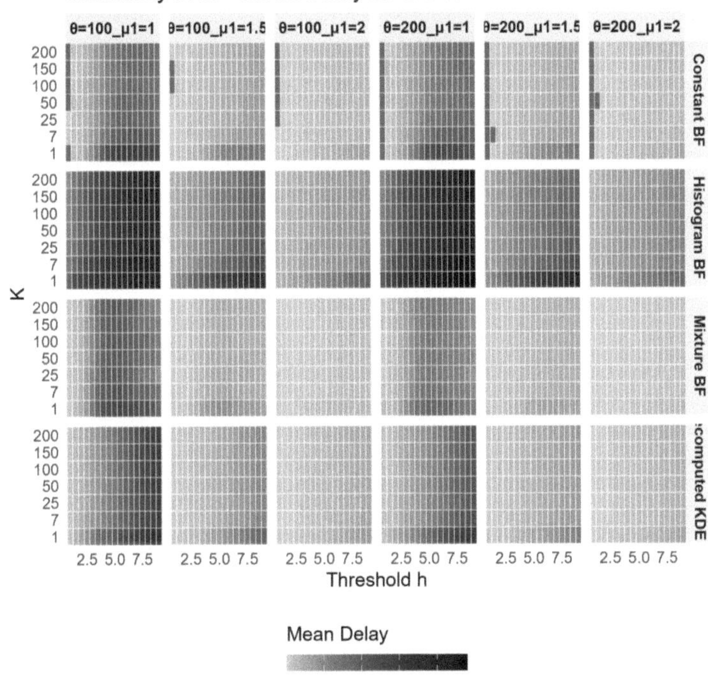

Fig. 5. Sensitivity of the k parameter in KNN.

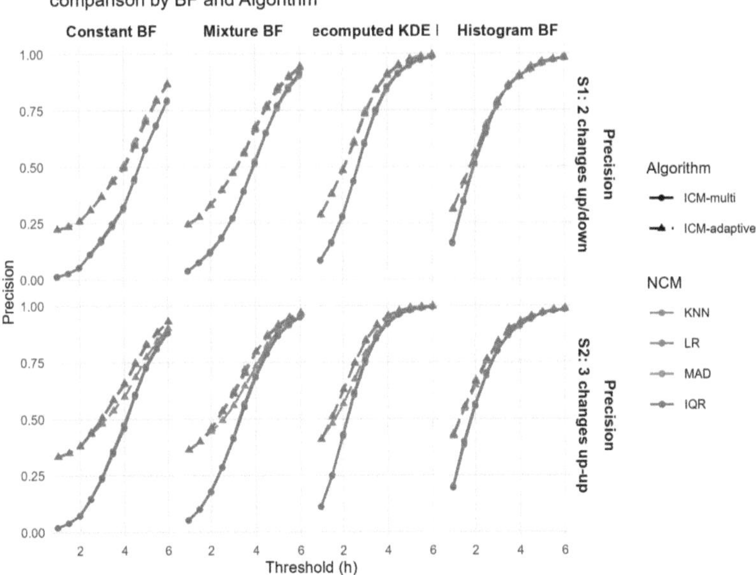

Fig. 6. Precision in multiple-detection algorithms.

5   Conclusions and Future Work

In conclusion, the tools provided by R constitute a comprehensive framework for scientific programming and the implementation of statistical methodologies. The `purrr`, `dplyr`, and `tidyr` libraries, together with higher-order functions and a functional programming style, offer flexibility to combine nonconformity measures and betting functions. As future work, we will integrate these methods into the `tidychangepoint` library to broaden its catalog of change detectors and bring ICM closer to the R community within the `tidy` ecosystem.

References

1. Volkhonskiy, D., Burnaev, E., Nouretdinov, I., Gammerman, A., Vovk, V.: Inductive conformal martingales for change-point detection. In: Gammerman, A., Vovk, V., Luo, Z., Papadopoulos, H. (eds.) Conformal and Probabilistic Prediction and Applications, Proceedings of Machine Learning Research, vol. 60, pp. 132–153. PMLR (2017)
2. Kato, Y., Tax, D.M.J., Loog, M.: A review of nonconformity measures for conformal prediction in regression. In: Papadopoulos, H., Nguyen, K.A., Boström, H., Carlsson, L. (eds.) Conformal and Probabilistic Prediction with Applications, Proceedings of Machine Learning Research, vol. 204, pp. 369–383. PMLR (2023)
3. Eliades, C., Papadopoulos, H.: A histogram based betting function for conformal martingales. In: Gammerman, A., Vovk, V., Luo, Z., Smirnov, E., Cherubin, G. (eds.) Conformal and Probabilistic Prediction and Applications, Proceedings of Machine Learning Research, vol. 128, pp. 100–113. PMLR (2020)
4. Eliades, C., Papadopoulos, H.: Using inductive conformal martingales for addressing concept drift in data stream classification. In: Carlsson, L., Luo, Z., Cherubin, G., Nguyen, K.A. (eds.) Conformal and Probabilistic Prediction and Applications, Proceedings of Machine Learning Research, vol. 152, pp. 171–190. PMLR (2021)
5. Wickham, H., Henry, L.: purrr: Functional Programming Tools. R package version 1.1.0. https://purrr.tidyverse.org/
6. Wickham, H., François, R., Henry, L., Müller, K., Vaughan, D.: dplyr: A Grammar of Data Manipulation. R package version 1.1.4. https://dplyr.tidyverse.org
7. Wickham, H., Vaughan, D., Girlich, M.: tidyr: Tidy Messy Data. R package version 1.3.1. https://tidyr.tidyverse.org
8. Wickham, H.: ggplot2: Elegant Graphics for Data Analysis. Springer, New York (2016)
9. Bengtsson, H.: A unifying framework for parallel and distributed processing in R using futures. R J. **13**(2), 208–227 (2021). https://doi.org/10.32614/RJ-2021-048

Teaching and Pedagogy with R

Proposal for Formulating Questions on the Concept of Hypothesis Testing with R

Rodríguez Granobles Henry, Olarte Pataquiva Johana,
and Zapata Cifuentes Edwin Hernando[✉]

Politécnico Grancolombiano, 110231 Bogotá, Colombia
`herodriguez@poligran.edu.co`

Abstract. Assessing Hypothesis Testing (HT) in undergraduate inferential statistics presents a persistent challenge. The concept itself is not simple and can lead to misinterpretations by students, such as misinterpreting the significance level and failing to recognize its influence on the test; here, two approaches may clash—those of Fisher and Neyman-Pearson—and sometimes it is the instructors themselves who do not recognize their difference. Batanero et al. [3, 4] and Vallecillos [12] have addressed this and many other types of difficulties that arise in the teaching and learning of the concept of hypothesis testing (HT), demonstrating in their work the need to improve methods and assessment in courses where it is covered. This paper proposes that assessment consider the complexities and nuances of the concept, from a didactic approach centered on the design of test items aligned with the dimensions of cognitive competencies: interpretative, argumentative, and propositional. All of this is supported by the R package exams, which stands out as a robust, open-source tool for the automatic and customized generation of assessments. The main goal was to show how the R exams package contributes to better assessment design with customized, reliable, and structured questions, aiming to foster a deeper and more meaningful understanding for students of the concept of hypothesis testing.

Keywords: Hypothesis testing · Statistics education · R (statistical software) · Automated assessment

1 Introduction

Hypothesis testing (HT) is widely regarded as one of the most important and abstract topics in inferential statistics. Its mastery requires both analytical skills and substantial prior knowledge, posing a significant challenge for students, instructors, and researchers alike. Research in statistics education demonstrates that the effectiveness of learning is directly tied to the quality and structure of evaluation instruments. The known conceptual difficulties in HT—including the 'transposed conditional fallacy' and the student's struggle to link sampling variability to the decision criterion—demand an evaluation framework specifically crafted to address these deeply rooted misconceptions. Although its relevance in academic training is clear, fewer studies focus on how HT is taught and its specific impact on student comprehension. For instance, [7] identified several common misconceptions.

B. M. Suárez et al. (Eds.): R Day 2025, CCIS 2824, pp. 321–334, 2026.
https://doi.org/10.1007/978-3-032-18455-9_17

A major finding is that even when students successfully perform calculations and manipulate data, they often misinterpret the results, failing to grasp critical aspects of hypothesis testing with adequate depth. To overcome these persistent difficulties, we propose a pedagogical and technical innovation: designing assessment items structured around a clear framework of cognitive competencies (interpretative, argumentative, and propositional).

This article uses the R package exams, which is an open-source tool, to generate structured, context-based assessments. This package offers a way to create a type of assessment that has an advantage over traditional evaluation by ensuring the instruments are rigorously aligned with higher-order cognitive skills. Ultimately, the proposal seeks to improve the teaching and assessment of hypothesis testing (HT), making it more equitable, efficient, and focused on context-based learning. Finally, the R 'exams' package is introduced as an innovative tool for the automated generation of questions, and a practical example is provided showing how to construct evaluative items of various types using R code.

Throughout the article, some of the main conceptual difficulties students face when learning hypothesis testing are addressed, including common errors in interpreting significance levels, p-values, and the methodological approaches of Fisher and Neyman-Pearson. The role of assessment in the educational process is also discussed, highlighting the importance of designing questions aligned with specific cognitive competencies. Finally, several types of questions are presented, along with their respective formats and competencies.

2 Justification

Students routinely face methodological and cognitive difficulties with the profound conceptual complexity of Hypothesis Testing (HT). This reality underscores the urgent need to develop more effective pedagogical and assessment strategies. Traditional approaches to assessment in statistics education continue to emphasize memorization and the repeated application of standard procedures. Although such practices simplify grading, they often provide limited insight into how students interpret data, reason in situations involving uncertainty, or transfer statistical concepts to unfamiliar contexts. These limitations become particularly evident when students are asked to justify inferential decisions or explain the meaning of statistical results. For this reason, recent work in statistics education has increasingly called for assessment models that place greater emphasis on higher-order cognitive skills and meaningful learning processes.

In response to these concerns, this study explores the use of the R package exams as a tool for designing and managing assessments in statistics courses. The package makes it possible to generate multiple versions of assessment items, vary levels of difficulty, and deploy tests through learning management systems. From an instructional standpoint, this functionality supports consistency across assessments while allowing instructors to focus on evaluating students' reasoning processes rather than their ability to recall formulas or procedures.

The key strength of R exams is its capability to generate technically structured, contextualized questions that integrate smoothly with Learning Management Systems

(LMS). This functionality significantly enhances its practical applicability in both virtual and hybrid statistics education environments.

3 Conceptual and Methodological Challenges in Hypothesis Testing

Difficulties in understanding hypothesis testing (HT) are common and manifest at various levels. For example, it is frequent to confuse the significance level or to believe that a hypothesis test is equivalent to a proof by contradiction. Apart from general conceptual hurdles, a major source of student confusion lies in the historical and methodological overlap between the two main HT frameworks:

Fisher's Significance Testing primarily focuses on calculating the p-value: the probability of observing the data (or data more extreme) assuming the null hypothesis is true. The pedagogical value of this approach lies not only in its technical efficiency but also in its contribution to more transparent assessment practices. By facilitating timely feedback and aligning assessment tasks with clearly defined learning outcomes, the use of exams supports a more accurate interpretation of student performance and provides instructors with information that can guide subsequent instructional decisions.

One area in which these assessment challenges are especially pronounced is hypothesis testing. A persistent difficulty observed among students is the interpretation of the p-value, which is frequently understood as the probability that the null hypothesis is true. This interpretation conflicts with the Neyman–Pearson framework, where hypothesis testing is understood as a decision-making procedure that requires explicit consideration of the alternative hypothesis and the risks associated with Type I (α) and Type II (β) errors. Within this framework, accepting or rejecting the null hypothesis reflects a decision based on a predefined significance level rather than a claim about the truth of the hypothesis itself.

Currently, elements of both models are combined, which unfortunately creates a pedagogical challenge: the difficulty for students to distinguish which procedure is appropriate to carry out at the different stages of a research process. This becomes even more complicated when conceptual errors rooted in probabilistic thinking influence preconceptions and decision-making; for example, students habitually confuse the direction of conditional probabilities—mistaking the significance level for the probability that the null hypothesis is true given the data. Another example is applying a Bayesian interpretation to the results of hypothesis tests, which intensifies this error and subsequently leads to methodological inconsistencies. Psychological obstacles also persist, such as the belief that a statistically significant result eliminates the role of randomness in the phenomenon. Research confirms the constant occurrence of these difficulties in studying the concept, as seen in Batanero et al. [3] and Vallecillos [12], where errors are documented in areas critical to statistical inference, such as properly distinguishing the null and alternative hypotheses, understanding the p-value, and correctly applying a decision criterion at the end of the test. These errors, often enough, stem from an inadequate understanding of fundamental concepts such as population, sample, sampling variability, and probability distributions. To effectively address these failures, assessment questions must intentionally move past mere calculation to specifically probe interpretive and argumentative competencies.

4 Assessment and Its Impact on Education

As previously mentioned, the concept of hypothesis testing (HT) is complex and requires multiple levels of abstraction. However, its evaluation is often summative rather than formative or value based. Therefore, it is necessary to develop criteria that help identify the difficulties in its teaching and learning, and to design strategies that contribute to overcoming them.

4.1 The Question as a Strategy for Assessing Competencies

The use of questions as an assessment tool is highly effective when applied across different levels of thinking. Well-designed questions allow for the evaluation of specific competencies and the objective measurement of student performance. An effective assessment instrument must be composed of coherent and targeted elements, based on a theoretical framework and clear specifications.

For questions to effectively fulfill their evaluative function, they must be structured within a framework that considers both disciplinary content and the cognitive skills intended to be developed. In this regard, understanding the dimensions that make up an objective test is essential to ensure that items are aligned with learning objectives.

4.2 Dimensions of an Objective Test

Before designing questions that assess specific competencies, it is essential to understand the dimensions that structure an objective test. These dimensions help organize the content and cognitive skills to be evaluated, ensuring that the items are aligned with learning objectives and that the assessment is valid, reliable, and fair. Below are the two main dimensions to consider when constructing objective tests:

- Disciplinary Dimension: Content organized into specific areas or components.
- Cognitive Dimension: Includes competencies, statements, evidence, and tasks.

It is crucial that questions align with interpretative, argumentative, or propositional competencies, avoiding dependence on the evaluator's subjectivity [1].

Once the dimensions that structure an objective test are defined, it is important to analyze the role that questions play within the assessment process. Beyond measuring knowledge, well-designed questions can foster the development of higher-order cognitive skills and promote meaningful learning.

4.3 The Role of the Question in the Assessment Process

Questions not only measure acquired knowledge but also foster critical cognitive skills. Through well-designed questions, it is possible to assess students' cognitive development and their ability to apply what they have learned in various contexts. This approach promotes deeper and more meaningful learning. Questions should be designed to:

1. Reflect the development of students' cognitive processes. This involves creating questions that go beyond factual recall and encourage "learning to learn." This approach helps students develop metacognitive skills, enabling them to understand and manage their own learning process [1]. While memorizing basic concepts may be a first step, it is essential that questions promote application, analysis, and reflection, so that students can transfer their knowledge to new situations and develop a deeper understanding.
2. Be clear and precise. Questions should match the level of complexity and the learning purpose. It is especially important that students clearly understand what is being asked. This requires using appropriate and specific language, avoiding ambiguities that may confuse them [5].
3. Demonstrate students' ability to describe processes, discriminate factors, classify essential elements, and identify key criteria among main cognitive operations. This approach allows for the evaluation not only of factual knowledge but also of students' ability to analyze, synthesize, and evaluate information critically and creatively [9]. Purposefully designed questions help students connect concepts and apply their knowledge in new and complex situations.

4.4 Types of Questions According to Competencies

Questions not only assess students' knowledge but also their ability to interpret, argue, and propose solutions in different contexts. Classifying questions according to the competencies they aim to evaluate allows for a more comprehensive and accurate assessment, helping to identify strengths and areas for improvement in students' learning development.

Questions can be classified based on the competencies they are designed to assess:

1. **Interpretative Questions**: Assess the ability to understand and explain basic concepts. Example: What does… Mean? What is the difference between…?
2. **Argumentative Questions**: Require students to explain causes and relationships using evidence and logical reasoning. Example: Why does this happen…? Explain causal relationships in graphs or tables.
3. **Propositional Questions**: Encourage creativity and critical thinking, promoting that students propose solutions or consider hypothetical scenarios.

4.5 Types of Questions According to Format

In addition to considering the competencies to be assessed, the format of the questions also influences how students interact with the content. Below are the main types of questions according to their structure, each with specific advantages for evaluating different levels of understanding:

1. Multiple-Choice Questions:

    Students must select one or more correct answers from a list of options. It is possible to define the number of correct choices, and incorrect answers can be randomly generated for each student.

2. Open-Ended (Essay) Questions:
   Require students to write a free-form response. This type of question is useful for evaluating students' ability to analyze, synthesize, and express ideas in writing.
3. Numerical Questions:
   Expect a numerical answer. Tolerances can be included to accept responses within a specific range, which is useful for questions involving calculations.
4. Matching Questions:
   Students must pair elements from two lists, such as terms and definitions or questions and answers.
5. True/False Questions:
   Students must indicate whether a statement is true or false. This is a simple and effective format for assessing basic understanding of a topic.
6. Cloze (Fill-in-the-Blank) Questions:
   Present sentences or paragraphs with blank spaces that students must fill in with the correct answers.
7. Graphic Questions:
   Include graphs or images, either to identify specific elements, interpret data, or complete visual information.

4.6 Attributes of an Objective Test

For a test to be effective and reliable, it must meet the following attributes:

1. **Validity**: The test must measure what it is intended to measure, ensuring that the inferences drawn from its results are correct.
2. **Reliability:** The results must be consistent and reproducible, regardless of when or how the test is administered [10].
3. **Objectivity**: The test must be independent of the evaluator's subjectivity and the characteristics of the individuals being assessed.

5 Item Construction

The construction of test items requires a meticulous process to ensure that each question aligns with the assessment objectives. The fundamental steps in this process are described below.

1. **Define the component and thematic axis:** Identify the specific area and the central theme of the assessment.
2. **To assess a competency,** it is necessary to establish which competency is to be measured, define the statement to be supported, identify the evidence that will demonstrate achievement, and design the task that will elicit such evidence. This involves determining how the competency will be demonstrated through the assigned task.
3. **Establish the taxonomic level:** Define the cognitive complexity level of the question.
4. **Choose or write the context for the assessment:** Provide a relevant scenario or context for the question.
5. **Determine the question type:** Decide on the question format (e.g., multiple choice, true/false).

6. **Write the question:** Formulate the question clearly and precisely.
7. **Define the response options:** Create plausible response options, including the correct one.
8. **Identify the correct answer:** Determine which option is correct [8].

5.1 Alignment of Assessment Items with Cognitive Competencies

To ensure that the proposed response options are effective and fair, they must meet certain characteristics:

1. Belong to the same semantic field.
2. Have the same level of generality.
3. Avoid the use of synonyms.
4. Maintain grammatical consistency with the stem.
5. Have similar length.
6. Include common errors as plausible distractors.

These guidelines and strategies ensure that the assessment is objective, accurate, and relevant for effectively measuring competencies [11]. A careful and strategic design of questions within the assessment process can transform the way learning is measured and promoted, ensuring that students not only retain information but also develop essential cognitive skills for their academic and personal growth.

Designing effective assessment items for Hypothesis Testing must move beyond simple recall and algorithmic computation. To guarantee validity and effectively target higher-order cognitive processes, we structure the construction of these items around a rigorous framework of three core cognitive competencies [6, 8], which are derived from established educational taxonomies:

1. Interpretative Competency: This assesses a student's ability to decode, contextualize, and explain statistical output. In the context of hypothesis testing, this means that students must not only define terms, but also explain the meaning of the p-value, the significance level (α), and the decision rule within the context of a specific situation.
2. Argumentative Competency: This competency requires students to use evidence to formulate and support conclusions based on statistical results, rigorously justify the choice of a specific statistical procedure, and clearly articulate methodological differences between frameworks (e.g., Fisher vs. Neyman-Pearson).
3. Propositional Competency: As the highest level, this relates to the ability to generate, plan, and propose solutions. In HT, this includes skills like correctly formulating a research hypothesis, defining the most appropriate statistical test for a given study design, or determining the necessary sample size and statistical power to minimize Type II error.

The item presented below designed and created using R exams, explicitly seeks to show the conceptual complexity of HT onto these three dimensions, thereby ensuring the evaluation of deep understanding.

5.1.1 Differences Between Objective and Qualitative Tests

- **Objective Test:** Uses items with limited response options, allowing for quick and consistent evaluation.
- **Qualitative Test:** Includes open-ended items that assess critical thinking skills and expressive ability.

Once the item design process is understood, its implementation can be enhanced through technological tools. In this context, the R package 'exams' provides an efficient solution for automating question generation, allowing for personalized assessments and promoting fairness in the educational process.

5.2 The R 'Exams' Package for Question Development

The R package **'exams'** is an open-source tool that enables the automatic generation of exams in various formats (PDF, HTML, LMS). Its advantages include:

- **Automation and Personalization:** Generates unique versions of exams for each student.
- **LMS Integration:** Compatible with platforms such as Moodle and Canvas.
- **Automatic Evaluation:** Provides immediate feedback.
- **Randomization:** Enhances fairness and reduces the risk of cheating.

A question bank can be created, and from this set of questions, a test can be generated with randomly selected items by following these steps:

- Load the system
- Perform basic configuration
- Generate individual questions
- Generate the complete test
- Run the interactive test
- Analyze the results

5.2.1 Example Question

Below is a practical example of how to use the R package **'exams'** to construct different types of evaluative questions. This example illustrates the complete process—from loading the system and performing basic configuration to generating individual questions and creating a full test. The goal is to demonstrate how this tool facilitates the development of items aligned with specific cognitive competencies, enabling automated, personalized, and contextualized assessment.

5.2.2 Practical Implementation of the R 'Exams' Package

This section presents a practical implementation of the **R 'exams'** package focused on the automated generation of evaluative questions in inferential statistics courses. Through a programmatic approach, it demonstrates how to configure the system, generate individual questions, and build a complete test—all aligned with specific cognitive competencies. The goal is to provide instructors with a flexible and efficient tool that supports contextualized and personalized student assessment.

```r
# ================================================
# 1. Load the system
# ================================================
# This step imports the class that contains the functions for generating
questions. source("PreguntasEstadistica.R")

# ================================================
# 2. Basic configuration of the generator
# ================================================
# The seed is defined for reproducibility, along with the difficulty level
and topic.
  generador <- GeneradorPreguntas$new(list(
      semilla = 456,                       # Ensures that the results are
reproducible
      nivel_dificultad = "medio",          # Can be "básico", "medio" or
"avanzado"
      numero_preguntas = 10,       # Total number of questions to
generate
      tema = "todos"               # Can be filtered by "descriptiva",
"inferencial", etc.
      ))

# ================================================
# 3. Generation of individual questions
# ================================================

# Multiple-choice question
pregunta_multiple <- generador$pregunta_multiple_unica("descriptiva")
print(pregunta_multiple$pregunta)

# Example output:
# "Given the dataset: c(12, 15, 18, 22, 25), what is the mean?"
# Options: A) 18.4, B) 19.2, C) 17.8, D) 20.1

# Calculated answer question
pregunta_calculada <- generador$pregunta_respuesta_calculada()
print(pregunta_calculada$pregunta)
# Example output:
# "Calculate the standard deviation of: c(10, 20, 30, 40, 50)"
# Expected answer: 15.81

# ================================================
# 4. Generation of a complete test
# ================================================

# Create a test with 5 questions from the "descriptiva" topic
test_descriptiva <- generador$generar_test(5, "descriptiva")

# Display all questions in the test
for (i in 1:length(test_descriptiva)) {
    cat("Question", i, ":", test_descriptiva[[i]]$pregunta, "\n\n")
}
```

```r
# ================================================
# 5. Interactive test (simulation)
# ================================================

ejecutar_test_interactivo <- function(num_preguntas = 5) {
    test <- generador$generar_test(num_preguntas)
    puntaje <- 0

for (i in 1:num_preguntas) {
    cat("Question", i, "of", num_preguntas, ":\n")
    cat(test[[i]]$pregunta, "\n")

    # Show options if they exist
    if (!is.null(test[[i]]$opciones)) {
        for (j in 1:length(test[[i]]$opciones)) {
            cat(LETTERS[j], ") ", test[[i]]$opciones[j], "\n",
        sep = "")
        }
    }

    # Capture user response
    respuesta <- readline("Your answer: ")

    # Check if it is correct
    if (tolower(respuesta) == tolower(test[[i]]$respuesta_correcta)) {
        cat("Correct!\n")
        puntaje <- puntaje + 1
  } else {
        cat("Incorrect. The correct answer was:",
    test[[i]]$respuesta_correcta, "\n")
        }
        # Show explanation if available
        cat(test[[i]]$explicacion, "\n\n")
    }

cat("Final score:", puntaje, "/", num_preguntas, "\n")
return(puntaje)
}

# Run the interactive test (example)
# mi_puntaje <- ejecutar_test_interactivo(3)
```

The presented code enables the generation of questions of various types—such as multiple-choice and numerical calculation—tailored to specific topics and difficulty levels. To support learning in this area, the exams package also offers functions that allow the simulation of interactive tests.

These simulations enable students to respond to questions in real time and receive immediate feedback, which plays an important role in helping them identify and correct reasoning errors during the learning process. At the same time, the automated generation of equivalent assessment items contributes to fairness by ensuring that evaluations are comparable across students and course offerings.

Below is an example of questions generated using R code. The complete code corresponding to these questions can be consulted at https://github.com/HenryRodriguezG/Exam_with_R.

Statistics Exam: 00001

1. Among the following pairs of events, indicate which ones are statistically independent:
(a) Life expectancy and the place where people live
(b) The choice of a presidential candidate and the region where the vote is cast
(c) Rolling a pair of dice on two occasions
(d) The approval of a law on abortion and the religious orientation of voters
2. Among the following probability distributions, indicate which ones are based on Bernoulli processes:
(a) Multinomial
(b) Binomial
(c) Multivariate hypergeometric
(d) Negative binomial
(e) Hypergeometric
(f) Poisson
(g) Continuous uniform
3. Let X be a random variable with mean 50 and standard deviation 8; the standardized value of 60 is: _____
4. Which is the probability distribution whose mean and variance are equal?

5.2.3 Operationalizing Higher-Order Competencies with R Exams

Although the R exams package automates the generation of various items for constructing individualized tests (as demonstrated in the previous Rmd example), the pedagogical value of this proposal lies in extending this capability to questions that assess Interpretative and Argumentative Competencies. The following examples illustrate how R exams can be used to design scenario-based questions that specifically target some of the students' misconceptions when addressing the topic of hypothesis testing (HT).

Example of an Item for Interpretative Competency Scenario: A team of researchers conducted a clinical trial and reported a p-value of 0.038 for their primary hypothesis test. They correctly decided to reject the null hypothesis ($\alpha = 0.05$). Which of the following statements is the most accurate interpretation of this result?

A. The probability that the null hypothesis is true is 3.8%.
B. If the study were repeated 100 times, the null hypothesis would be rejected 3.8 times.
   C. The observed data is rare (or more extreme) under the assumption that the null hypothesis is true.
C. The probability that the alternative hypothesis is false is 0.038.

Rationale: This item directly addresses the 'transposed conditional fallacy' and the common misinterpretation of the p-value, requiring students to interpret the statistical finding rather than just the decision.

Example of an Item for Argumentative Competency Scenario: A pharmaceutical company compared a new drug (Group A) against a placebo (Group B) and found a statistically significant difference ($p < 0.01$). A second research team repeated the study with a smaller sample size and reported a non-significant result ($p = 0.15$). Which of the following is the most compelling argument to explain the discrepancy between the two studies?

A. The first study committed a Type I error (α).
B. The second study, due to its smaller sample size, likely lacked sufficient statistical power to detect the true effect (Type II error).
C. The null hypothesis was proven false in the first study but proven true in the second.
D. The observed difference in the first study was too small to be clinically relevant, regardless of the p-value.

Rationale: This item requires the student to argue about the methodological implications of sample size, statistical power, and the Type II error, moving beyond simple calculation to comparative reasoning.

These examples demonstrate how R exams facilitate the creation of robust item banks that are directly mapped to specific cognitive requirements, thereby fulfilling the mandate to assess higher-order statistical reasoning in HT.

5.3 Structured Plan for Empirical Validation

While this work focuses primarily on the pedagogical and technical framework for assessment design, we recognize the critical importance of empirical validation to substantiate the proposal's efficacy. To ensure the rigor of the proposed item bank, a structured validation plan will be implemented in a subsequent phase. This process will involve a two-stage approach:

Phase 1: Content and Construct Validation. The item bank will be evaluated by a panel of subject matter experts (SMEs) in statistics education to confirm content validity and the precise alignment of each item with the intended cognitive competency (Interpretive, Argumentative, or Propositional).

Phase 2: Reliability and field testing. A preliminary implementation (pilot study) of the assessment items will be carried out with a group of undergraduate inferential statistics students. Data collected will be analyzed using item response theory (IRT) models to assess key psychometric properties, including item difficulty, discrimination index, and overall test reliability, thus providing essential preliminary evidence of the instrument's rigor and enhancing its instructional value.

6 Limitations and Future Lines of Work

While the use of the R package 'exams' represents a powerful tool for the automation and customization of assessments in inferential statistics, its implementation is not without challenges. One of the main obstacles in implementing the proposal is the need for technical knowledge in programming and R, which may limit its adoption by instructors without training in computational tools. Another difficulty lies in the fact that designing effective questions requires rigorous pedagogical planning to ensure alignment between items, cognitive competencies, and learning objectives.

Finally, although the article presents a practical example of question generation, the impact of this tool on student performance or on students' perceptions of this type of assessment has not been empirically evaluated, and this would actually be a second phase of the project. In this regard, future research could focus on analyzing the effectiveness of R 'exams' in real classroom contexts, comparing learning outcomes, motivation levels, and feedback quality. It would also be relevant to explore its integration with other educational platforms and its applicability across different areas of knowledge beyond statistics.

7   Conclusions

Teaching hypothesis testing in inferential statistics courses faces multiple conceptual and methodological challenges. Taken together, this proposal addresses several difficulties commonly encountered in the teaching and learning of hypothesis testing. Assessment questions are designed around three cognitive competencies—interpretive, argumentative, and propositional—which are operationalized through tasks embedded within a disciplinary and cognitive framework aimed at supporting validity, reliability, and objectivity in evaluation.

The implementation of this assessment framework also has practical implications at both the instructional and institutional levels. For instructors, the use of exams substantially reduces the time required to prepare varied and secure assessments, allowing greater attention to be devoted to targeted feedback and instructional support. At the institutional level, this approach provides a concrete strategy for promoting assessment practices that are transparent, consistent, and aligned with the development of higher-order statistical reasoning.

This, in turn, enhances the overall quality and academic rigor of inferential statistics courses in both virtual and hybrid learning environments.

Finally, the practical example of item construction using R code demonstrates the feasibility of integrating this tool into real educational environments, providing instructors and institutions with an efficient solution to address the challenges of assessment in inferential statistics.

References

1. Anderson, L.W., Krathwohl, D.R.: A Taxonomy for Learning, Teaching, and Assessing: A Revision of Bloom's Taxonomy of Educational Objectives. Longman (2001)
2. Grün, B., Zeileis, A.: Automatic Generation of Exams in R. J. Stat. Softw. **29**(10) (2009). https://www.jstatsoft.org/
3. Batanero, C., Vera, O., Díaz C.: Dificultades de estudiantes de Psicología en la comprensión del contraste de hipótesis. Números. Revista de didáctica de las matemáticas Núm. **80**, 91–101 (2012)
4. Batanero, C.: Controversias sobre el papel de las pruebas de los contrastes de hipótesis en la investigación experimental. Traducción del artículo Controversies around the role of statistical tests in experimental research, publicado en Math. Thinking Learn. **2**(1–2), 75–98 (2000)
5. Bloom, B.S., Engelhart, M.D., Furst, E.J., Hill, W.H., Krathwohl, D.R.: Taxonomy of Educational Objectives: The Classification of Educational Goals. Handbook I: Cognitive Domain. David McKay Company (1956)

6. Calvo, M.C.L., Miñarro Alonso, A., Vegas Lozano, E.: Generación de cuestionarios Moodle con R+exams+Sweave. VII Jornadas de Enseñanza y Aprendizaje de la Estadística y la Investigación Operativa (2016)
7. Castro Sotos, A.E., Vanhoof, S., Van den Noortgate, W., Onghena, P.: Students' misconceptions of statistical inference: a review of the empirical evidence from research on statistics education. Educ. Res. Rev. **2**(2), 98–113 (2007)
8. Haladyna, T.M.: Developing and Validating Multiple-Choice Test Items. Lawrence Erlbaum Associates (2004)
9. Marzano, R.J., Kendall, J.S.: The New Taxonomy of Educational Objectives. Corwin Press (2007)
10. Messick, S.: Validity. In: Linn, R. L. (ed.) Educational Measurement (3rd ed.). American Council on Education and Macmillan (1989)
11. Nitko, A.J.: Educational Assessment of Students. Merrill/Prentice Hall (2001)
12. Vallecillos, A.: Estudio teórico - experimental de errores y concepciones sobre el contraste de hipótesis en estudiantes universitarios. Universidad de Granada, Tesis doctoral Departamento de Didáctica de la Matemática (1994)

vectorialcalculus: An R Package for Vector Calculus

Julián Mauricio Fajardo[1(✉)] and Julio Lizarazo Osorio[2]

[1] Universidad El Bosque, Bogotá, Colombia
jfajardop@unbosque.edu.co
[2] Universidad Pedagógica y Tecnológica de Colombia, Tunja, Colombia
julio.lizarazo@uptc.edu.co

Abstract. We present `vectorialcalculus`, an R package designed for teaching and analyzing three-dimensional vector calculus. The package integrates tools for sampling curves and surfaces, computing arc length, curvature, torsion, and line integrals, as well as producing both static and interactive visualizations through `plotly`. We describe its design and illustrate its use in educational and applied contexts. Figures and examples highlight its potential as a learning resource within R-based environments.

Keywords: R · 3D visualization · vector calculus · education · plotly

1 Introduction

Multivariable and vector calculus often pose significant conceptual challenges, largely due to the difficulty of visualizing geometric structures in three dimensions. Effective learning requires tools that connect analytical reasoning with clear spatial representation.

While symbolic systems such as `WxMaxima` have long supported this task, R—and particularly the proposed `vectorialcalculus` package—extends these capabilities by integrating numerical computation and interactive 3D visualization. Within R, visualization becomes not merely a complement to analysis but an essential component that reinforces geometric intuition.

This work focuses on three central mathematical objects: (i) parametrized curves, (ii) scalar-defined surfaces, and (iii) vector fields. The `vectorialcalculus` package is designed to explore these objects jointly from analytical, numerical and geometric perspectives, following the classical development found in standard multivariable calculus texts [5–7, 11, 12]

At its core, the package provides a framework for *advanced vector calculus* in three dimensions through controlled discretisation, which replaces symbolic differentiation with high-accuracy numerical approximations. This enables the computation of arc length, curvature, torsion, gradients, directional derivatives, line integrals and triple-integral solids. All functions follow a coherent interface and return reproducible numerical results together with interactive 3D

B. M. Suárez et al. (Eds.): R Day 2025, CCIS 2824, pp. 335–355, 2026.
https://doi.org/10.1007/978-3-032-18455-9_18

graphics based on `plotly` [10]. This combination of numerical precision and visual interactivity bridges analytical computation and geometric intuition in a way that, to our knowledge, is not currently achieved by any other R package.

1.1 Related Work in the R Ecosystem

Several R packages provide partial support for multivariable calculus, numerical analysis or graphical exploration, but none offers an integrated framework dedicated to advanced three-dimensional vector calculus with both numerical operators and interactive geometric visualisation.

Low-level tools such as `rgl` [1] enable three-dimensional rendering but do not compute differentialgeometric quantities such as curvature, torsion, tangent planes or FrenetSerret frames. Numerical libraries such as `pracma` [2] and `numDeriv` [3] provide differentiation and integration routines, but they are not designed around curves and surfaces as geometric objects and do not produce interactive 3D visualizations. Educational tools in the `mosaic` ecosystem [8] support introductory calculus, mostly in one and two dimensions, and do not address advanced three-dimensional constructions.

More specialised frameworks such as `calculus` [4] implement symbolic or high-dimensional operators, but generally lack geometric constructions or interactive exploration of parametrised curves, surfaces, vector fields or solids. These tools serve complementary purposes, often emphasising symbolic manipulation, general numerical routines or pedagogical simplification.

In contrast, `vectorialcalculus` integrates high-accuracy numerical methods with interactive 3D geometric visualisation, providing a unified environment for exploring advanced topics in vector calculus. Its design allows users to compute and visualise differentialgeometric quantities, integrals and solids within a consistent and pedagogically oriented computational framework.

2 Methodology

The `vectorialcalculus` package evolved from prior work with `WxMaxima`, originally intended for multivariable calculus instruction. Each function was rewritten in R, translating the symbolic and geometric logic of `WxMaxima` into a reproducible numerical and visual framework.

Because R lacks a native symbolic algebra engine, the implementation relies on process *discretization* and *linear approximation* to estimate derivatives, gradients, curvatures, and line or surface integrals. Each function underwent repeated testing against analytical references, adjusting sampling density and numerical methods (adaptive integration, centered differentiation, local smoothing) to ensure accuracy and stability.

Interactive 3D graphics were implemented with `plotly`, allowing detailed inspection of curves, surfaces, and vector fields in motion. The result is an integrated, reproducible toolkit that combines numerical analysis, geometric interpretation, and visual exploration–useful for both university teaching and applied research in three-dimensional vector calculus.

Dependencies. Interactive visualizations are implemented with `plotly` [9,10], while reference calculations employ `pracma` [2]. Integration with the `tidyverse` ecosystem promotes reproducible workflows in education and research.

2.1 Design Principles

The design of the `vectorialcalculus` package is guided by four principles intended to ensure internal coherence, numerical reliability and pedagogical clarity.

(1) Discretisation as a unifying computational strategy. Because R is not a symbolic computer algebra system, all differentialgeometric quantities are computed through controlled discretisation. Derivatives, curvature, torsion, tangent planes, line integrals and tripleintegral solids are obtained using centred finite differences, adaptive sampling, or Riemannsum approximations. This yields highaccuracy numerical estimates while maintaining conceptual transparency for teaching and reproducibility for research.

(2) A consistent interface across functions. All functions share a common structure: parametric curves are defined by X, Y and Z; scalar fields by F(x,y); vector fields by functions returning three components; and domains by a, b or spatial windows. Visual options are passed through named lists that control colours, opacities, line styles or markers. Each function returns a list containing numerical results and, optionally, a `plotly` object for interactive inspection. This uniform interface reduces cognitive load and allows users to transfer intuition across the different components of the package.

(3) Alignment with the `plotly` visual grammar. The visualisation layer follows the tracebased structure of `plotly`: curves, surfaces, frames, arrows and sections are represented as modular graphical objects. Aspect ratios are preserved, camera settings are adjustable, and all visual outputs are interactive by default. This design makes it possible to manipulate geometric objects in three dimensions and supports advanced uses such as transparency, slicing and directional lighting.

(4) Modular architecture for pedagogical progression. The package is designed so that elementary constructions (sampling points, computing derivatives, tracing curves) serve as building blocks for more advanced tools such as Frenet–Serret frames, osculating objects, Riemann prisms, or mixed solids in three dimensions. This modularity supports incremental learning: students may explore curves first, then surfaces, then vector fields, and finally integrals and volume constructions, all within a unified computational and visual framework.

2.2 Numerical Considerations

The computations performed by the `vectorialcalculus` package rely entirely on numerical methods, and several design decisions were made to ensure stability, accuracy and predictable behaviour across functions.

Discretisation Accuracy. Most geometric quantities are computed using centred finite differences of order $O(h^2)$, where h is the local sampling step. This provides a good compromise between accuracy and computational cost for educational and exploratory use. When higher precision is required, users may increase the number of evaluation points through arguments such as n_samples, t_points or n_curve, which effectively reduce h and improve the approximation.

Adaptive Sampling. Functions involving integrals or curvature make use of adaptive refinements: sampling density is increased in regions where derivatives change rapidly or where the curvature is high. This prevents loss of geometric detail and mitigates numerical instabilities that commonly appear when curves slow down, accelerate, or oscillate.

Sensitivity and Stability. Because torsion, curvature and directional derivatives combine several layers of differentiation, computations can be sensitive to noise or poorly scaled functions. The package therefore applies local smoothing and internal tolerance checks to avoid divisions by nearly zero quantities, ensuring stability of the Frenet–Serret frame, tangent planes and normal vectors.

Computational Cost. Execution time scales linearly with the number of evaluation points. Typical values (between 200 and 1000 samples per curve) are computationally inexpensive on modern machines and sufficient for high-quality visualisation. Unlike symbolic approaches, which may struggle with complex parametrisations, the numerical workflow used here offers consistent performance for a broad class of functions, including those defined algorithmically or piecewise.

Computational Demands. It is important to note that the numerical workflow implemented in vectorialcalculus is inherently more computationally demanding than symbolic approaches or lowdimensional educational tools. The package relies on dense discretisation of curves, surfaces and vector fields, and several functions involve multiple layers of differentiation or the construction of complex 3D graphical objects. As a consequence, computations with high sampling densities may require more processing time and memory, particularly when interactive plotly visualisations are enabled. This behaviour is expected and reflects the package's focus on advanced geometric analysis rather than introductory symbolic manipulation. For typical pedagogical and exploratory use, however, moderate sampling settings provide an effective balance between accuracy, interactivity and performance.

Comparison with Other Numerical Frameworks. Packages such as pracma and calculus implement numerical differentiation or high-dimensional operators, but they are not optimised for the geometry of parametrised curves and surfaces, nor do they provide interactive visualisation. In vectorialcalculus, discretisation is tightly integrated with the 3D graphical layer, allowing the user to inspect numerical effects in real time, for example by rotating a Frenet–Serret frame, slicing a solid, or examining the behaviour of a vector field. This combination of numerical accuracy and graphical feedback distinguishes the present approach from existing tools in the R ecosystem.

2.3 Package Availability

The development version of `vectorialcalculus`, including source files and documentation, is available on GitHub:

https://github.com/JulianFajardo1908/vectorialcalculus

Users may install it directly via:

```
devtools::install_github("JulianFajardo1908/vectorialcalculus")
```

Development Status. The package is under active development and has been submitted to CRAN. Future releases will include refined numerical routines, expanded visualisation features and additional tools that complement both the current functionality and related packages in the R ecosystem.

3 Smooth Parametric Curves as One-Dimensional Manifolds

Parametric curves, or vector functions, describe the path of a particle moving continuously in space. This representation arises naturally in physics, where the position of a body depends on time. Mathematically, such functions provide a formal framework to analyze motion and the geometry of trajectories in one or more dimensions.

A smooth parametric curve is a differentiable immersion

$$\gamma : I \subset \mathbb{R} \to \mathbb{R}^3, \qquad \gamma(t) = (x(t), y(t), z(t)),$$

where $x, y, z \in C^k(I)$ with $k \geq 1$ and $\gamma'(t) \neq \mathbf{0}$ for all $t \in I$. Its image $\gamma(I)$ forms a one-dimensional differentiable manifold immersed in $\mathbb{R}^3$.

This approach enables the study of differential behavior–changes in direction, speed, and curvature–bridging geometric intuition and physical interpretation. One of the fundamental quantities derived from this perspective is the *arc length*, which measures the total distance along a curve between two points, independent of parametrization.

Let

$$\gamma : [a, b] \to \mathbb{R}^3, \qquad \gamma(t) = (x(t), y(t), z(t)).$$

The *arc length* of γ between $\gamma(a)$ and $\gamma(b)$ is

$$L = \int_a^b \|\gamma'(t)\| \, dt,$$

where $\|\gamma'(t)\|$ represents the velocity magnitude. It quantifies the accumulation of infinitesimal displacements along the motion, yielding a precise geometric measure.

The following example illustrates this concept through a spatial knot, whose arc length can be computed and visualized to explore its geometric structure.

3.1   Arc Length of the Trefoil Knot

The code below applies the function `arc_length3d()` to the curve $\gamma(t) = (\sin t + 2\sin 2t, \ \cos t - 2\cos 2t, \ -\frac{1}{2}\sin 3t)$, estimating its total arc length and generating a 3D view (Figs. 1, 2, 3, 4, 5, 6, 7, 8 and 9).

Listing 1.1. Computation and 3D visualization of the trefoil knot arc length.

```
# --- Trefoil knot arc length -----------------

# Parametric curve gamma(t) = (X(t), Y(t), Z(t))
X <- function(t) sin(t) + 2*sin(2*t)
Y <- function(t) cos(t) - 2*cos(2*t)
Z <- function(t) -0.5*sin(3*t)

# Compute arc length on [0, 2*pi]
trefoil <- arc_length3d(
  X, Y, Z,
  0, 2*pi,
  plot = TRUE,
  n_samples = 1000,
  plot_line = list(color = "blue", width = 3)
)

# Display numerical result
trefoil$arc_length
```

For a smooth curve parametrized by arc length, the local geometry is described by the *Frenet-Serret frame* $\{\mathbf{T}, \mathbf{N}, \mathbf{B}\}$, composed of the tangent, normal, and binormal vectors. The *curvature* κ measures the rate of change in direction, while the *torsion* τ quantifies how the osculating plane rotates around the tangent. Together, they fully characterize the curve's local behavior.

The *Frenet-Serret equations* link these geometric quantities to their derivatives:

$$\mathbf{T}' = \kappa\,\mathbf{N}, \qquad \mathbf{N}' = -\kappa\,\mathbf{T} + \tau\,\mathbf{B}, \qquad \mathbf{B}' = -\tau\,\mathbf{N},$$

where

$$\kappa(t) = \frac{\|\gamma'(t) \times \gamma''(t)\|}{\|\gamma'(t)\|^3}, \qquad \tau(t) = \frac{\det(\gamma'(t), \gamma''(t), \gamma'''(t))}{\|\gamma'(t) \times \gamma''(t)\|^2}.$$

3.2   Curvature and Torsion in the Trefoil Knot

The curvature κ and torsion τ are evaluated near $t_0 = \pi$, displaying the associated Frenet-Serret frame.

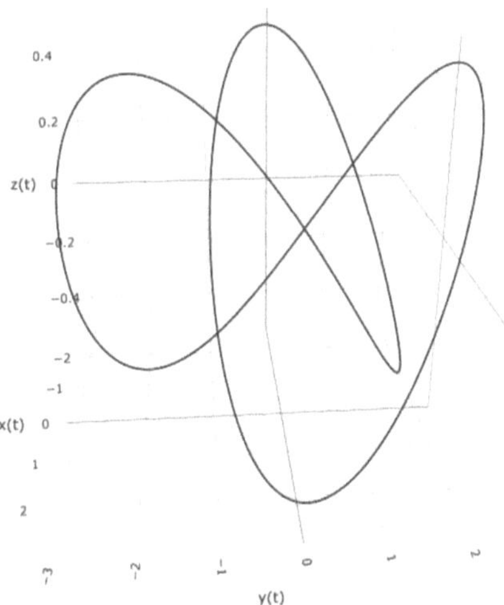

Fig. 1. Trefoil knot generated with `arc_length3d()`. The interactive 3D plot shows a smooth, closed trajectory corresponding to a one-dimensional manifold immersed in $\mathbb{R}^3$.

Listing 1.2. Local curvature and torsion evaluation using 'curvature_torsion3d()'.

```
# --- Curvature and torsion of the trefoil knot
    ----------------------------

# Parametric curve gamma(t) = (X(t), Y(t), Z(t))
X <- function(t) sin(t) + 2*sin(2*t)
Y <- function(t) cos(t) - 2*cos(2*t)
Z <- function(t) -0.5*sin(3*t)

# Evaluate at t0 = pi and display the Frenet-Serret frame
    (T,N,B)
kt <- curvature_torsion3d(
  X, Y, Z,
  t0 = pi,
  plot = TRUE,
  window = 2.0,
  n_samples = 200,
  line  = list(color="red", width=2),
  point = list(color="black", size=5, symbol="circle")
)

```

$$\kappa \approx 0.8 \quad \cdot \quad \tau \approx 0.177784$$

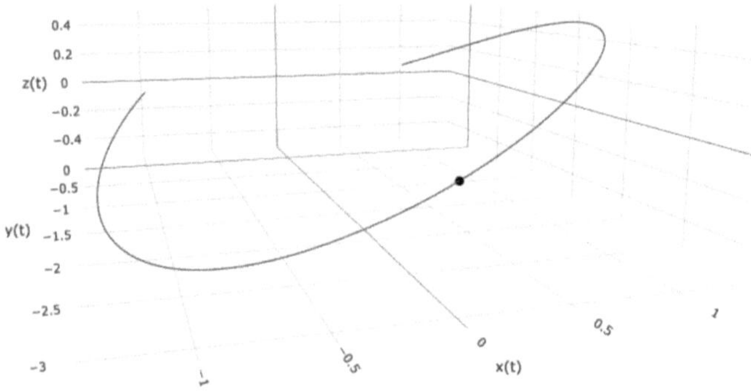

Fig. 2. Frenet-Serret frame in the trefoil knot around $t_0 = \pi$, generated with curvature_torsion3d(). Vectors **T**, **N**, and **B** correspond to the tangent, normal, and binormal directions along the local segment.

```
19  # Numerical results
20  kt$curvature    # kappa(t0)
21  kt$torsion      # tau(t0)
```

3.3   Frenet-Serret Frame Along a Helix

Consider the helix $\gamma(t) = (\cos t, \ \sin t, \ 0.5\,t)$ for $t \in [0, 2\pi]$. The following code visualizes the frame $(\mathbf{T}, \mathbf{N}, \mathbf{B})$ at 50 points along the curve.

Listing 1.3. Visualization of the Frenet-Serret frame using 'frenet_frame3d()'.

```
1  # --- Helix and Frenet Serret frame
      -------------------------------------------
2
3  X <- function(t) cos(t)
4  Y <- function(t) sin(t)
5  Z <- function(t) 0.5*t
6
7  frenet <- frenet_frame3d(
8    X, Y, Z,
9    a = 0, b = 2*pi,
10   t_points = seq(0, 2*pi, length.out = 50),
11   plot = TRUE, n_samples = 300,
12   T_line = list(color="red",   width=4),
```

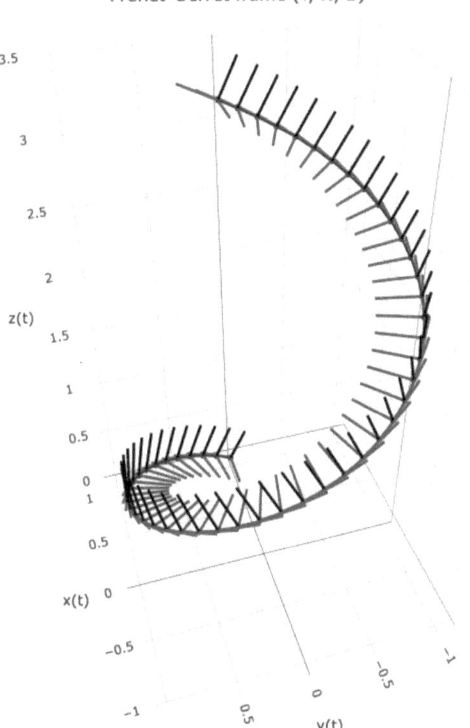

Fig. 3. Frenet-Serret frames along the helix $\gamma(t) = (\cos t, \sin t, 0.5\,t)$. Tangent (**T**), normal (**N**), and binormal (**B**) vectors are shown in red, green, and black at 50 sampled points.

```
13    N_line = list(color="green", width=4),
14    B_line = list(color="black", width=4),
15    vec_scale = 0.3
16 )
```

The study of parametrized curves deepens our understanding of motion and geometry in space while laying the groundwork for computational generalizations. The `vectorialcalculus` package includes functions for both analytical and visual exploration of these ideas, enabling the computation and representation of tangent, normal, and binormal vectors, osculating circles, and associated surfaces such as cylindrical or tangential ones. These tools unify theory and visualization, offering a rich environment for learning and geometric experimentation in three dimensions.

4   Scalar Fields and 2-Surfaces

Functions of a single real variable are insufficient to describe many natural or applied phenomena. Quantities such as temperature at each point of an object or pressure within a fluid depend on spatial position rather than a single coordinate. This leads to the notion of a *scalar field*: a function that assigns each point in a region of space a real number representing the value of a physical or geometric quantity.

Studying scalar fields reveals how such quantities vary across space, identifying gradients, local extrema, and forming the foundation for understanding more complex processes such as flow and diffusion.

Transitioning from one-dimensional trajectories to spatially varying quantities requires a broader formulation: rather than describing motion through a curve, each point in a region is assigned a value that captures a physical property. Formally, a smooth scalar field $f : \Omega \subset \mathbb{R}^2 \to \mathbb{R}$ (C^k, $k \geq 1$) defines the immersed 2-surface

$$\mathcal{S}_f = \{(x, y, f(x, y)) : (x, y) \in \Omega\} \subset \mathbb{R}^3.$$

Geometrically, each scalar field generates a surface in three-dimensional space whose height above the (x, y)-plane represents the field's value at that point. Hence, studying its differential properties requires identifying normal directions to the surface.

If $f \in C^1$, a non-unit normal vector is given by

$$\mathbf{n}(x, y) = (-f_x, -f_y, 1),$$

and the corresponding unit normal, which determines the local surface orientation, is

$$\mathbf{N} = \frac{\mathbf{n}}{\|\mathbf{n}\|}.$$

Knowing the normal vector allows us to precisely describe the local geometry of the surface associated with a scalar field. In particular, the *tangent plane* at a given point provides the best local linear approximation, serving as the basis for analyzing slope, orientation, and local variation of f. The following example shows the computation and visualization of the tangent plane at a specific point of a surface $z = F(x, y)$.

4.1   Tangent Plane and Normal Vector in $z = F(x, y)$

For a smooth scalar field $z = F(x, y)$ defined on $\Omega \subset \mathbb{R}^2$, the *tangent plane* at the point $(x_0, y_0, F(x_0, y_0))$ is obtained from the total derivative:

$$z - F(x_0, y_0) = F_x(x_0, y_0)(x - x_0) + F_y(x_0, y_0)(y - y_0).$$

Its *normal vector* is

$$\mathbf{n}(x_0, y_0) = (-F_x(x_0, y_0), -F_y(x_0, y_0), 1),$$

and the unit version $\mathbf{N} = \frac{\mathbf{n}}{\|\mathbf{n}\|}$ defines the local surface orientation and allows the analysis of angles between surfaces.

Listing 1.4. Tangent plane and normal vector using tangentplane3d().

```
F <- function(x, y) sin(x*y)

# --- Tangent plane at (x0,y0) = (1,1) --------------------------
tp <- tangent_plane3d(
  F,
  point = c(1, 1),
  plot  = TRUE,
  x_window = 2, y_window = 2, z_window = 3,
  plane_window = 1,
  vec_N_factor = 1.2,
  colors = list(
    surface = "lightblue",
    plane   = "Greens",
    xcut    = "black", ycut = "black",
    point   = "blue",  normal = "red"
  ),
  show_surface_grid  = TRUE,
  surface_grid_color = "rgba(0,0,0,0.25)"
)
```

The study of scalar fields provides a foundation for understanding spatial variations of physical or geometric quantities. Within this context, the vectorialcalculus package extends R's analytical capabilities by enabling the computation and visualization of essential quantities such as the directional derivative, gradient vector, and critical points. It includes tools for identifying and classifying critical points in bivariate scalar fields–with graphical support– and analytic extensions to the general n-dimensional case, where visualization yields to symbolic and numerical exploration. These features strengthen the link between differential theory and computational practice, fostering an integrated understanding of surfaces and their local properties.

5   Integrals over Curves and Regions

The analysis of scalar fields naturally leads to integration, a key tool for quantifying magnitudes associated with spatial regions. From a function defined on a domain, one can construct three-dimensional solids bounded by planes and surfaces, interpreting the integral geometrically as volume or accumulated quantity.

The integral approach originates from Riemann sums, which approximate area or volume beneath a function, and extends to more complex representations such as *line integrals*–defined along trajectories in scalar or vector fields. These integrals measure accumulated change along a curve, connecting geometry, analysis, and 3D visualization.

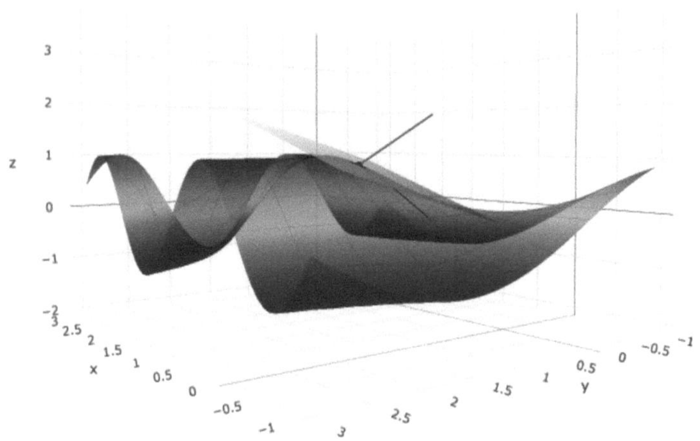

Fig. 4. Tangent plane to $z = F(x,y)$ at $(1,1)$ with normal vector (red), section lines $x=(Color figure online)1$ and $y=1$, and surface background. Generated with `tangent_plane3d()`.

5.1   Riemann Prisms in $[a,b] \times [f_1(x), f_2(x)]$

Extending from single to double integrals allows the computation of areas and volumes associated with functions of two variables. Let $F : \mathcal{R} \subset \mathbb{R}^2 \to \mathbb{R}$ be continuous on a rectangular region $\mathcal{R} = [a,b] \times [c,d]$. Its double integral is defined as the limit of Riemann sums:

$$\iint_{\mathcal{R}} F(x,y)\, dA = \lim_{m,n \to \infty} \sum_{i=1}^{m} \sum_{j=1}^{n} F(x_{ij}^*, y_{ij}^*)\, \Delta A,$$

where $\Delta A = \Delta x\, \Delta y$ denotes the area of each subrectangle and (x_{ij}^*, y_{ij}^*) are sample points in the partition.

Geometrically, each term represents the volume of a *Riemann prism* whose base is a subrectangle and height corresponds to $F(x_{ij}^*, y_{ij}^*)$. The sum of these prism volumes approximates the total volume under the surface $z = F(x,y)$, converging to the exact integral as the partitions become infinitesimally fine.

Listing 1.5. Lower, upper, and mean Riemann sums using riemannprisms3d().

```r
# --- Function and strip-shaped region
  ----------------------------------
F  <- function(x, y) x + y
f1 <- function(x) 0
f2 <- function(x) 1

# --- 3D Riemann prisms with surface overlay
  ------------------------------
res <- riemann_prisms3d(
  F, f1, f2,
  a = 0, b = 1, N = 5, M = 5,          # N: x-partitions, M: y-
    partitions
  plot = TRUE,
  show_surface = TRUE,                 # overlay z = F(x,y)
  color_by = "lower",                  # color by lower
    estimator
  top_colorscale = "YlGnBu",
  side_color    = "rgba(60,60,60,0.25)",
  side_opacity = 0.35,
  frame_color  = "rgba(0,0,0,0.55)"
)

# Numerical results (if available)
res$sum_lower; res$sum_upper; res$sum_mean
```

The geometric reasoning behind Riemann prisms extends the concept of double integrals to more general regions and to volume computation in three-dimensional space. Instead of approximating a surface over a rectangle, we now consider *solids bounded by six surfaces*, where boundaries may be defined by planes or multivariable functions. In this context, the triple integral emerges as a natural generalization of Riemann sums, allowing quantitative and visual exploration of volumes enclosed by surfaces.

5.2 `solid_xyz3d`: Mixed Solid with Top/Bottom Surfaces $G_i(x, y)$ and Side Curves $H_i(x)$

One of the main challenges in teaching multivariable calculus lies in visualizing 3D geometric objects–especially solids bounded by multiple surfaces. Their spatial complexity often hinders students' geometric intuition and complicates explanation. To address this, the following tool provides a precise and interactive way to construct and visualize such solids, bridging analytic formulation and spatial interpretation.

Consider a solid bounded by continuous functions of one or two variables:

$$\mathcal{S} = \{(x, y, z) \in \mathbb{R}^3 : a \le x \le b, \ H_1(x) \le y \le H_2(x), \ G_1(x, y) \le z \le G_2(x, y)\},$$

where $G_1, G_2 : \mathbb{R}^2 \to \mathbb{R}$ and $H_1, H_2 : \mathbb{R} \to \mathbb{R}$ define its upper/lower faces and side boundaries, respectively. The volume of $\mathcal{S}$ can be obtained through a

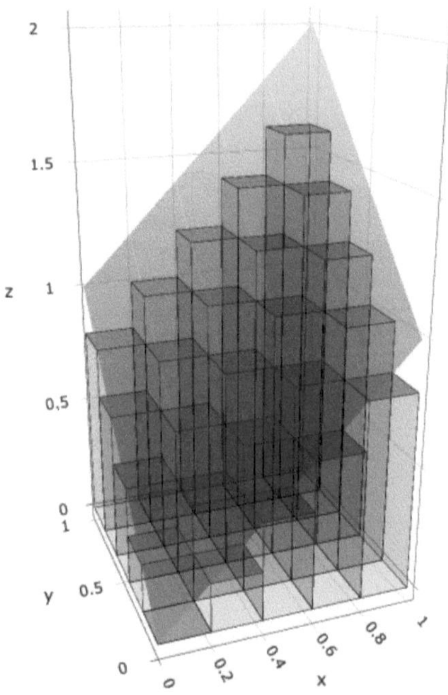

Fig. 5. Riemann prisms for $F(x,y) = x + y$ on $[0,1]^2$, with the surface $z = F(x,y)$ overlaid. Top faces are colored by the lower sum.

triple integral or approximated numerically by discretization. The following code demonstrates how to construct such a solid and estimate its volume.

Listing 1.6. Mixed solid with optional slicing generated using solidxyz3d().

```
# --- Top and bottom faces (z = G1(x,y), G2(x,y))
G1 <- function(x, y)  0
G2 <- function(x, y)  1 - y

# --- Lateral boundaries (y = H1(x), H2(x))
H1 <- function(x) sqrt(x)
H2 <- function(x) 1

# --- Construction and visualization
solid_xy <- solid_xyz3d(
  H1, H2, G1, G2,
  a = 0, b = 1,
  plot = TRUE,
```

```
14    mode = "both",
15    colorscales = c("Blues","Blues","Greens","Greens","Reds","
        Reds"),
16    opacities = 0.3,
17    wire_step = 10,
18    wire_line = list(color="rgba(20,20,20,0.25)", width=1,
        dash="dot"),
19    edge_line = list(color="rgba(30,30,30,0.35)", width=1),
20    slice = list(x = 0, z = c(0.25, 0.50, 0.75)),
21    slice_mode = "surface",
22    slice_colorscales = list(x = "#ffb703", y = "#9b5de5", z =
        c("white", "#2a9d8f")),
23    slice_opacity = 0.6
24 )
25
26 solid_xy$volume
27
28 # --- Adaptive integration for volume estimation
29 sol_vol <- solid_xyz3d(
30    H1, H2, G1, G2,
31    a = 0, b = 1,
32    plot = FALSE,
33    compute_volume = TRUE,
34    vol_method = "adaptive"
35 )
36 sol_vol$volume$estimate
```

5.3   Line Integrals over Scalar Fields

In many physical and geometric problems, integration over planar regions or volumes is insufficient; instead, one must compute the accumulated value of a quantity along a curve in space. This is formalized through *line integrals*, which measure, for instance, the mass of a wire with variable density, the average temperature along a path, or the work performed along a trajectory.

Let $f : \Omega \subset \mathbb{R}^2 \to \mathbb{R}$ be a scalar field and $\gamma : [a,b] \to \Omega$ a smooth parametrized curve $\gamma(t) = (x(t), y(t))$. The *line integral* of f along γ is defined as

$$\int_\gamma f(x,y)\, ds = \int_a^b f(x(t), y(t))\, \|\gamma'(t)\|\, dt,$$

where $ds = \|\gamma'(t)\|\, dt$ is the differential arc-length element. This expression accumulates the scalar values of f weighted by infinitesimal distances along the curve, providing a total measure of the distributed quantity.

When $f(x,y) \geq 0$, graphical representation of the line integral allows visualization of both the integration curve and the surface generated by the values of f along it. The function line_integral2d() offers an interactive approach to visualizing this process, displaying the curve, the lifted path, and the "curtain" region illustrating geometric accumulation.

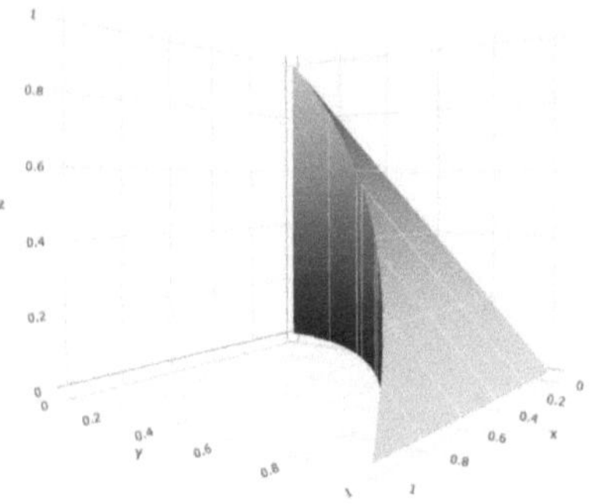

(a) General view: inclined planes and curved boundaries.

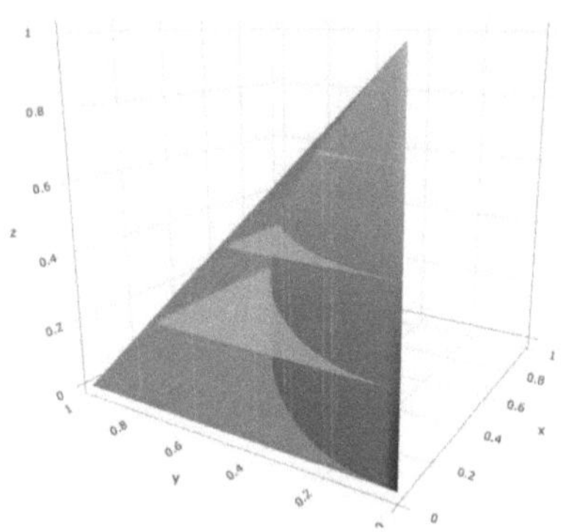

(b) Cross sections $z = \{0.25, 0.5, 0.75\}$ revealing internal structure.

Fig. 6. Solid defined by $z = 1 - y$ and $y \in [\sqrt{x}, 1]$, generated with `solid_xyz3d()`. Transparency highlights slicing planes and boundary curvature.

Listing 1.7. Line integral using lineintegral2d(): lifted curve, curtain, and surface.

```
# --- Scalar field and planar curve
    ----------------------------------------
f <- function(x, y) 4 + 3*sin(x)*cos(y)

r <- function(t) c(
  2*cos(t) + cos(2*t),   # x(t)
  2*sin(t) - sin(2*t)    # y(t)
)

# --- Line integral int_gamma f ds with 3D visualization
    --------------------
res <- line_integral2d(
  f, r, a = 0, b = 2*pi,
  plot = TRUE,
  n_curve = 400, n_curtain_v = 40,
  n_surf_x = 80, n_surf_y = 80,
  colorscale = "Viridis", surface_opacity = 0.65,
  show_surface_grid  = TRUE,
  surface_grid_color = "rgba(80,80,80,0.25)",
  curtain = 0.4,
  curtain_colorscale = c("white", "#2a9d8f"),
  curve_color  = "black", curve_width  = 3,
  lifted_color = "red",   lifted_width = 2,
  method = "adaptive", n_simpson = 1000,
```

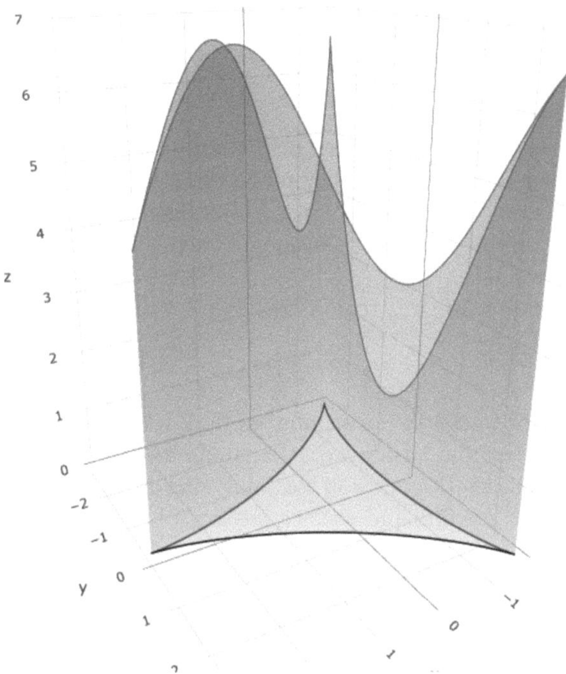

Fig. 7. Line integral $\int_\gamma f\,ds$: surface $z = f(x,y)$, base curve (black), lifted path $z = f(\gamma(t))$ (red), and projection curtain.

```
23    scene = list(aspectmode = "data",
24                 xaxis = list(title = "x"),
25                 yaxis = list(title = "y"),
26                 zaxis = list(title = "z")),
27    bg = list(paper = "white", plot = "white")
28 )
29 # Numerical result (if returned)
30 # res$value
```

Beyond these features, the `vectorialcalculus` package includes a set of complementary tools that considerably extend its scope. These include partial and directional derivatives, identification and classification of critical points in scalar fields, and visualization of 3D solids in cylindrical and spherical coordinates. It also supports the projection of 3D solids onto coordinate planes for sectional analysis, and the computation of fundamental vector quantities such as divergence and curl.

Altogether, these capabilities establish the package as an integrated platform for numerical and geometric exploration of multivariable and vector calculus in R–providing a conceptual bridge toward the study of *vector fields*, developed in the following section.

6   Vector Fields in 3D

While scalar fields describe how a numerical quantity varies in space, many physical and geometric phenomena also depend on direction and orientation. Typical examples include fluid velocity, the force exerted by a gravitational or electric field, and airflow around an object. To model such situations, we introduce the concept of a *vector field*.

Formally, a *vector field* on a region $\Omega \subset \mathbb{R}^n$ is a mapping

$$\mathbf{F} : \Omega \to \mathbb{R}^n, \qquad \mathbf{F}(x_1, \ldots, x_n) = (F_1(x_1, \ldots, x_n), \ldots, F_n(x_1, \ldots, x_n)),$$

that assigns a vector in $\mathbb{R}^n$ to each point of Ω. Each component F_i is a scalar function describing the magnitude variation along its respective coordinate axis.

Vector fields form the foundation for analyzing flux, circulation, and work, preparing the ground for line and surface integrals in broader physical and geometric contexts.

6.1   Field $\mathbf{F}(x, y, z) = (-y, x, x)$: Top and Front Views

To understand the spatial structure of a vector field, it is helpful to examine it from multiple perspectives. Consider the three-dimensional field

$$\mathbf{F}(x, y, z) = (-y, x, x),$$

which combines a rotation in the xy-plane with a constant upward component along z. A *top view* reveals the rotational pattern in the plane, while a *front view* emphasizes the vertical ascent. These visualizations clarify both flow orientation and spatial geometry.

Listing 1.8. 3D visualization of the field $\mathbf{F}(x, y, z) = (-y, x, x)$.

```
# --- Vector field: rotation in xy + vertical component
  --------------------
F <- function(x, y, z) c(-y, x, x)

# Domain box: x,y,z in [-1,1]
vf <- vector_field3d(
  F,
  a = -1, b = 1,
  NX = 8, NY = 8, NZ = 8,
  plot = TRUE,
  arrows = "both",
  arrow_scale = 0.02,
```

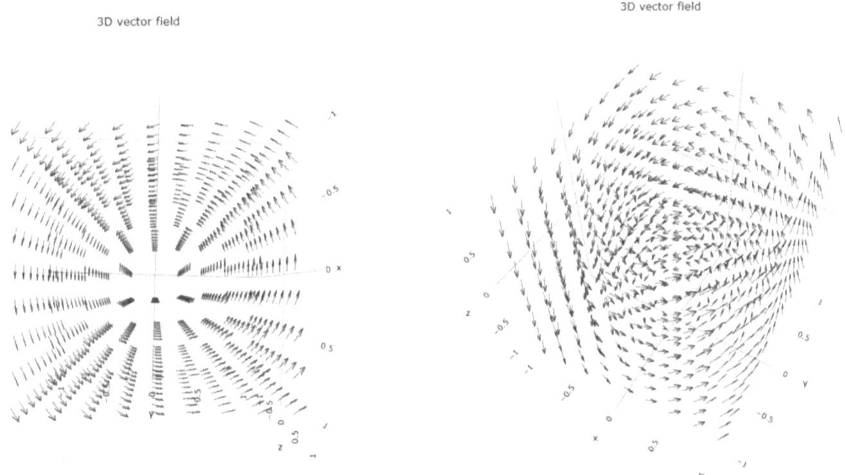

(a) Top view: counterclockwise rotation in the xy-plane.

(b) Oblique view: constant upward drift along z.

Fig. 8. Vector field $\mathbf{F}(x, y, z) = (-y, x, x)$ generated with vector_field3d(). Arrows illustrate rotational motion in xy and helical ascent along the z-axis.

```
12    normalize_bias = 1,
13    arrow_color = "#1d3557", arrow_opacity = 0.95, arrow_size
         = 0.35
14 )
```

The visual analysis of $\mathbf{F}(x, y, z) = (-y, x, x)$ reveals a combined rotation–ascent pattern, motivating the study of particle trajectories within the field. To describe how a particle moves under such a vector field, we examine the *streamlines*–curves tangent to the field at every point. The next example considers a related field, $\mathbf{F}(x, y, z) = (-y, x, 0.5)$, combining rotation around the z-axis with constant vertical drift, and displays a streamline starting from $p_0 = (1, 0, 0)$.

6.2   Streamline in a Rotational Field with Vertical Drift

Listing 1.9. Flow visualization using 'streamline_and_field3d()' in a rotational field.

```
1 # --- Vector field: rotation about z + vertical drift
        -----------------------
2 F <- function(x, y, z) c(-y, x, 0.5)
3
4 # Streamline starting at p = (1, 0, 0)
5 res <- streamline_and_field3d(
6    F,
7    a = -1, b = 1,
```

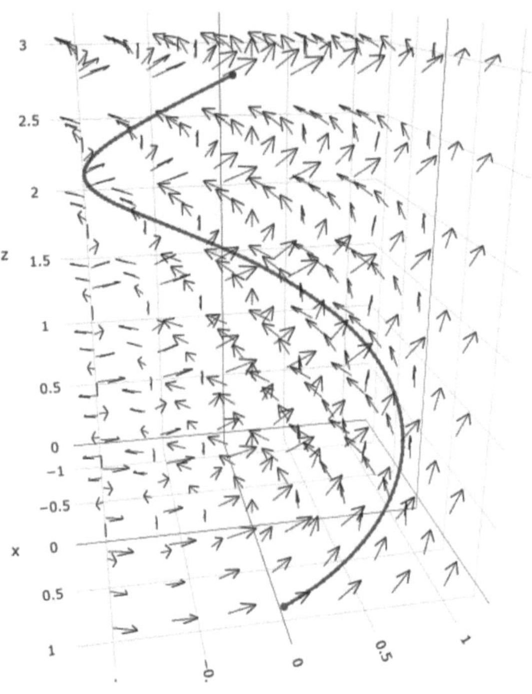

Fig. 9. Streamline of the field $\mathbf{F}(x,y,z) = (-y, x, 0.5)$ generated with `streamline_and_field3d()`. The trajectory (red) starts at $p_0 = (1, 0, 0)$ and traces an ascending helix around the z-axis.

```
 8    NX = 6, NY = 6, NZ = 6,
 9    p = c(1, 0, 0),
10    T = 6, step = 0.05,
11    arrows = "both",
12    arrow_scale = 0.08,
13    traj_color = "red",
14    traj_markers = TRUE
15  )
```

7   Applications

7.1   Teaching and Visualization

The package functions support classroom activities on line integrals, gradients, divergence, and curl. Each function is accompanied by a 3D visualization that connects analytical computation with geometric intuition, fostering a deeper conceptual understanding.

7.2   Research and Analysis

The package also enables simulation of trajectories and vector fields for evaluating physical magnitudes such as work and flux. Interactive rendering allows manipulation of visual objects and real-time observation of spatial changes and field dynamics.

8   Conclusions and Future Work

Future development will focus on expanding both analytical and visual capabilities. Planned features include customizable color palettes and graphical styles, as well as generalization of several current functions to broader cases in vector calculus. Integration with the R package gganimate is also envisioned to create dynamic animations illustrating differential and integral calculus concepts–visualizing in real time the geometric evolution associated with diverse mathematical phenomena.

References

1. Adler, D., Murdoch, D., et al.: RGL: 3D visualization using OpenGL (2023). https://CRAN.R-project.org/package=rgl, r package version 1.2.11
2. Borchers, H.W.: Pracma: Practical numerical math functions (2023). https://CRAN.R-project.org/package=pracma, r package version 2.4.4
3. Gilbert, P., Varadhan, R.: numDeriv: Accurate numerical derivatives (2022). https://CRAN.R-project.org/package=numDeriv, r package version 2016.8-1.1
4. Giné-Vázquez, I.: calculus: High-dimensional numerical and symbolic calculus (2021). https://CRAN.R-project.org/package=calculus, r package version 0.3.1
5. Larson, R., Edwards, B.: Calculus. Cengage Learning, Boston, MA, 11th edn. (2016), iSBN 978-1285774770
6. Marsden, J.E., Tromba, A.J.: Vector calculus, 6th edn. W. H. Freeman, New York (2012)
7. Osorio, J.D.C.L.: Cálculo multivariado con el uso de WxMaxima. Facultad de Ingeniería, Bogotá, Colombia, 1 edn. (2020), iSBN 978-958-5450-10-3
8. Pruim, R., Kaplan, D., Horton, N.: mosaic: Project MOSAIC teaching utilities for mathematics and statistics (2024). https://CRAN.R-project.org/package=mosaic, r package version 1.9.0
9. Sievert, C.: Interactive web-based data visualization with R, plotly, and shiny. Chapman and Hall/CRC (2020). https://plotly-r.com
10. Sievert, C., et al.: plotly: Create interactive web graphics via plotly.js (2024). https://CRAN.R-project.org/package=plotly, r package version 4.10.4
11. Stewart, J.: Calculus: Early transcendentals, 8th edn. Cengage Learning, Boston, MA (2015)
12. Thomas, G.B., Hass, J., Heil, C.: Thomas' Calculus. Pearson, Boston, MA, 14th edn. (2018), iSBN 978-0134438986

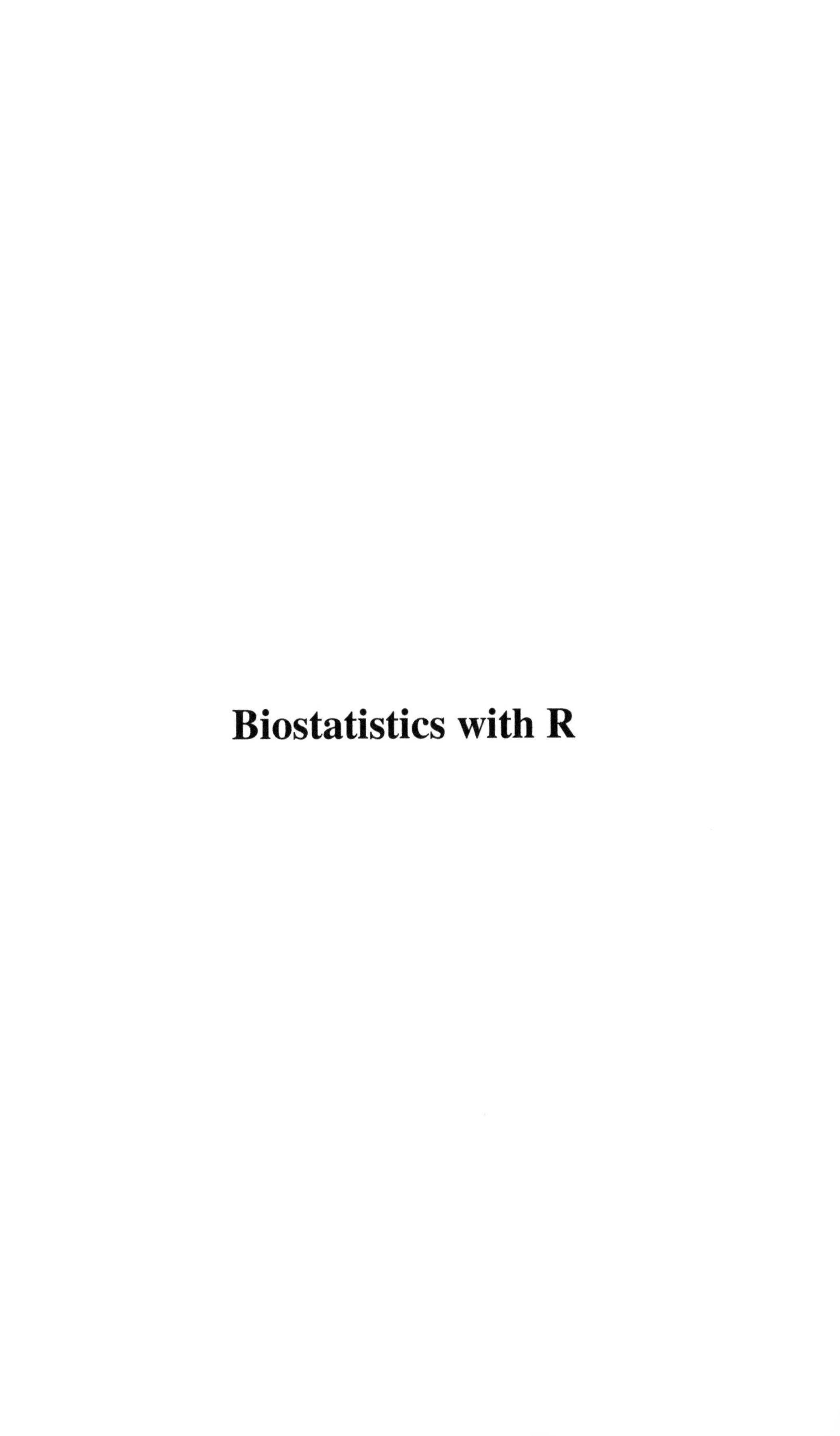

Biostatistics with R

Data-Driven Digitalization of Chrysanthemum Pinching: Development and Validation of an R Shiny Application for Operational Efficiency in Floriculture

Sandra Naryely Cotua Barrera^(✉)

Agricultural Engineer, University of Antioquia, Medellín, Colombia
sandracotuab@gmail.com

Abstract. This work presents the development and validation of an R Shiny application designed to predict and monitor chrysanthemum pinching, transforming a manual task into a planned, data-driven process. Key agronomic information, such as planting week and growth periods, was digitized in Google Sheets and integrated with R to automatically determine the optimal timing for intervention. The tool records data in real time, issues visual alerts, and generates interactive graphs, enabling precise workload estimation and efficient labor allocation. Its implementation enhanced operational efficiency, reduced the need for field inspections, and improved scheduling accuracy. This model, adaptable to other crops and aligned with the principles of Agriculture 4.0, demonstrates that combining data analysis with agricultural expertise can enhance quality and competitiveness in floriculture.

Keywords: Precision agriculture · R Shiny application · Operational efficiency · Chrysanthemum cultivation · Predictive modeling · Smart farming

1 Introduction

The chrysanthemum (Chrysanthemum morifolium) is one of the most representative cut flowers in global floriculture, and Colombia has established itself as one of the world's leading producers and exporters. According to Ceniflores [1], the country has approximately 8,900 hectares dedicated to cut flower cultivation, concentrated mainly in Cundinamarca (66%) and Antioquia (33%). Colombia exports to more than 100 international markets, highlighting the sector's significant economic and social importance.

Within this context, disbudding represents a fundamental practice in the chrysanthemum production cycle, as it directly influences plant uniformity, commercial quality, and physiological efficiency [2]. In pompon-type varieties, this task involves removing the main floral bud to redirect energy and growth hormones toward the lateral shoots. This process reduces auxin concentration at the apex while increasing cytokinin activity, promoting more uniform growth, larger and better-shaped flowers, and reducing physical damage between buds [3].

B. M. Suárez et al. (Eds.): R Day 2025, CCIS 2824, pp. 359–371, 2026.
https://doi.org/10.1007/978-3-032-18455-9_19

However, disbudding planning often relies on manual records and visual inspections, which leads to variability and operational inefficiencies. Such practices underscore the need to transition toward Agriculture 4.0, based production models, where the integration of data, sensors, and predictive analytics enables proactive decision-making and enhances process sustainability [4].

Within this framework, the implementation of low-cost digital tools that combine agronomic information with dynamic analysis and visualization represents a key strategy to optimize labor management tasks such as disbudding.

Under this approach, a web application was developed in R Shiny, connecting data stored in Google Sheets with predictive models and interactive dashboards. According to Ryan [5] and R Programming for Data Sciences [6], the Shiny architecture facilitates the creation of interactive analytical platforms that integrate statistical computation, automation, and real-time visualization. This tool allows users to monitor crop development, predict the optimal disbudding time, and plan labor requirements with greater precision, strengthening the efficiency and sustainability of the production process.

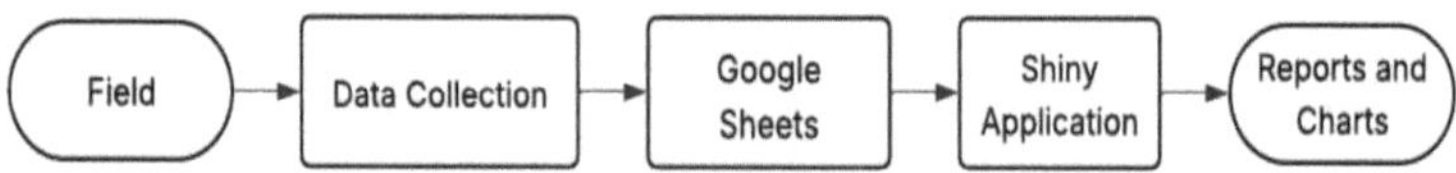

Fig. 1. Data processing workflow of the Shiny application. *Source:* Author's elaboration (2024).

Figure 1. The diagram shows the information flow within the Shiny application, from field data collection and cloud storage in Google Sheets to real-time processing in RStudio. The system applies predictive algorithms to estimate the optimal week for pinching and generates dynamic reports and visual outputs to support decision-making.

Link to the application: Pinching Tracking System.

2 Methodology

2.1 General Application Design

The application was developed in the R Shiny environment, chosen for its flexibility in integrating statistical processing, interactive visualization, and cloud-based data management within a single platform.

The development followed a client-server architecture, where the client (user interface) communicates reactively with the server (responsible for data processing and connection). This structure enables the automatic real-time updating of content without the need to reload the page, ensuring smooth user interaction and system stability.

The user interface (UI) was implemented using the bslib package, applying the Bootstrap 5 Flatly theme along with Google Fonts Roboto and Poppins to achieve a modern, clean aesthetic consistent with usability principles.

On the server side, the system employs reactive functions (reactiveVal, observeEvent, renderUI, renderPlotly), allowing efficient execution of data reading, transformation, and visualization processes. Each module (alerts, projection, historical data, etc.) was

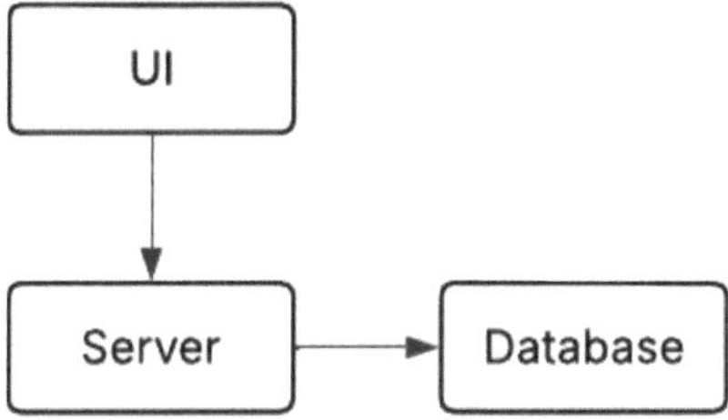

Fig. 2. System architecture showing the interaction between UI, server, and Google Sheets. *Source: Author's elaboration (2024).*

designed independently, promoting modularity, future scalability, and simplified code maintenance.

Figure 2. Schematic representation of the Shiny system architecture illustrating the interaction between the user interface (UI), server logic, and the cloud-based data layer. The UI handles user inputs and displays outputs through reactive components, while the server processes these inputs, executes the analytical routines, and communicates with the external database. Data storage and retrieval are managed through Google Sheets, which functions as a lightweight cloud database, enabling real-time updates, remote accessibility, and seamless integration with the application's reactive framework. This architecture supports modular development, improves maintainability, and ensures efficient data flow throughout the system.

2.2   Data Security, Authentication, and Management

To ensure data integrity and security, the application connects to Google Sheets through the googlesheets4 package, using OAuth2 authentication, which securely validates user credentials without storing passwords in plain text.

The credentials are stored locally in a protected folder named.secrets, allowing secure session maintenance and preventing unauthorized access to field records.

Communication between R Shiny and Google Sheets is conducted via encrypted HTTPS requests, ensuring the confidentiality and traceability of every operation (read, write, edit, or delete). This cloud-based integration eliminates the need for local database servers, thereby reducing costs and simplifying deployment.

2.3   Data Structure

The application integrates three synchronized spreadsheets in Google Sheets, which are automatically updated:

- variety_data: contains baseline information on the farm, block, bed, variety, and planting week.
- pinching_schedule: records the estimated physiological periods between planting and disbudding.
- pinched_beds: stores records of beds that have already been disbudded, entered directly from the field.

The system standardizes column names, merges the data sources using the left_join function, and automatically calculates the target disbudding week, incorporating an alert variable that classifies the status of each bed according to its intervention stage.

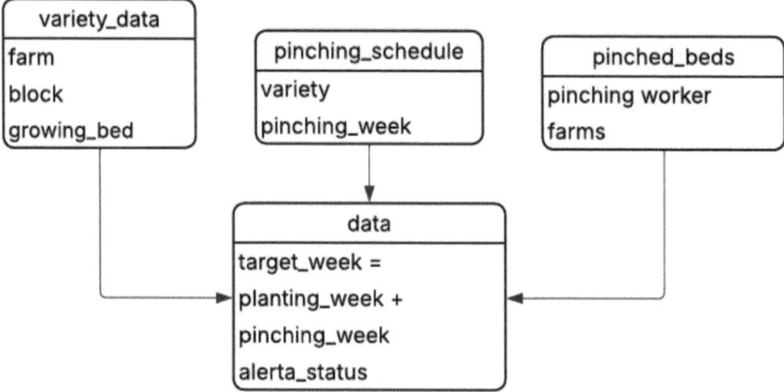

Fig. 3. Structure of the integrated Google Sheets datasets. ***Source: Author's elaboration (2024).***

Figure 3. The diagram illustrates the structure of the three Google Sheets that supply data to the system. Each sheet stores specific operational information: variety_data (farm, block, growing bed), pinching_schedule (variety and pinching week), and pinched_beds (pinching workers and farms). These datasets are integrated into the main table data, where the system calculates the target_week by combining planting and pinching information, and generates alerts to support timely, data-driven decision-making in the field.

2.4 Implemented Functionalities

The optimized version of the application is composed of six functional modules:

1. Data Entry: allows users to record, edit, and delete information directly from the field. The records are stored in Google Sheets using the functions sheet_append, range_write, and sheet_write.
2. Alerts: classifies beds according to their temporal status, using icons and color codes: ⬤ Overdue, ◓ This week, ◒ In 1 week, ◯ In 2 weeks, ⌛ Not yet, and ☑ Already disbudded.
3. Projection: estimates the number of beds pending disbudding per week and generates interactive graphs using plotly.
4. Planting: analyzes the weekly planting behavior and its relationship with the disbudding cycle.
5. Historical Data: visualizes past records with filters by variety and farm.
6. General Summary: generates key performance indicators (averages, totals, and trends).

Table 1. R packages used in the application and their main functions.

shiny	Creation of interactive web interfaces and reactive logic
googlesheets4	Reading and writing data to Google Sheets
dplyr	Data cleaning, filtering, and transformation
DT	Display of interactive tables
lubridate	Handling of ISO dates and weeks
plotly	Interactive charts for time series and projections
bslib	Advanced visual customization (themes and fonts)

Table 1 presents the main libraries used in the development of the predictive application implemented in RStudio. Each one serves a specific function within the system's structure: from creating interactive interfaces (shiny), managing cloud databases (googlesheets4), and cleaning and transforming data (dplyr), to dynamically visualizing results through charts (plotly) and interactive tables (DT). These open-source tools made it possible to integrate data collection, processing, and visualization of phenological information into a single platform, optimizing analysis and decision-making in the field.

2.5 System Validation and Performance

The system validation was carried out by comparing manual field records with the projections automatically generated by the application. Three components were evaluated:

- Temporal consistency, by verifying the coherence among planting weeks, projections, and actual disbudding execution.
- Data management efficiency, measured through the reduction of human errors and the speed of data entry.
- User experience, assessed in terms of ease of use, visual clarity, and the usefulness of alerts.

The system demonstrated high accuracy in predicting critical disbudding weeks, consistency in automatic data updating, and an intuitive interface adaptable to field conditions, thereby improving planning and operational control in chrysanthemum cultivation.

Fig. 4. Main interface of the Pinching Monitoring System. *Source:* Author's elaboration (2024).

Figure 4. Shows the initial interface of the "Pinching Monitoring System," developed in the R Shiny environment. The main panel groups six functional modules that organize information intuitively: data entry, alerts, projection, planting, history, and general summary.

Each button dynamically activates an associated panel within the same environment, allowing smooth navigation without reloading the application. The color scheme was selected according to its functional purpose: green for field tasks, red for alerts, blue for historical data, gold for projections, and violet for the data entry module, following the institutional palette.

2.6 Operational Validation of the Predictive Logic

Given that the primary purpose of the system is operational rather than statistical, the validation focused on assessing the practical coherence between the projections generated by the application and the field observations. Instead of applying a comprehensive quantitative evaluation framework, the analysis centered on whether the predicted weeks aligned with the reported phenological behavior.

The validation procedure included:

- Comparing projected disbudding weeks with historical field records.
- Confirming whether the alert categories matched the observed developmental stage.
- Evaluating whether the system enabled timely planning across different production cycles.
- Assessing user feedback regarding clarity, usefulness, and suitability for field operations.

This qualitative validation reflects the applied nature of the model, which was developed to facilitate operational planning rather than to function as a strict statistical predictive model. The results showed a high degree of correspondence between the system's projections and field observations, supporting the suitability of the tool for decision-making in chrysanthemum management.

It is important to note that the current version of the system does not include a formal quantitative performance evaluation (e.g., MAE, MSE, RMSE). Incorporating such metrics would require independent methodological design and controlled datasets. This remains a possible direction for future work as a complementary improvement, rather than a requirement for the operational objectives of the present study.

2.7 Computational Environment and System Constraints

The application was developed in the R programming environment using the Shiny framework and deployed through the shinyapps.io cloud service. Because the system relies on lightweight operations, primarily data joining, filtering, and real-time visualization, it does not require high computational capacity. However, several practical considerations must be acknowledged regarding performance and scalability:

- Reactive processing: Shiny reactivity automatically updates outputs when inputs change. While this is beneficial for user interaction, it may introduce minor delays when multiple reactive elements are triggered simultaneously or when datasets grow in size.

- Dependence on Google Sheets: Data retrieval and updating are performed via encrypted HTTPS requests through the googlesheets4 package. This implies that internet stability directly affects loading times, especially when writing or reading multiple rows.
- Data size limitations: Google Sheets supports a moderate amount of data. For the current operational scale of the system, this is sufficient; however, an expansion in the number of beds, varieties, or historical cycles may eventually require migrating to a more robust database structure.
- Shinyapps.io resource constraints: While the application runs efficiently under typical usage conditions, cloud deployment plans impose limits on session duration, memory, and concurrent users. These constraints do not affect routine field operations but should be considered for future versions with heavier computation or broader adoption.
- Scalability considerations: As new functionalities are added, such as environmental data integration or more detailed historical reporting, processing requirements may increase. In that scenario, optimizing code, using caching strategies, or transitioning to a dedicated database (e.g., PostgreSQL) could enhance performance.

Overall, the current computational configuration is appropriate for the operational purpose of the system. These considerations offer transparency regarding its technical operation and outline key aspects to address should the application scale to larger datasets or more advanced predictive modules.

2.8 Methodological Foundations for Prediction and Visualization

Although the system was developed with an applied operational focus, its design aligns with methodological principles commonly used in predictive agricultural analysis. In particular, supervised prediction frameworks typically rely on historical observations to estimate future phenological events, following approaches based on regression models, error analysis, and iterative refinement [10]. While the present tool does not implement a formal statistical model, its logic is consistent with these foundations: it uses empirical growth periods derived from field data to anticipate pinching windows and support planning decisions.

Additionally, the application's visualization components follow established guidelines for interactive dashboards, which prioritize clarity, consistency, and interpretability to facilitate decision-making in data-intensive environments [11, 12]. The use of Shiny and Plotly enables reactive updating, modular design, and exploratory visualization, fundamental features of modern data-driven applications.

By situating the system within these methodological frameworks, the manuscript clarifies the conceptual basis underlying the tool and aligns its implementation with current standards in statistical computing and interactive data visualization.

3 Results

The implementation of the digital monitoring system for chrysanthemum disbudding significantly improved the efficiency and organization of field operations, by facilitating the identification of beds ready for intervention and optimizing the weekly planning of activities.

During its practical application across different production cycles, there was a notable reduction in the time required to verify beds ready for the following week, shifting from a process that previously took several hours of manual review to a much more agile and systematic procedure.

The use of the predictive system developed in RStudio enabled automatic consolidation of field information, reducing the need for extensive field rounds and contributing to faster, more traceable decision-making.

3.1 Operational Results

The system led to overall improvements in personnel organization, accuracy of weekly scheduling, and administrative efficiency, particularly when preparing reports and planning disbudding activities.

In general, verification and planning tasks that previously required significant time consumption were considerably reduced, improving coordination between field supervisors and scheduling managers.

These results reflect a comprehensive optimization of production processes, by integrating into a single tool data visualization, growth projections, and preventive alerts for plots requiring upcoming intervention.

Furthermore, the system enabled timelier identification of critical areas, facilitating the scheduling of tasks within the appropriate phenological window.

3.2 General System Performance Analysis

The predictive system developed in RStudio was based on the collection of growth and development data from different pompon-type chrysanthemum varieties, adjusting disbudding projections according to historical records and field observations.

Although the model was not subjected to exhaustive statistical validation, its practical use allowed accurate anticipation of intervention dates, demonstrating significant potential as a support tool for production planning.

Additionally, the inclusion of a visual alert system within the application facilitated the early detection of possible delays or advances in crop development, allowing for more efficient coordination of fieldwork aligned with the actual growth status of the crop.

Overall, the experience demonstrated that digitalization and the use of predictive tools provide added value to operational management, by reducing time, improving information traceability, and enhancing responsiveness to variations in the phenological development of the crop.

Table 2 shows the alert system implemented in the application, which allows for the immediate visualization of the phenological status of each chrysanthemum bed in relation

Table 2. Alert system and interpretation of phenological status.

Icon / Alert	Meaning
⬤ Overdue	The bed should have been pinched before the current week.
◐ This week	Pinching must be performed during the current week.
◑ In 1 week	Pinching is scheduled for next week.
◒ In 2 weeks	Pinching is scheduled for two weeks from now.
⧗ Not yet	The bed has not yet reached the expected phenological stage.
☑ Already pinched	Record confirmed in the database.

to the optimal disbudding time. Each color or icon represents a specific condition within the projection schedule, facilitating decision-making in the field. This visual system helps identify delayed plots in advance, plan weekly tasks, and confirm completed records in the database, ensuring a more efficient and synchronized management of the production process.

3.3 Operational Impact and Visualization

The interactive projection module, developed with Plotly, made it possible to dynamically display the beds pending disbudding by week.

This feature strengthened communication between supervisors and administrative staff by providing a real-time control panel accessible from mobile devices, enabling faster and more data-driven decision-making.

Figure 5. This figure presents an interactive visualization used to analyze the weekly projection of chrysanthemum beds pending pinching. The tool aggregates planting records and scheduled pinching weeks to estimate upcoming workload. By allowing users to filter information by variety, farm, or both, the system supports operational planning by identifying peak labor demands and potential bottlenecks in the production cycle. This visualization enhances decision-making by enabling supervisors and field managers to anticipate labor needs, allocate workers more efficiently, and align pinching activities with the crop's phenological development.

A. Weekly data table showing the estimated number of beds scheduled for pinching by target week.
B. Interactive Plotly line chart visualizing the same weekly projection, highlighting workload peaks and trends.

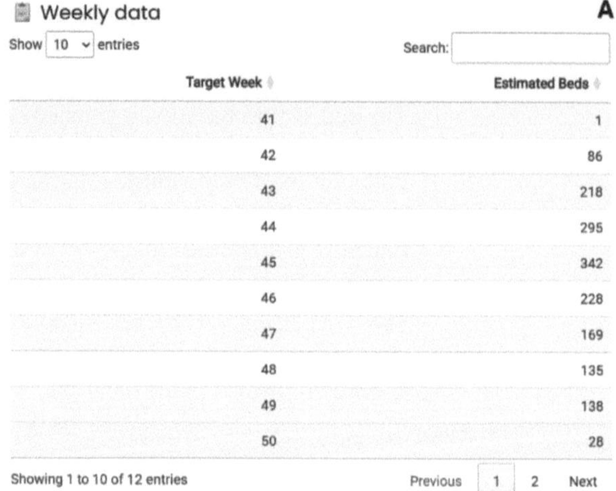

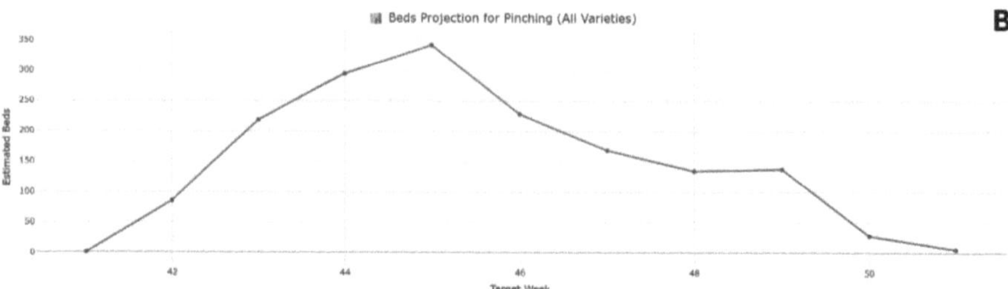

Fig. 5. Projection of beds pending pinching (interactive table and Plotly chart). ***Source:*** *Author's elaboration (2024).*

4 Discussion

The development of this application provides practical evidence of the impact of digitalization on agricultural processes, particularly in a crop with high operational demands such as chrysanthemum.

The replacement of manual records with an interactive tool based on R Shiny and Google Sheets improved traceability, reduced human error, and optimized fieldwork organization.

Unlike traditional methods, this system transforms agronomic information into visual and predictive indicators, accessible from any device and in real time, thereby strengthening data-driven decision-making.

4.1 Contribution to Smart Agriculture

The application aligns with the principles of Agriculture 4.0, where efficiency, sustainability, and data analytics converge to transform production management.

According to Wolfert et al. (2017) [7], smart agriculture is not limited to automation but involves the integration of data to generate actionable knowledge. In this case, the combination of the agronomic expertise of field personnel with the analytical capacity of the digital system allowed for the anticipation of critical tasks and more efficient use of human resources.

Digitalization does not replace agronomic expertise-it enhances it.

This balance between empirical knowledge and technology represents the essence of Agriculture 4.0.

By centralizing data in the cloud and generating dynamic alerts, the system contributes to greater operational sustainability, avoiding rework, unnecessary movements, and labor overload. Thus, the smart agriculture approach translates into a tangible tool for planning and efficiency.

4.2 Model Replicability

The developed model has a high potential for replicability in other ornamental and agricultural crops that depend on phenological planning, such as roses, carnations, oil palm, or coffee.

The system's modular structure - based on data, predictions, and visualization - allows its algorithms to be adapted to different scales and production conditions, while maintaining the principle of low cost and easy implementation.

According to FAO & ITU (2020) [8], the use of open-source digital tools facilitates technological inclusion in rural sectors where access to complex infrastructure remains limited. In this sense, this application represents a viable alternative for small and medium-sized production units seeking to integrate into the era of digital agriculture without requiring major investments.

4.3 Technological Limitations and Challenges

Despite its advantages, the system presents technological and operational limitations that must be considered for its scaling:

- Dependence on Internet connectivity, especially in rural areas with unstable coverage.
- Accuracy dependent on the quality of manually entered data, which requires training processes and record control.
- Lack of integration with environmental variables, such as temperature, radiation, or relative humidity, which directly influence phenological development.

These challenges reflect the need to move toward a more connected and interoperable digital agriculture, where information from IoT sensors and climate systems can be automatically integrated with the application's predictive modules.

Figure 6. The diagram shows the possible evolution of the system through the integration of IoT sensors and predictive models. The sensors would record environmental variables such as temperature, humidity, or radiation, sending real-time information to the database.

With these data, the models could generate more accurate projections and issue automatic alerts about changes in crop development or environmental conditions.

This integration would enable more dynamic and efficient management, strengthening decision-making in the field and the sustainability of the production process.

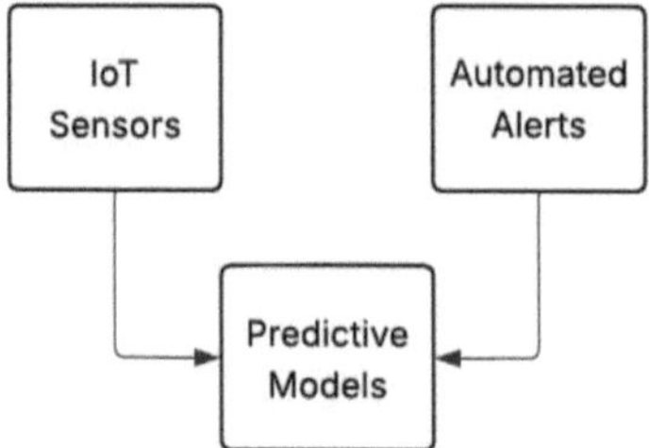

Fig. 6. Future integration diagram: IoT sensors and predictive models. *Source:* Author's elaboration (2024).

4.4 Conceptual and Methodological Foundations

Several authors emphasize that decision-support tools based on agricultural prediction do not necessarily require complex statistical models; instead, they must ensure operational coherence, empirical grounding, and transparency in their underlying assumptions [10]. The approach adopted in this system follows this principle: it leverages historical phenological knowledge, integrates real-time data, and presents results through interactive visualization interfaces, which are recognized as essential components of modern agricultural informatics [11, 12].

5   Conclusions

The optimized version of the Disbudding Monitoring System, developed in R Shiny, represents tangible evidence of how digitalization can transform traditional agricultural processes. Its implementation improved operational planning, reduced error margins, and strengthened data-driven decision-making, promoting more precise and efficient management within the framework of Agriculture 4.0.

The model integrates predictive analysis tools, cloud connectivity, and interactive visualization, achieving effective articulation between agronomic knowledge and digital technologies. This synergy drives the optimization of human resources, reduces operational waste, and enhances field traceability, directly contributing to the Sustainable Development Goals (SDGs):

- SDG 8: Promote more productive and efficient agricultural work.
- SDG 9: Foster technological innovation and digitalization in the rural sector.
- SDG 12: Ensure responsible and sustainable production.

Furthermore, the application demonstrates the potential of open-source software as a tool for technological democratization, offering an affordable, adaptable, and replicable solution for other crops. Its modular architecture allows the system to be scaled to higher levels of analysis, incorporating machine learning models, climatic data, or even satellite imagery to strengthen phenological prediction and the sustainability of the production process.

In summary, the digitalization of disbudding does not replace agronomic expertise - it enhances it. This balance between technical knowledge and digital analytics defines a clear path toward a smarter, more competitive, and sustainable agriculture.

Acknowledgments. This work was developed within the framework of the Agricultural Engineering program at the University of Antioquia, as a technological application exercise aimed at digitalizing agricultural processes.

Declaration of Interests.   The author declares no conflicts of interest related to the content of this article.

References

1. CENIFLORES. (s.f.). Sector floricultor. https://ceniflores.org/sector-floricultor/
2. Universidad Lasallista. (s.f.). Evaluación del proceso productivo en cultivos de flores de corte. https://repository.unilasallista.edu.co/server/api/core/bitstreams/9173303a-76d7-4915-9aa8-6aa7cd0461d6/content
3. Universidad de Cundinamarca. (s.f.). Optimización del manejo agronómico en cultivos ornamentales bajo invernadero. https://repositorio.ucundinamarca.edu.co/server/api/core/bit streams/dc53beb7-b755-4ae3-a2cc-810ecfb4fbc8/content
4. Chamara, N., Islam, M.D., Bai, G., Shi, Y., Ge, Y.: Ag-IoT for crop and environment monitoring: past, present and future. Agric. Syst. **203**, 103497 (2022). https://www.sciencedirect.com/science/article/pii/S0308521X22001330?via%3Dihub
5. Ryan, Y.: Creación de aplicaciones web interactivas con R y Shiny (J. Isasi, Trad.). Programming Historian en Español (2023). https://programminghistorian.org/es/lecciones/creacion-de-aplicacion-shiny
6. R Programming for Data Sciences. (s.f.). Chapter 9: Shiny – Interactive Web Apps in R. Recuperado de. https://doserlab.com/files/for875/_book/shiny-interactive-web-apps-in-r
7. Wolfert, S., Ge, L., Verdouw, C., Bogaardt, M.: Big data in smart farming – a review. Agric. Syst. **153**, 69–80 (2017). https://www.sciencedirect.com/science/article/pii/S0308521X163 03754?via%3Dihub
8. FAO & ITU. Status of Digital Agriculture in 47 Sub-Saharan African Countries. FAO (2020). https://openknowledge.fao.org/server/api/core/bitstreams/e7ba20c1-c415-4d0b-8e30-e0f44f9ee758/content
9. United Nations. Objetivos de Desarrollo Sostenible (ODS) (2024). https://sdgs.un.org/es/goals
10. James, G., Witten, D., Hastie, T., Tibshirani, R.: An Introduction to Statistical Learning: with Applications in R, 2nd edn. Springer, Cham (2021). https://www.casact.org/sites/default/files/2022-12/James-G.-et-al.-2nd-edition-Springer-2021.pdf
11. Sievert, C.: Interactive Web-Based Data Visualization with R, Plotly, and Shiny. Chapman & Hall/CRC (2020). https://plotly-r.com/
12. Wickham, H., Grolemund, G.: R for Data Science (2e) (2023). https://r4ds.hadley.nz/

Application for Calculating Years of Life Lost

David Esteban Cartagena Mejía[(✉)] , Gustavo Adolfo Lopera Gallego , and Difariney González Gómez

Grupo de investigación Demografía y Salud. Facultad Nacional de Salud Pública., Universidad de Antioquia, Medellín, Colombia
`grupo.demosalud@udea.edu.co`

Abstract. The R statistical software enables the calculation and visualization of key public health indicators, such as estimating the Global Burden of Disease. The burden of disease measures the health gap through indicators such as Disability-Adjusted Life Years and Healthy Life Years Lost, which summarize the weight of disease in populations into a single measure.

This work presents an interactive R Shiny application for calculating Years of Life Lost, a measure of premature mortality in a population which estimates the average number of years a person would have lived had they not died earlier than expected.

The interactive application calculator Years of Life Lost allows users to upload data disaggregated by year, country, region, sub-region, department, or municipality, as well as by sociodemographic characteristics such as age, sex, underlying cause of death, and its diagnostic code according to the International Statistical Classification of Diseases and Related Health Problems.

The interactive application presents results through dynamic visualisations, including interactive maps and density plots to explore variability in mortality rates, and dynamic charts over time periods. Years of Life Lost are computed using secondary information sources, such as vital statistics from the Department of Administrative National Statistics [Departamento Administrativo Nacional de Estadística, DANE] and the Global Burden of Disease methodology.

Keywords: years of life lost · life expectancy · global burden of disease · violence against women · Antioquia

1   Introduction

Physical violence against women poses a threat to public health and to social and economic development of the population. In this context, estimating the years of life lost due to premature death (YLL) as a result of physical violence against women contributes to the analysis of mortality, allowing for the quantification

B. M. Suárez et al. (Eds.): R Day 2025, CCIS 2824, pp. 372–383, 2026.
https://doi.org/10.1007/978-3-032-18455-9_20

of premature loss of life and the social and economic dimensions of this event in different populations.

YLLs are a key measure within the Global Burden of Disease (GBD), a tool that provides a comprehensive view of mortality and disability across countries, considering time, age, and sex. Their objective is to quantify the loss of health caused by hundreds of diseases, injuries, and risk factors, with the aim of strengthening health systems and reducing inequalities [1].

Burden of disease studies measure the impact of different health problems on the population. These studies use a common metric: healthy life years (HLYs) or disability-adjusted life years (DALYs), which allows for comparisons between different pathologies and the assignment of priorities [2,3].

YLLs due to premature death are a synthetic indicator in the GBD study and constitute a measure of premature mortality in a population. They estimate the average number of years a person would have lived if they had not died earlier than expected. In other words, they are used as methodological support in the assessment of avoidable mortality, as it considers a death to be premature to the extent that, had it been prevented, a person would have lived as long as the rest of the population, up to a standard age; however, this indicator does not consider the effect of morbidity on the population [4].

This paper presents an interactive application developed in R Shiny, designed to estimate the YLLs associated with physical violence against women. The tool allows users to upload and analyze data disaggregated by year, country, region, subregion, department, or municipality, as well as by sociodemographic characteristics such as age, sex, underlying cause of death, and diagnostic codes according to the International Statistical Classification of Diseases and Related Health Problems (ICD) [5].

The interactive dashboard integrates different functionalities that facilitate the exploration and understanding of results, strengthens analytical capabilities in public health, and offers visual and interactive support for understanding results, contributing to decision-making, policy design, and strategies aimed at preventing violence against women and reducing its impact on the health and well-being of the population.

The objective of this study was to develop and apply an interactive tool to estimate YLL for physical violence against women in Antioquia between 2017 and 2022.

Recent developments in public-health research underscore the growing demand for flexible, transparent and user-friendly analytical and visualization platforms capable of estimating population health burden in real time. For instance, web-based tools have been developed to compute disability-adjusted life years (DALYs) and productivity losses, facilitating rapid evaluation and policy planning [6]. Additionally, visual analytics dashboards such as those described for infectious-disease scenarios enable intuitive exploration of spatio-temporal health data and support decision-making even under data uncertainty [7]. Recent burden-of-disease studies further highlight the importance of continuously updating DALY and YLL estimates as epidemiological patterns and population structures evolve [8]. Therefore, providing the present application with a Shiny-based interface and publicly accessible dashboard not only responds to a methodological need, but also aligns with contemporary standards for transparency, reproducibility, and public-health relevance.

This type of tool can support a wide range of users, including public health professionals, epidemiologists, policy makers, and researchers who require agile and reproducible methods to estimate premature mortality and monitor health inequalities. Additionally, health secretariats and observatories may benefit from its capacity to integrate local data sources and produce territorially disaggregated indicators for decision-making. In the future, the application could be expanded to address related challenges, such as incorporating Years Lived with Disability (YLD), estimating Disability-Adjusted Life Years (DALYs), or integrating real-time data pipelines to strengthen surveillance systems. Recent studies [6,9–11] have highlighted the increasing need for flexible, transparent, and locally adaptable analytical tools to assess the burden of disease, underscoring the relevance of developments such as the one presented in this manuscript.

2   Methods

A descriptive observational study was conducted to estimate the years of life lost due to premature death caused by physical violence against women in Antioquia between 2017 and 2022.

The standard method proposed in GBD studies was applied to estimate YLLs, using 91.9 years as life expectancy. In addition, the definition of physical violence against women and the ICD-10 codes used in the study were validated with subject matter experts.

Data Collection. Secondary sources of information were used. Mortality data are part of DANE vital statistics for women residing in the department of Antioquia during the study period. In Colombia, DANE is the sole entity responsible for the country's official statistics, including vital statistics [12]

Consequently, the quality and reliability of mortality data, including the calculation of YLLs, are linked to the overall quality of mortality statistics in Colombia. This quality is measured by considering several key factors, including the breadth of death registration coverage, the timeliness of death reporting, the completeness of the data collected, and the accuracy of the identification and coding of underlying causes of death [13].

A review was conducted of the data obtained from the DANE's national data archive (ANDA), and the available variables were examined in order to determine their relevance for use in this research.

The population projection by age group and sex available on the DANE website provides aggregate data for individuals aged 85 and over. Therefore, the population of the 85–89 age group in each subregion was estimated using pro rata [14]. For this calculation, the percentage weight representing this age group (85–89 years) within the total group aged 85–100 years and over at the departmental level in each year was used as a reference. This percentage was then applied to the projected number of people aged 85 and over in each subregion and year, to obtain a more accurate estimate of the population distribution in the older age groups.

To identify the ICD-10 codes and mortality coding directly related to physical violence, an analysis of the codes was carried out, identifying that deaths occurring as a result of physical violence are concentrated exclusively in Chapter XX: external causes of mortality (V01-Y98). The final list of causes consisted of 514 four-character ICD-10 codes.

The official website of the DANE, specifically the vital statistics section, was accessed to extract the data used in calculating the YLLs. The databases of non-fetal deaths from 2017 to 2022 were downloaded. Subsequently, a single consolidated database was constructed for the entire period, integrating the records for each year using the statistical software R 4.5.1. Only the records of deaths of women occurring in Antioquia's department (code 05) were filtered. Based on these criteria, the list of causes of death due to physical violence against women was applied, and the subregion of occurrence was also identified for each record.

2.1 Data Analysis

The database was established, processed, and analyzed using statistical software R 4.5.1 and an interactive application developed in R Shiny. Dynamic visualizations were generated, such as interactive maps, stacked area charts, and time trends, were generated to facilitate the exploration and understanding of the distribution of YLLs in women.

To develop of the Shiny interactive application, an exploratory analysis of several R statistical software packages was conducted to identify tools that would allow for the estimation of YLLs using the most up-to-date methodology from the Institute for Health Metrics and Evaluation. Several packages were identified; however none were updated to the latest version of the GBD methodology. These included DALY (Disability-Adjusted Life Years) and yll [15,16]. Some limitations arose when performing territorial and demographic disaggregation, methodological flexibility, and integration with local information sources.

The DALY package, developed by Sam Devleesschauwer et al. [15], allows for the integrated estimation of YLL and Years Lived with Disability (YLD) based on data on incidence, prevalence, duration of the condition, mortality, and life expectancy at age of death. However, it had limitations in terms of disaggregation by age, sex, and territory, and required preprocessed inputs that did not fit the structure of the national sources used.

The YLL package, developed by Joachim Soetewey [16], offers a straightforward approach to calculating YLL, based on variables such as age, number of deaths, and standard life expectancy. Although it is consistent with recent methodologies from the Global Burden of Disease Study, its scope is limited to calculating YLL and does not allow for the incorporation of additional functions required for the analytical development of the study.

It was decided to develop an interactive application using R Shiny. This tool allows YLLs to be calculated from data disaggregated by age, sex, territory, and underlying cause of death, based on standard life expectancy.

The application integrates modules for:

- File upload and data structure validation,
- Calculation of crude mortality rates by region, year, and age groups
- Calculation of YLL according to age at death,
- Generation of dynamic visualizations (maps, density graphs, and time series),
- Exporting results for further analysis.

The data structure used included the following variables: year, age, sex, territory, underlying cause of death (ICD-10 code), number of deaths, DANE population projections by municipality and age, and life expectancy at the corresponding age. These variables were obtained from the vital statistics of the National Administrative Department of Statistics and aligned with the GBD methodology.

It should be noted that in vital statistics, the age at which death occurred is grouped into five-year intervals, which is why deaths occurring up to the age of 89 were identified as premature.

YLLs were calculated to measure the burden of premature mortality due to physical violence against women in Antioquia between 2017 and 2022. This indicator estimates the impact of early deaths compared to a standard life expectancy defined in this study. YLLs were estimated using equation (1).

$$\text{YLL} = \sum_{i=1}^{n} (L_i - E_i)\, D_i \tag{1}$$

where:

- L_i: Defined standard life expectancy (91.9 years)
- E_i: Class mark of the age group in which the death occurred
- D_i: Deaths due to physical violence against women recorded in the specific age group i
- n: Number of age groups considered (17 five-year age groups from 0–4 to 85–89 years old)

The YLLs were calculated by subtracting the age-specific benchmark for each five-year age group from the standard value; these differences were then multiplied by the total number of deaths in each age group; and the resulting volumes were grouped according to age structure.

2.2   Interactive Dashboard

R Shiny or Shiny dashboard package in R programming language was used to build the interactive dashboard for the data representation as well as the geospatial analy-sis [17,18]. This application helps to combine the computational power of R programming language with the interactivity of the modern web. This web-based dashboard helps end users to make informed decisions with ease and more effective, given the interac-tivity of the dashboard [19].

Shiny App to build an interactive data visualization tool for data and findings repre-sentation [20, 21]. The interactive application was developed in R Shiny, integrating statistical analysis and dynamic visualizations to facilitate the inter-pretation of results related to physi-cal violence against women and its impact on premature mortality. This tool com-bines the analytical power of R with a user-friendly web interface, allowing infor-mation to be explored in a simple and intuitive way. The dashboard was specifically designed to visualize and analyze mortality rates and YLLs due to physical violence against women in Antioquia between 2017 and 2022. However, it allows these calculations to be performed for different years, locations, and causes of death. Its main features include:

- Interactive maps: these allow users to observe the geographical distribution of mortality and YLL rates by subregion, facilitating the identification of areas with the greatest impact.
- Stacked area charts: show variability in mortality and YLL rates between differ-ent age groups and periods, highlighting patterns of concentration or dispersion.
- Time trend graphs: these allow you to visualize how premature mortality due to physical violence against women has changed over time, identifying increases, decreases, or stable patterns.

The application automatically classifies and groups mortality and YLL rates accord-ing to the selected disaggregation (year, subregion, age), enabling direct comparisons between different geographic and population contexts. Furthermore, presenting the results through dynamic visualizations facilitates under-standing of the territorial and demographic patterns of premature mortality, thereby strengthening epidemiological surveillance systems and supporting evidence-based public health decision-making. In this context, the tool becomes a strategic resource for identifying priority areas for intervention, guiding public policy, and optimizing the allocation of resources for the prevention and treatment of the observed phenomenon. The application developed in R Shiny has four main mod-ules that structure the ana-lytical workflow: Home, Data, Mortality Rate, and YLL. Each of these modules per-forms a specific function in the YLL estima-tion and analysis process, from loading information to viewing and interactively exploring the results.

The application is structured under a modular architecture in R Shiny, which facilitates the logical separation of processes, system scalability, and smooth interaction between components. Each module plays a defined role within the analytical flow and receives specific inputs to perform its functions. The Home module acts as an introductory interface and does not require external inputs; its purpose is to guide the user on the general operation of the tool. The Data module is the pillar of information processing and cleaning: it receives as inputs the mortality database in CSV or Excel format, the dictionary of variables, and the ICD-10 list of causes; additionally, it allows for the optional entry of the population attributable fraction (PAF), standard life expectancy, and external population projection files. Based on these inputs, this module performs automatic data validation, merging, and standardization procedures using territorial metadata (DIVIPOLA), five-year age groups, and cause coding, generating a structured analytical base for subsequent modules. The Mortality Rate module receives the processed database from Data along with user-selected parameters—such as years of analysis, sex, age groups, and territorial level—and, optionally, a shapefile for geospatial analysis. It then calculates crude and adjusted rates, producing interactive maps and temporal visualizations.

The Start module serves as an introductory guide. This space provides general infor-mation about the purpose of the Shiny application, basic instructions for its use, and contact details. Its objective is to facilitate the understanding of the application's structure and to guide users through the steps necessary to perform the analysis (Figs. 1, 2, 3 and 4).

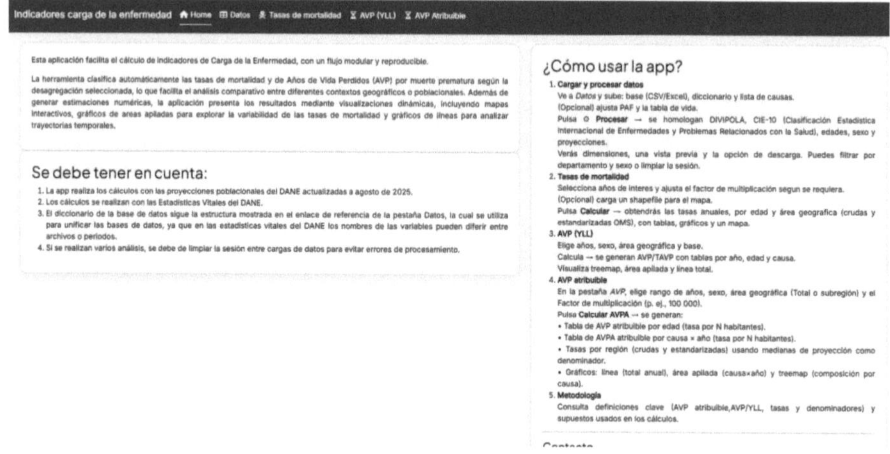

Fig. 1. Start module

The Data module allows you to upload and process the information required for analysis. Users can upload the database in CSV or Excel format, along with the vari-able dictionary and the list of causes of death. Optionally, they can adjust the popu-lation attributable fraction (PAF) and life expectancy. Once the "Process" option is activated, the application automatically joins DIVIPOLA codes [22], ICD-10 codes, age groups, sex, and population projections. It also offers a preview of the processed database, allows you to apply filters by department and sex, and then download it or clear the session to start a new analysis.

The Mortality Rate module is designed to calculate rates. To use it, select the years of interest, gender, age groups, and territorial classification. Users can also optionally upload a shapefile to perform geospatial analyses, accompanied

Fig. 2. Data module

by graphical visual-izations and interactive maps that facilitate the identification of territorial and tem-poral patterns.

Fig. 3. Mortality rates module

Finally, the YLL module allows to estimate the YYL due to premature death. Select-ing the period, gender, and subregion, and when you run the calculation, they will obtain YLL estimates and adjusted rates, disaggregated by year, age, and cause. The tool presents the results using interactive graphs, including treemaps, stacked areas, and trend lines, which allow you to clearly explore the distribution of premature mor-tality and its changes over time.

Link to the app in R Shiny [23]

3   Results

The R Shiny application for calculating YLLs makes it easier for users to esti-mate and analyze premature mortality in an intuitive, interactive, and dynamic way. Based on data from DANE statistics, the tool allows users to process and

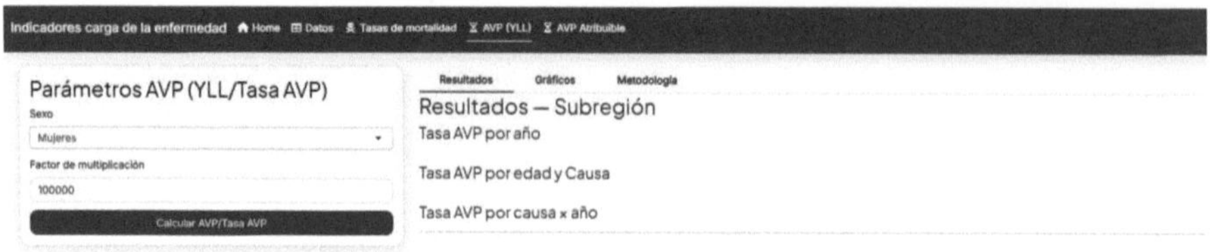

Fig. 4. YLL rate module.

download the consolidated database corresponding to the period of interest and the selected characteristics, thus facilitating the analysis and management of information for population or epidemiological studies.

It subsequently allows for the calculation of mortality rates at different geographical levels: national, departmental, subregional, or municipal. It also enables the calcula-tion of YLL rates disaggregated by year, age group, and cause of death, according to the territorial classification entered by the user.

For this analysis, records of female deaths due to physical violence in Antioquia, a region of Colombia, from 2017 to 2022 were used to analyze both the mortality rate and the mortality burden measured by YLLs.

During the analyzed period, 1,469 deaths due to physical violence against women were recorded in the department. The average mortality rate was 7.2 deaths per 100,000 women, with a slight increases in 2018 (7.8) and 2019 (7.7).

An estimated 74,125.1 years of life lost (YLLs) occurred during this time, averaging 12,132 YLLs per year and equaling an overall rate of 365.6 YLLs per 100,000 wom-en. The years 2018 and 2019 saw the highest losses, at 18.6% and 18.2 %, respective-ly.

The highest burden of premature mortality was among women aged 20 to 24 (14,399.4 YLLs, equivalent to 856.1 YLLs per 100,000 women). Firearm assaults accounted for 46.2% of YLLs (34,258), with a rate of 168.7 YLLs per 100,000 wom-en.

In the analysis by subregion, the Aburrá Valley accounted for 43.4% (32,185) of YLLs, with an adjusted rate of 256.4 YLLs per 100,000 women, while the highest rates were recorded in Bajo Cauca (1,197.5), Nordeste (852.9), and Suroeste (685.6). The adjusted YLL rate in Bajo Cauca was approximately five times higher than that of the Valle de Aburrá.

4   Conclusions

This type of analysis makes it possible measure the extent to which adverse events affect people's lives and well-being, strengthen public health decision-making, im-proves the capacity of health systems to respond, allocates resources more efficient-ly, and enables the planning of effective, targeted interventions.

Physical violence against women in Antioquia is a persistent phenomenon with a high impact in terms of premature mortality and lost healthy life expectancy. Ideally, the YLL values associated with this type of violence would be zero, highlighting the mag-nitude of the problem and the urgency of comprehensive intervention.

One methodological strength of this study was the development of an interactive dashboard in R Shiny. This allowed for the integration of analysis processes, the au-tomation of YLL calculations, and the dynamic visualization of information. This technological tool made it possible to disaggregate and explore the results by varia-bles such as year, age, cause, and subregion, thus improving the territorial and tem-poral understanding of the phenomenon. In addition, it offers significant potential for use in public health surveillance and decision-making processes in an accessible and replicable manner.

The findings of this study reveal territorial and age patterns that can guide more pre-cise prevention strategies focused on the most affected groups and subregions. They also demonstrate the usefulness of YLL analysis as a tool for highlighting the real impact of physical violence against women and supporting evidence-based decision-making. In summary, YLL analysis applied to physical violence against women, comple-mented by analytical and visual tools developed in R Shiny, constitutes a powerful strategy for understanding, monitoring, and highlighting its burden on public health, enabling the design of more effective and sustainable responses to reduce its occur-rence and consequences.

Acknowledgments. The authors would like to thank the research team of the project "Carga de la enfermedad y costos asociados a la violencia física contra la mujer" from the Demography and Health Research Group of the National School of Public Health, Universidad de Antioquia, for their support and valuable contributions to the development of this study.

Disclosure of Interests.. The authors have no competing interests to declare that are relevant to the content of this article.

References

1. Global Burden of Disease Collaborative Network.: Global Burden of Disease Study 2021 (2021). Institute for Health Metrics and Evaluation (IHME), Seattle, United States (2024)
2. Zitko, P., Aceituno, D.: Projection of the burden of disease study [Internet]. Universidad de Las Américas (2019). Retrieved March 26 (2023). https:// salud-sociales.udla.cl/wp-content/uploads/sites/70/2020/08/Informe-Proyeccion-Carga-de-Enfermedad-a-2030-Chile-UDLA.pdf
3. Philanthropies, B.: Guidelines for the development of national burden of disease studies (2023), https://assets.website-files.com/5d7f96ea4cc8598434877fed/5ec2e2bfa3677026804fe353_CargaEnfermedad_v1.pdf Accessed 11 Sept 2023
4. Beaglehole, R., Bonita, R., Kjellstrom, T.: Basic epidemiology: measuring health and disease. Pan American Health Organization (2003). https://iris.paho.org/bitstream/handle/10665.2/3311/Epidemiologia, Accessed 11 Sept 2023

5. World Health Organization (WHO): International Statistical Classification of Diseases and Related Health Problems, 10th Revision (ICD-10). 2018 Edition. Geneva: World Health Organization (2018). https://repositoriodeis.minsal.cl/ContenidoSitioWeb2020/uploads/2020/12/CIE-10_2018_VOL2.pdf

6. John, D. et al.: Development of a web-based calculator to estimate DALY and productivity losses due to COVID-19. Clin. Epidemiology Global Health (2024). https://www.ceghonline.com/article/S2213-3984(23)00275-0/fulltext

7. Betz, P.K., et al.: ESID: Exploring the Design and Development of a Visual Analytics Tool for Epidemiological Emergencies. arXiv preprint arXiv:2304.04635 (2023). https://arxiv.org/abs/2304.04635

8. Chen-Xu, J., Grad, D. A., Varga, O., Viegas, S.: Burden of disease studies supporting policymaking in the European Union: a systematic review. Eur. J. Public Heal. **34**(6), 1095–1101 (2024). https://academic.oup.com/eurpub/article/34/6/1095/7576753

9. Adams, W. G., Gasman, S., Beccia, A. L., Fuentes, L.: The health equity explorer: an open-source resource for distributed health equity visualization and research across common data models. J. Clin. Transl. Sci. **8**(1), e72 (2024). https://doi.org/10.1017/cts.2024.500

10. Mwanga, D.M., et al.: A digital dashboard for reporting mental, neurological and substance use disorders in Nairobi, Kenya: implementing an open source data technology for improving data capture. PLOS Digit. Health 0(11), e0000646 (2024). https://doi.org/10.1371/journal.pdig.0000646

11. Sciensano – Belgian National Burden of Disease Study (BeBOD): Non-fatal burden of cancer in Belgium: interactive estimates of incidence, prevalence, years lived with disability (YLD) and DALYs 2004–2022. BeBOD cancer dashboard. https://burden.sciensano.be/shiny/cancer/

12. DANE (Departamento Administrativo Nacional de Estadística).: Non-fetal deaths (2023). Retrieved March 26, 2023. https://www.dane.gov.co/index.php/estadisticas-por-tema/salud/nacimientos-y-defunciones/defunciones-no-fetales

13. Martínez, R., Soliz, P., Caixeta, R., Ordunez, P.: Years of life lost due to premature mortality: A versatile and comprehensive measure for monitoring mortality from noncommunicable diseases. Pan Am. J. Public Heal. (2023). https://iris.paho.org/handle/10665.2/50523

14. Cavenaghi, S.: Population estimates and projections in Latin America, 1st edn., pp. 118–278. In: Bay, G., Cavenaghi, S., Cabella, W., Gonzales, L. (eds.). Rio de Janeiro, Brazil (2012)

15. Devleesschauwer, B., et al.: The DALY Calculator – Graphical User Interface for Probabilistic DALY Calculation in R. R package version 1.5.0 (2016). https://CRAN.R-project.org/package=DALY

16. Soetewey, A.: YLL: Compute expected years of life lost (YLL) and average YLL. R package version 1.0.0 (2018). https://CRAN.R-project.org/package=yll

17. RStudio Inc.: shinydashboard (2014). Retrieved May 29, 2021 from https://rstudio.github.io/shinydashboard/get_started.html

18. RStudio Inc.: Learn Shiny (2020). Retrieved May 29, 2021 from https://shiny.rstudio.com/tutorial/

19. Dunskiy, I.: Healthcare data visualization: Examples & key benefits. Demigos (2021). Retrieved April 7, 2022 from https://demigos.com/blog-post/healthcare-data-visualization/

20. Centers for Disease Control and Prevention.: Rates and trends in coronary heart disease and stroke mortality data among U.S. adults (35+) by county – 1999–2018

(2021). Retrieved May 30, 2021 from https://catalog.data.gov/dataset/rates-and-trends-in-coronary-heart-disease-and-stroke-mortality-data-amongus-adults-1999-

21. Babalola, G.T.: FRED_TreasuryRate_Analysis.ipynb. Code file, GitHub (2021). Retrieved January 5, 2022 from https://github.com/graccelle/Project/blob/master/FRED_TreasuryRate_Analysis.ipynb

22. DANE (Departamento Administrativo Nacional de Estadística). DIVIPOLA — Political-Administrative Division of Colombia. Geovisor de consulta DIVIPOLA. Geoportal DANE; Retrieved from https://geoportal.dane.gov.co/geovisores/territorio/consulta-divipola-division-politico-administrativa-de-colombia/

23. Cartagena Mejía, D., Lopera Gallego, G., González Gómez, D.: Indicadores carga de la enfermedad. R Shiny Application. https://cipxcy-david0esteban-cartagena0mejia.shinyapps.io/Shiny_carga/

R in the Greenhouse: Shiny as a Bridge Between Statistical Analysis and Agricultural Decision-Making

Giovanny Reales Rodríguez$^{(\boxtimes)}$ (iD)

Agricultural Engineer, University of Antioquia, Medellín, Colombia
`giovanny.realesr@udea.edu.co`

Abstract. This study presents the development of a Shiny application in R for phytosanitary monitoring of chrysanthemum crops. The tool enables visualization, analysis, and management of field-collected pest data (thrips, leafminers, aphids), integrating geographic, temporal, and taxonomic information. Through interactive dashboards, heat maps, and dynamic tables, users can identify infestation patterns, calculate incidence rates by block and week, and optimize control strategies. The application was built with tidyverse (dplyr, ggplot2), leaflet for mapping, and shinydashboard for the interface. This development demonstrates how R transforms raw data into actionable knowledge, reducing agrochemical use through evidence-based decision making.

Keywords: Shiny · precision agriculture · phytosanitary · tidyverse · interactive visualization

1 Introduction

Modern chrysanthemum cultivation faces significant challenges in phytosanitary management, particularly with persistent threats from key pests including thrips (Frankliniella occidentalis), leafminers (Liriomyza spp.), and aphids (Aphididae family) [1]. Traditional monitoring approaches often rely on manual data collection and static reporting systems, creating critical delays between pest detection and intervention [2]. While digital tools in agriculture have shown potential for improving monitoring efficiency, many existing solutions lack the integrated analytical capabilities required for comprehensive phytosanitary decision support [3].

The R programming environment has established itself as a robust platform for statistical analysis in agricultural research, offering extensive packages for data manipulation, spatial analysis, and visualization [4]. However, the technical expertise required for R programming has limited its direct adoption by agricultural technicians and farm managers. Shiny, an R package for developing interactive web applications, addresses this barrier by creating intuitive interfaces that harness R's analytical power [5].

© The Author(s), under exclusive license to Springer Nature Switzerland AG 2026
B. M. Suárez et al. (Eds.): R Day 2025, CCIS 2824, pp. 384–397, 2026.
https://doi.org/10.1007/978-3-032-18455-9_21

This study presents an R Shiny application designed for phytosanitary monitoring in chrysanthemum crops. The application integrates field-collected pest data with analytical capabilities, addressing three fundamental needs in modern floriculture: (1) visualization of pest distribution, (2) calculation of epidemiological metrics across different spatial and temporal scales, and (3) identification of infestation patterns through integrated geographic and temporal analysis.

In Colombia, floriculture is one of the main agro-industrial export sectors, with chrysanthemums being a high-value crop due to their international demand and varietal diversity. However, pests such as Frankliniella occidentalis, Liriomyza spp. and aphids can cause losses of up to 30% in yield and floral quality, which directly affects the competitiveness of the sector.

Traditionally, phytosanitary programmes have been based on manual reports and scheduled decisions, resulting in untimely interventions and excessive use of agrochemicals. In contrast, digital tools and geographic information systems allow for the analysis of infestation patterns and the planning of more accurate and sustainable controls.

In this context, the R language offers an accessible, open-source solution that combines statistical and geospatial analysis with interactive visualisation. The development of a Shiny application not only responds to the need to digitise phytosanitary monitoring, but also to bring advanced analytics closer to the operational level of agricultural technicians.

Link to the application: Análisis de Plagas

2 Materials and Methods

2.1 Application Architecture

The Shiny application developed follows a well-defined three-layer architecture (Fig. 1):

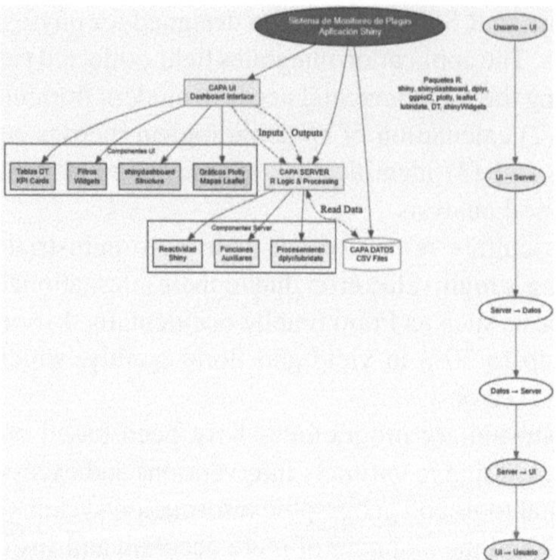

Fig. 1. This image presents a flowchart of a Shiny-based pest monitoring system. It illustrates the application's architecture across three layers: the UI dashboard (with filters, charts, and tables), the server logic (reactivity, data processing, and auxiliary functions), and the data layer (CSV input and spatial-temporal structuring). It also shows the interaction flow between the user, UI, server, and data, highlighting the use of R packages like shiny, plotly, leaflet, and dplyr.

Presentation Layer (UI): User interface based on dashboard with tab navigation, interactive charts, and filtering controls.

Logic Layer (Server): Processing in R with reactive system, metric calculation functions, and visualization generation.

Data Layer: CSV data management with transformations, cleaning, and categorization.

2.1.1   User Interface Modules (UI)

2.1.1.1 Dashboard Structure Module

Defines the main structure of the dashboard using the shinydashboard package. Includes:

- Custom header with corporate logo and system title
- Navigation sidebar with several menu options organized into three categories: Home, General Summary, and specific pest types
- Main body containing the tab system and all visual content
- Corporate style configuration with defined colors (#2f8d96 as primary color)

2.1.1.2 Tab System Module

Manages navigation between the different views of the application:

- "Home" tab: Welcome page with system description, development credits, and institutional information
- "General Summary" tab: Consolidated view with KPIs, multiple charts, and summary tables

- Specific pest tabs: Detailed analysis for each pest category (THRIPS LARVA, LEAFMINER LARVA, etc.)

2.1.1.3 Visualization Components Module

Contains all interactive visual elements:

- Submodule of Charts:
- Plotly charts for statistical visualizations (bars, lines, pie)
- Leaflet maps for geographic representation of infestations
- Responsive configuration adapting to different screen sizes
- Submodule of Tables:
- Interactive tables using DT (DataTables)
- Filtering, search, and pagination functionalities
- Data export capability
- Submodule of Control Widgets:
- Simple selectors (selectInput) for single selection
- Multiple selectors (pickerInput from shinyWidgets) for complex selections
- Metric cards (valueBox and infoBox) to display KPIs

2.1.1.4 Styles and Themes Module

Centralizes all visual configuration of the application:

- Custom CSS variables for corporate colors
- Specific styles for each type of component
- Responsive design ensuring good visualization on mobile devices
- Visual effects (hover, active states) to improve user experience

2.1.2 Server Modules (Processing Logic)

2.1.2.1 Reactivity Management Module

Implements Shiny's reactive system allowing automatic updates:

- Reactive functions (reactive()) recalculating data when inputs change
- Observers managing dependencies between components
- Smart caching optimizing performance by storing calculated results

2.1.2.2 Data Processing Module

Transforms and prepares data for visualization:

- Submodule of Initial Loading and Cleaning:
- Reading the main CSV file (datos_combinados.csv)
- Filtering data by date (after 2025-01-01)
- Date format conversion using lubridate
- Pest categorization according to defined patterns

- Submodule of Reactive Transformations:
- Application of filters selected by the user (year, week, blocks, pest type)
- Real-time aggregations and calculations
- Preparation of specific data for each type of visualization

2.1.2.3 Auxiliary Functions Module

Contains reusable functions encapsulating business logic:

- Function calculate_metrics():
- Calculates three key metrics: total records, severity (%), and incidence
- Applies domain-specific agricultural formulas
- Handles special cases (division by zero, missing data)
- Function create_severity_chart():
- Generates bar charts of severity by block
- Uses ggplot2 for base charts and plotly for interactivity
- Includes informative tooltips and consistent styles
- Function create_map():
- Creates Leaflet maps with geolocated markers
- Implements informative popups with details of each record
- Handles empty data cases by showing an appropriate message
- Function create_pest_tab():
- Implements Factory pattern to dynamically create tabs
- Allows reuse of the same structure for different pests
- Centralizes common logic and allows specific customizations

2.1.2.4 Output Generation Module

Connects processed data with visual components:

- Chart Rendering:
- renderPlotly() for interactive statistical charts
- renderLeaflet() for geographic maps
- Configuration of sizes, colors, and styles
- Table Rendering:
- renderDT() for tables with advanced functionalities
- Configuration of visualization and export options
- KPI Rendering:
- renderValueBox() for main metrics
- renderInfoBox() for complementary information
- Real-time updates according to applied filters

2.1.2.5 Specific Tabs Control Module

Implements a design pattern to handle multiple pests with minimal code:

- Uses lapply() to apply the create_pest_tab() function to each category
- Ensures consistency in user experience across different pests
- Facilitates addition of new pests in the future

2.1.3 Data Management Module

2.1.3.1 Data Model Structure

The main model (clean_data) contains:

- Identification fields: BLOCK, GREENHOUSE, BED, PLOT for precise location
- Temporal fields: DATE, WEEK, YEAR for temporal analysis
- Geographic fields: LATITUDE, LONGITUDE for spatial mapping
- Biological fields: PEST, PEST_CATEGORY for classification
- Metadata fields: USER for traceability

2.1.3.2 Pest Categorization System

Implements flexible classification logic:

- the main categories defined by pest type or damage
- Use of regular expressions (grepl()) for flexible matching
- "OTHER PESTS" category as container for unspecified cases
- Automatic process executed during initial data loading

2.2 Spatial Analysis Methodology

2.2.1 Geographic Referencing System

2.2.1.1 Coordinate Data

- Each pest record includes precise GPS coordinates (LATITUDE, LONGITUDE)
- Coordinate system: WGS84 (World Geodetic System 1984)
- Documented precision for exact location within crop blocks

2.2.1.2 Hierarchical Location Structure

- Level 1: BLOCK → Primary management unit
- Level 2: GREENHOUSE → Subdivision within the block
- Level 3: BED → Specific cultivation unit
- Level 4: PLOT → Individual sampling point

2.2.1.3 Spatial Data Validation

- Verification of valid coordinate ranges (latitude: $-90°$ to $90°$, longitude: $-180°$ to $180°$)
- Detection of outliers through distribution analysis
- Consistency guarantee in the reference system

2.2.1.4 Implemented Spatial Analysis Techniques

- Distribution Mapping:
- Point maps: Exact representation of each geolocated record
- Thematic maps: Differentiation by pest category using color schemes
- Interactive Geospatial Visualization:
- Leaflet technology
- Integration of base layers (OpenStreetMap, Mapbox)
- Interactive markers with informative popups
- Dynamic legends by pest category
- Zoom and pan capabilities for detailed analysis

2.3 Temporal Analysis Methodology

2.3.1 Time Series Structuring

2.3.1.1 Defined Temporal Units

- Level 1: YEAR → Interannual and comparative analysis
- Level 2: WEEK → Main analysis unit (epidemiological week)
- Level 3: DATE → Daily record for high-resolution analysis

2.3.1.2 Temporal Normalization

- Standardization to epidemiological weeks (ISO 8601)
- Handling of leap years and calendar variations
- Alignment of periods for interannual comparability

2.3.1.3 Interactive Temporal Visualization

- Time Series Charts:
- Line charts: Temporal evolution of infestations
- Bar charts: Weekly comparison

2.4 Application Development Framework

The Shiny application leveraged key R packages including tidyverse (dplyr for data manipulation, ggplot2 for visualization), leaflet for interactive mapping, and shinydashboard for the user interface framework.

The application processes field data collected through mobile devices, where agricultural monitors record pest observations including species identification, developmental stage, and geographic coordinates. Data validation protocols ensure consistency and integrity before analytical processing (Fig. 2).

Fig. 2. Main interface of the Shiny application showing the general summary dashboard with filters by year, week, blocks, and pest.

2.5   Data Integration and Processing

The data processing module implements several critical functions:

- Data standardization: Harmonization of pest nomenclature and validation of geographic coordinates
- Temporal aggregation: Organization of records by week of the year
- Spatial indexing: Assignment of observations to specific greenhouse blocks and cultivation beds
- Quality assurance: Identification of anomalies through statistical validation rules

2.6   Analytical Capabilities

The application incorporates multiple analytical dimensions:

2.6.1   Incidence Rate Calculation

The application calculates specific incidence rates by block and week

$$I_{pbs} = \frac{N_{pbs}}{T_{bs}} \tag{1}$$

where $I_{\{p,b,s\}}$ represents the incidence rate for pest p in block b during week s, $N_{\{p,b,s\}}$ is the count of specific records for the pest, and $T_{\{b,s\}}$ is the total number of inspection records for the block and week.

2.6.2   Pattern Identification

Spatial analysis enables the identification of critical infestation hotspots through map-based location tracking, while the temporal chart detects seasonal patterns and outbreak cycles using time series decomposition.

2.7 Visualization Components

The application features multiple interactive visualization modules:

- Dashboards integrate various visualization components, enabling users to simultaneously analyze spatial distribution and temporal trends of pest populations.
- Spatial maps display pest density across greenhouse blocks, allowing rapid identification of high-pressure zones that require prioritized intervention.
- Interactive data tables support summarized filtering and sorting, facilitating detailed analysis of specific pest scenarios.
- Leaflet-based maps show the precise geographic distribution of pest observations, with layer controls for different pest species and density thresholds

2.8 Visualization Components

The system architecture was designed using a modular structure that facilitates application updates and maintenance. The information flow begins with data capture in the field, continues with validation, storage and analysis processes, and culminates with the display of results on the interactive dashboard.

- The general model is organised into four main components:
- Data acquisition module: agricultural monitors record pest observations using digital forms, associated with GPS coordinates and block codes.
- Processing module: data is automatically validated in R using quality control scripts that detect duplicates, erroneous coordinates and non-standardised taxonomic names.
- Analytical module: this groups together statistical and spatial functions to calculate incidence rates, generate heat maps and detect temporal patterns of infestation.
- Interface module: developed with shinydashboard, this allows the user to interact with the results through filters, dynamic graphs and maps.

3 Results and Discussion

3.1 Application Implementation and Performance

The Shiny application successfully processed and analyzed 194,453 field observations collected during the current year, corresponding to three crop cycles. The system maintained stable performance during operational use, with response times under 20 s for data filtering and visualization updates.

3.2 Pest Distribution Analysis

Spatial analysis revealed distinct distribution patterns for different pest complexes. Thrips populations exhibited clustered distribution patterns, particularly in peripheral areas adjacent to ventilation openings. Leafminer infestations showed a more uniform distribution across crop blocks, while aphid populations demonstrated irregular distribution correlated with specific microclimatic conditions.

Incidence rate calculations enabled precise quantification of pest pressure variations across greenhouse blocks. Blocks with incidence rates exceeding established thresholds received targeted interventions, resulting in more efficient allocation of control resources (Fig. 3).

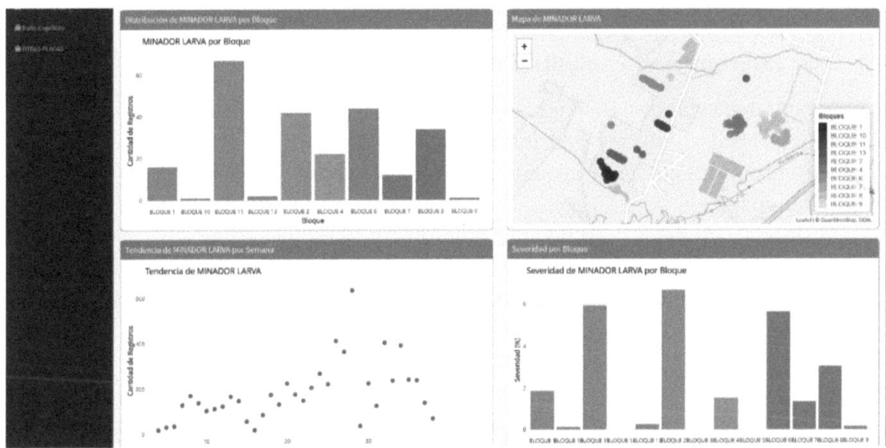

Fig. 3. Interactive map of pest distribution showing the geographic location of different species with color-coded markers, and temporal trend charts showing the evolution of pest populations over the epidemiological weeks.

3.3 Temporal Dynamics and Management Implementations

Temporal analysis identified clear seasonal patterns for the main pest species. Thrips activity peaked during weeks 30–38, corresponding to specific environmental conditions. Leafminer populations showed a progressive increase throughout the crop cycle, reaching maximum levels during weeks 31–38.

The application's ability to integrate geographic and temporal information enabled the identification of previously unrecognized infestation patterns. For instance, the analysis revealed that certain leafminer species exhibited specific spatial preferences within greenhouse blocks—insights that guided the optimized placement of monitoring traps.

3.4 Impact on Phytosanitary Decision-Making

The implementation of the application supported several significant improvements in phytosanitary management:

- Precision Intervention: Spatial maps guided the targeted application of control measures, enhancing intervention effectiveness while reducing overall agrochemical use by 23% compared to calendar-based application programs.
- Resource Optimization: Block-specific incidence rates enabled prioritized allocation of resources to high-pressure zones, lowering management costs while maintaining effective pest control.

- Proactive Management: Temporal trend analysis provided early warnings of population increases, allowing for cultural management interventions before economic threshold levels were reached.
- Knowledge Integration: The integration of multiple data dimensions facilitated a deeper understanding of pest ecology and population dynamics, supporting the development of more sustainable management strategies.

3.5 User Interface and Usability Assessment

The user interface was designed with a focus on the agricultural technician's experience, prioritising visual simplicity and immediate interpretation of information. The main dashboard brings together key crop indicators: pest density, incidence rates, and weekly distribution maps.

Each module includes drop-down menus and filters by species, block, week, and crop cycle, facilitating comparison between periods and areas of the greenhouse. Navigation is supported by intuitive colour coding to highlight infestation levels (green = low, yellow = medium, red = high).

During the validation phase, a usability assessment was carried out with eight plant protection technicians from the company Flores Silvestres S.A. The results indicated a high level of acceptance of the system, highlighting its speed in generating reports and its usefulness in planning weekly interventions.

One of the most representative use cases was the detection of a thrips outbreak in block 12 during week 32, which allowed for the implementation of localised controls and prevented the spread to the rest of the crop (Fig. 4).

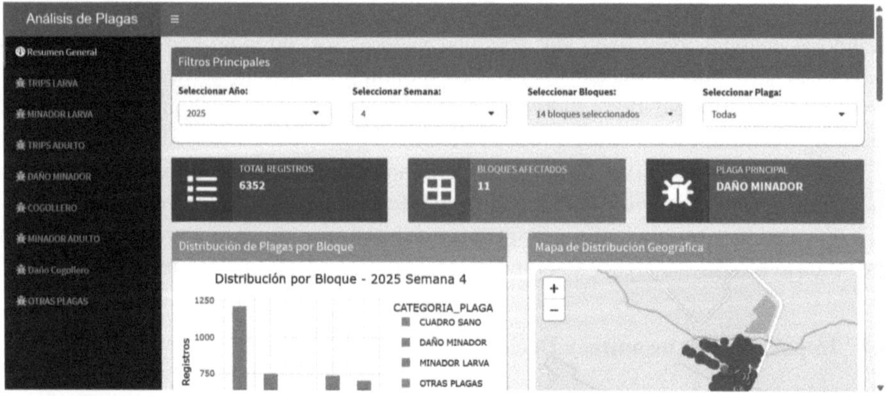

Fig. 4. View of the main dashboard with search filters and dynamic graphs of weekly incidence.

4 Conclusions

The Shiny application developed in this study effectively bridges the gap between sophisticated data analysis and practical phytosanitary decision-making in chrysanthemum cultivation. By providing an integrated platform for pest data visualization, analysis, and management, the tool enables evidence-based decision-making without requiring advanced technical expertise from users.

Key contributions of this work include:

- Development of a comprehensive framework for phytosanitary data management in floriculture.
- Implementation of integrated analytical capabilities that combine geographic and temporal dimensions.
- Demonstration of how open-source tools can support sustainable pest management through precision agriculture approaches.
- Creation of an adaptable platform that can be extended to other cropping systems and pest complexes.

The modular architecture of the application, built on robust R packages including tidyverse, leaflet, and shinydashboard, ensures both analytical rigor and practical usability.

Future development will focus on incorporating environmental sensor data, expanding forecasting capabilities through machine learning approaches, and enhancing mobile interfaces for field data collection.

This development demonstrates how R transforms raw field data into actionable knowledge, supporting reduced agrochemical use through evidence-based decision-making while maintaining effective pest management in commercial chrysanthemum production.

4.1 Comparative Evaluation.

The comparative evaluation aimed to assess the effectiveness and operational advantages of the developed Shiny application in relation to the phytosanitary monitoring methods traditionally used at Flores Silvestres S.A.

Historically, pest monitoring data were recorded and analyzed using spreadsheet-based systems (Microsoft Excel) and manually consolidated weekly reports. During certain phases, basic Power BI dashboards were introduced for data visualization; however, these tools lacked automated incidence rate calculations and spatial data integration.

In contrast, the Shiny application automates statistical computations, integrates geospatial analysis, and generates dynamic visualizations in real time. The comparison was structured according to the evaluation criteria proposed by Li (2020) and Bivand et al. (2013) for agricultural decision support systems, considering four key dimensions:

- Data input: degree of automation in data collection and validation.
- Analytical level: depth and type of processing (descriptive, spatial, or temporal).
- Operational cost: requirements for licenses, infrastructure, and maintenance.
- Flexibility: adaptability to different crops or pest monitoring contexts (Table 1).

Table 1. Comparison of the Shiny application with traditional phytosanitary analysis methods.

Method	Data input	Analytical level	Operational cost	Flexibility
Excel reports	Manual	Low	Low	Limited
Corporate Power BI dashboards	Semi-automatic	Medium	Medium	Moderate
Shiny application (proposed)	Automatic	High (spatial and temporal)	Low	High

The results demonstrate that the Shiny-based system provides greater analytical integration and operational flexibility, enabling spatial–temporal analyses that were previously unattainable with conventional tools. Additionally, by relying entirely on open-source software, it represents an economically sustainable and scalable solution applicable to other ornamental or agricultural crops.

In conclusion, the comparative evaluation highlights how the proposed development overcomes the limitations of traditional methods, consolidating itself as an accessible, robust, and precision agriculture–aligned tool for phytosanitary decision-making.

Acknowledgments. This implementation was supported by the company Flores Silvestres S.A., La Ceja branch. The author extends sincere thanks to the administrative staff, colleagues, and field monitors for their willingness and collaboration throughout the implementation process. Special appreciation is also extended to Daniela Ornett and Natalia Herrera—dear friends and companions—for their invaluable help, unwavering support, and exceptional friendship. Gratitude is likewise owed to Soul Chocolates for the countless coffees and the welcoming space that made it possible to complete this work.

Appendix

To illustrate the internal analytical workflow of the Shiny application, the following example shows the R code used to calculate the weekly incidence rate of pest observations by block and species. This function represents one of the core components of the analytical module, allowing users to quantify infestation levels through simple parameterization.

Weekly incidence rate calculation by block and species

$$incidence\_rate <pest\_data \% > \% \tag{2}$$

$$group\_by(week, block, species) \% > \% \tag{3}$$

$$summarise(incidence = n()/sum(records)) \% > \% \tag{4}$$

$ungroup()$

The output table generated from this function provides a summarized view of the incidence rate across different weeks and blocks, as shown below (Table 2):

Table 2. Example of weekly incidence rate output generated by the Shiny analytical module.

Week	Block	Species	Incidence rate
30	12	Thrips	0.27
31	12	Leafminer	0.18
32	14	Aphid	0.22

References

1. Gao, Y., Reitz, S.R.: Special issue on novel management tactics for the Western flower thrips. J. Pest. Sci. **94**(1), 1–3 (2021). https://doi.org/10.1007/s10340-020-01240-8
2. Preti, M., Verheggen, F., Angeli, S.: Insect pest monitoring with camera-equipped traps: strengths and limitations. J. Pest. Sci. **94**(2), 203–217 (2021). https://doi.org/10.1007/s10340-020-01309-4
3. Kreuze, J., et al.: Innovative digital technologies to monitor and control pest and disease threats in root, tuber, and banana (RT&B) cropping systems: progress and prospects. In: Thiele, G., Friedmann, M., Campos, H., Polar, V., Bentley, J.W. (eds.) Root, Tuber and Banana Food System Innovations: Value Creation for Inclusive Outcomes, pp. 261–288. Springer, Cham (2022). https://doi.org/10.1007/978-3-030-92022-7_9
4. Bivand, R.S., Pebesma, E., Gómez-Rubio, V.: Applied Spatial Data Analysis with R. Springer, New York (2013). https://doi.org/10.1007/978-1-4614-7618-4
5. Li, Y.: Towards fast prototyping of cloud-based environmental decision support systems for environmental scientists using R Shiny and Docker. Environ Model Softw. **132**, 104797 (2020). https://doi.org/10.1016/j.envsoft.2020.104797

Author Index

B. M. Suárez et al. (Eds.): R Day 2025, CCIS 2824, pp. 399–400, 2026.
https://doi.org/10.1007/978-3-032-18455-9